D1778744

commutative algebra and algebraic geometry

PURE AND APPLIED MATHEMATICS

A Program of Monographs, Textbooks, and Lecture Notes

EXECUTIVE EDITORS

Earl J. Taft
Rutgers University
New Brunswick, New Jersey

Zuhair Nashed
University of Delaware
Newark, Delaware

EDITORIAL BOARD

M. S. Baouendi
University of California, San Diego

Jane Cronin
Rutgers University

Jack K. Hale
Georgia Institute of Technology

S. Kobayashi
University of California, Berkeley

Marvin Marcus
University of California, Santa Barbara

W. S. Massey
Yale University

Anil Nerode
Cornell University

Donald Passman
University of Wisconsin, Madison

Fred S. Roberts
Rutgers University

Gian-Carlo Rota
Massachusetts Institute of Technology

David L. Russell
Virginia Polytechnic Institute and State University

Walter Schempp
Universität Siegen

Mark Teply
University of Wisconsin, Milwaukee

LECTURE NOTES IN PURE AND APPLIED MATHEMATICS

1. *N. Jacobson,* Exceptional Lie Algebras
2. *L.-Å. Lindahl and F. Poulsen,* Thin Sets in Harmonic Analysis
3. *I. Satake,* Classification Theory of Semi-Simple Algebraic Groups
4. *F. Hirzebruch et al., Differentiable Manifolds and Quadratic Forms*
5. *I. Chavel,* Riemannian Symmetric Spaces of Rank One
6. *R. B. Burckel,* Characterization of C(X) Among Its Subalgebras
7. *B. R. McDonald et al.,* Ring Theory
8. *Y.-T. Siu,* Techniques of Extension on Analytic Objects
9. *S. R. Caradus et al.,* Calkin Algebras and Algebras of Operators on Banach Spaces
10. *E. O. Roxin et al.,* Differential Games and Control Theory
11. *M. Orzech and C. Small,* The Brauer Group of Commutative Rings
12. *S. Thomier,* Topology and Its Applications
13. *J. M. Lopez and K. A. Ross,* Sidon Sets
14. *W. W. Comfort and S. Negrepontis,* Continuous Pseudometrics
15. *K. McKennon and J. M. Robertson,* Locally Convex Spaces
16. *M. Carmeli and S. Malin,* Representations of the Rotation and Lorentz Groups
17. *G. B. Seligman,* Rational Methods in Lie Algebras
18. *D. G. de Figueiredo,* Functional Analysis
19. *L. Cesari et al.,* Nonlinear Functional Analysis and Differential Equations
20. *J. J. Schäffer,* Geometry of Spheres in Normed Spaces
21. *K. Yano and M. Kon,* Anti-Invariant Submanifolds
22. *W. V. Vasconcelos,* The Rings of Dimension Two
23. *R. E. Chandler,* Hausdorff Compactifications
24. *S. P. Franklin and B. V. S. Thomas,* Topology
25. *S. K. Jain,* Ring Theory
26. *B. R. McDonald and R. A. Morris,* Ring Theory II
27. *R. B. Mura and A. Rhemtulla,* Orderable Groups
28. *J. R. Graef,* Stability of Dynamical Systems
29. *H.-C. Wang,* Homogeneous Branch Algebras
30. *E. O. Roxin et al.,* Differential Games and Control Theory II
31. *R. D. Porter,* Introduction to Fibre Bundles
32. *M. Altman,* Contractors and Contractor Directions Theory and Applications
33. *J. S. Golan,* Decomposition and Dimension in Module Categories
34. *G. Fairweather,* Finite Element Galerkin Methods for Differential Equations
35. *J. D. Sally,* Numbers of Generators of Ideals in Local Rings
36. *S. S. Miller,* Complex Analysis
37. *R. Gordon,* Representation Theory of Algebras
38. *M. Goto and F. D. Grosshans,* Semisimple Lie Algebras
39. *A. I. Arruda et al.,* Mathematical Logic
40. *F. Van Oystaeyen,* Ring Theory
41. *F. Van Oystaeyen and A. Verschoren,* Reflectors and Localization
42. *M. Satyanarayana,* Positively Ordered Semigroups
43. *D. L Russell,* Mathematics of Finite-Dimensional Control Systems
44. *P.-T. Liu and E. Roxin,* Differential Games and Control Theory III
45. *A. Geramita and J. Seberry,* Orthogonal Designs
46. *J. Cigler, V. Losert, and P. Michor,* Banach Modules and Functors on Categories of Banach Spaces
47. *P.-T. Liu and J. G. Sutinen,* Control Theory in Mathematical Economics
48. *C. Byrnes,* Partial Differential Equations and Geometry
49. *G. Klambauer,* Problems and Propositions in Analysis
50. *J. Knopfmacher,* Analytic Arithmetic of Algebraic Function Fields
51. *F. Van Oystaeyen,* Ring Theory
52. *B. Kadem,* Binary Time Series
53. *J. Barros-Neto and R. A. Artino,* Hypoelliptic Boundary-Value Problems
54. *R. L. Sternberg et al.,* Nonlinear Partial Differential Equations in Engineering and Applied Science
55. *B. R. McDonald,* Ring Theory and Algebra III
56. *J. S. Golan,* Structure Sheaves Over a Noncommutative Ring
57. *T. V. Narayana et al.,* Combinatorics, Representation Theory and Statistical Methods in Groups
58. *T. A. Burton,* Modeling and Differential Equations in Biology
59. *K. H. Kim and F. W. Roush,* Introduction to Mathematical Consensus Theory
60. *J. Banas and K. Goebel,* Measures of Noncompactness in Banach Spaces

61. *O. A. Nielson,* Direct Integral Theory
62. *J. E. Smith et al.,* Ordered Groups
63. *J. Cronin,* Mathematics of Cell Electrophysiology
64. *J. W. Brewer,* Power Series Over Commutative Rings
65. *P. K. Kamthan and M. Gupta,* Sequence Spaces and Series
66. *T. G. McLaughlin,* Regressive Sets and the Theory of Isols
67. *T. L. Herdman et al.,* Integral and Functional Differential Equations
68. *R. Draper,* Commutative Algebra
69. *W. G. McKay and J. Patera,* Tables of Dimensions, Indices, and Branching Rules for Representations of Simple Lie Algebras
70. *R. L. Devaney and Z. H. Nitecki,* Classical Mechanics and Dynamical Systems
71. *J. Van Geel,* Places and Valuations in Noncommutative Ring Theory
72. *C. Faith,* Injective Modules and Injective Quotient Rings
73. *A. Fiacco,* Mathematical Programming with Data Perturbations I
74. *P. Schultz et al.,* Algebraic Structures and Applications
75. *L Bican et al.,* Rings, Modules, and Preradicals
76. *D. C. Kay and M. Breen,* Convexity and Related Combinatorial Geometry
77. *P. Fletcher and W. F. Lindgren,* Quasi-Uniform Spaces
78. *C.-C. Yang,* Factorization Theory of Meromorphic Functions
79. *O. Taussky,* Ternary Quadratic Forms and Norms
80. *S. P. Singh and J. H. Burry,* Nonlinear Analysis and Applications
81. *K. B. Hannsgen et al.,* Volterra and Functional Differential Equations
82. *N. L. Johnson et al.,* Finite Geometries
83. *G. I. Zapata,* Functional Analysis, Holomorphy, and Approximation Theory
84. *S. Greco and G. Valla,* Commutative Algebra
85. *A. V. Fiacco,* Mathematical Programming with Data Perturbations II
86. *J.-B. Hiriart-Urruty et al.,* Optimization
87. *A. Figa Talamanca and M. A. Picardello,* Harmonic Analysis on Free Groups
88. *M. Harada,* Factor Categories with Applications to Direct Decomposition of Modules
89. *V. I. Istrăţescu,* Strict Convexity and Complex Strict Convexity
90. *V. Lakshmikantham,* Trends in Theory and Practice of Nonlinear Differential Equations
91. *H. L. Manocha and J. B. Srivastava,* Algebra and Its Applications
92. *D. V. Chudnovsky and G. V. Chudnovsky,* Classical and Quantum Models and Arithmetic Problems
93. *J. W. Longley,* Least Squares Computations Using Orthogonalization Methods
94. *L. P. de Alcantara,* Mathematical Logic and Formal Systems
95. *C. E. Aull,* Rings of Continuous Functions
96. *R. Chuaqui,* Analysis, Geometry, and Probability
97. *L. Fuchs and L. Salce,* Modules Over Valuation Domains
98. *P. Fischer and W. R. Smith,* Chaos, Fractals, and Dynamics
99. *W. B. Powell and C. Tsinakis,* Ordered Algebraic Structures
100. *G. M. Rassias and T. M. Rassias,* Differential Geometry, Calculus of Variations, and Their Applications
101. *R.-E. Hoffmann and K. H. Hofmann,* Continuous Lattices and Their Applications
102. *J. H. Lightbourne III and S. M. Rankin III,* Physical Mathematics and Nonlinear Partial Differential Equations
103. *C. A. Baker and L. M. Batten,* Finite Geometrics
104. *J. W. Brewer et al.,* Linear Systems Over Commutative Rings
105. *C. McCrory and T. Shifrin,* Geometry and Topology
106. *D. W. Kueke et al.,* Mathematical Logic and Theoretical Computer Science
107. *B.-L. Lin and S. Simons,* Nonlinear and Convex Analysis
108. *S. J. Lee,* Operator Methods for Optimal Control Problems
109. *V. Lakshmikantham,* Nonlinear Analysis and Applications
110. *S. F. McCormick,* Multigrid Methods
111. *M. C. Tangora,* Computers in Algebra
112. *D. V. Chudnovsky and G. V. Chudnovsky,* Search Theory
113. *D. V. Chudnovsky and R. D. Jenks,* Computer Algebra
114. *M. C. Tangora,* Computers in Geometry and Topology
115. *P. Nelson et al.,* Transport Theory, Invariant Imbedding, and Integral Equations
116. *P. Clément et al.,* Semigroup Theory and Applications
117. *J. Vinuesa,* Orthogonal Polynomials and Their Applications
118. *C. M. Dafermos et al.,* Differential Equations
119. *E. O. Roxin,* Modern Optimal Control
120. *J. C. Díaz,* Mathematics for Large Scale Computing
121. *P. S. Milojević,* Nonlinear Functional Analysis
122. *C. Sadosky,* Analysis and Partial Differential Equations

123. *R. M. Shortt,* General Topology and Applications
124. *R. Wong,* Asymptotic and Computational Analysis
125. *D. V. Chudnovsky and R. D. Jenks,* Computers in Mathematics
126. *W. D. Wallis et al.,* Combinatorial Designs and Applications
127. *S. Elaydi,* Differential Equations
128. *G. Chen et al.,* Distributed Parameter Control Systems
129. *W. N. Everitt,* Inequalities
130. *H. G. Kaper and M. Garbey,* Asymptotic Analysis and the Numerical Solution of Partial Differential Equations
131. *O. Arino et al.,* Mathematical Population Dynamics
132. *S. Coen,* Geometry and Complex Variables
133. *J. A. Goldstein et al.,* Differential Equations with Applications in Biology, Physics, and Engineering
134. *S. J. Andima et al.,* General Topology and Applications
135. *P Clément et al.,* Semigroup Theory and Evolution Equations
136. *K. Jarosz,* Function Spaces
137. *J. M. Bayod et al.*, *p*-adic Functional Analysis
138. *G. A. Anastassiou*, Approximation Theory
139. *R. S. Rees*, Graphs, Matrices, and Designs
140. *G. Abrams et al.*, Methods in Module Theory
141. *G. L. Mullen and P. J.-S. Shiue*, Finite Fields, Coding Theory, and Advances in Communications and Computing
142. *M. C. Joshi and A. V. Balakrishnan*, Mathematical Theory of Control
143. *G. Komatsu and Y. Sakane*, Complex Geometry
144. *I. J. Bakelman*, Geometric Analysis and Nonlinear Partial Differential Equations
145. *T. Mabuchi and S. Mukai*, Einstein Metrics and Yang–Mills Connections
146. *L. Fuchs and R. Göbel*, Abelian Groups
147. *A. D. Pollington and W. Moran*, Number Theory with an Emphasis on the Markoff Spectrum
148. *G. Dore et al.*, Differential Equations in Banach Spaces
149. *T. West*, Continuum Theory and Dynamical Systems
150. *K. D. Bierstedt et al.*, Functional Analysis
151. *K. G. Fischer et al.*, Computational Algebra
152. *K. D. Elworthy et al.*, Differential Equations, Dynamical Systems, and Control Science
153. *P.-J. Cahen, et al.*, Commutative Ring Theory
154. *S. C. Cooper and W. J. Thron*, Continued Fractions and Orthogonal Functions
155. *P. Clément and G. Lumer*, Evolution Equations, Control Theory, and Biomathematics
156. *M. Gyllenberg and L. Persson*, Analysis, Algebra, and Computers in Mathematical Research
157. *W. O. Bray et al.*, Fourier Analysis
158. *J. Bergen and S. Montgomery*, Advances in Hopf Algebras
159. *A. R. Magid*, Rings, Extensions, and Cohomology
160. *N. H. Pavel*, Optimal Control of Differential Equations
161. *M. Ikawa*, Spectral and Scattering Theory
162. *X. Liu and D. Siegel*, Comparison Methods and Stability Theory
163. *J.-P. Zolésio*, Boundary Control and Variation
164. *M. Křížek et al.*, Finite Element Methods
165. *G. Da Prato and L. Tubaro*, Control of Partial Differential Equations
166. *E. Ballico*, Projective Geometry with Applications
167. *M. Costabel et al.*, Boundary Value Problems and Integral Equations in Nonsmooth Domains
168. *G. Ferreyra, G. R. Goldstein, and F. Neubrander*, Evolution Equations
169. *S. Huggett*, Twistor Theory
170. *H. Cook et al.*, Continua
171. *D. F. Anderson and D. E. Dobbs*, Zero-Dimensional Commutative Rings
172. *K. Jarosz,* Function Spaces
173. *V. Ancona et al.*, Complex Analysis and Geometry
174. *E. Casas*, Control of Partial Differential Equations and Applications
175. *N. Kalton et al.*, Interaction Between Functional Analysis, Harmonic Analysis, and Probability
176. *Z. Deng et al.*, Differential Equations and Control Theory
177. *P. Marcellini et al.* Partial Differential Equations and Applications
178. *A. Kartsatos*, Theory and Applications of Nonlinear Operators of Accretive and Monotone Type
179. *M. Maruyama*, Moduli of Vector Bundles
180. *A. Ursini and P. Aglianò*, Logic and Algebra
181. *X. H. Cao et al.,* Rings, Groups, and Algebras
182. *D. Arnold and R. M. Rangaswamy,* Abelian Groups and Modules
183. *S. R. Chakravarthy and A. S. Alfa*, Matrix-Analytic Methods in Stochastic Models
184. *J. E. Andersen et al.*, Geometry and Physics
185. *P.-J. Cahen et al.*, Commutative Ring Theory

186. *J. A. Goldstein et al.*, Stochastic Processes and Functional Analysis
187. *A. Sorbi*, Complexity, Logic, and Recursion Theory
188. *G. Da Prato and J.-P. Zolésio*, Partial Differential Equation Methods in Control and Shape Analysis
189. *D. D. Anderson*, Factorization in Integral Domains
190. *N. L. Johnson*, Mostly Finite Geometries
191. *D. Hinton and P. W. Schaefer*, Spectral Theory and Computational Methods of Sturm–Liouville Problems
192. *W. H. Schikhof et al.*, *p*-adic Functional Analysis
193. *S. Sertöz*, Algebraic Geometry
194. *G. Caristi and E. Mitidieri*, Reaction Diffusion Systems
195. *A. V. Fiacco*, Mathematical Programming with Data Perturbations
196. *M. Křížek et al.*, Finite Element Methods: Superconvergence, Post-Processing, and A Posteriori Estimates
197. *S. Caenepeel and A. Verschoren*, Rings, Hopf Algebras, and Brauer Groups
198. *V. Drensky et al.*, Methods in Ring Theory
199. *W. B. Jones and A. Sri Ranga*, Orthogonal Functions, Moment Theory, and Continued Fractions
200. *P. E. Newstead*, Algebraic Geometry
201. *D. Dikranjan and L. Salce*, Abelian Groups, Module Theory, and Topology
202. *Z. Chen et al.*, Advances in Computational Mathematics
203. *X. Caicedo and C. H. Montenegro*, Models, Algebras, and Proofs
204. *C. Y. Yıldırım and S. A. Stepanov*, Number Theory and Its Applications
205. *D. E. Dobbs et al.*, Advances in Commutative Ring Theory
206. *F. Van Oystaeyen*, Commutative Algebra and Algebraic Geometry

Additional Volumes in Preparation

commutative algebra and algebraic geometry

proceedings of the Ferrara meeting in honor of Mario Fiorentini

edited by

Freddy Van Oystaeyen
University of Antwerp
Wilrijk, Belgium

MARCEL DEKKER, INC. NEW YORK • BASEL

Library of Congress Cataloging-in-Publication Data

Commutative algebra and algebraic geometry: proceedings of the Ferrara meeting in honor of Mario Fiorentini. / edited by Freddy Van Oystaeyen.
p. cm.— (Lecture notes in pure and applied mathematics ; v. 206)
ISBN 0-8247-1990-5 (alk. paper)
1. Commutative algebra—Congresses. 2. Geometry, Algebraic—Congresses. I. Fiorentini, Mario. II. Oystaeyen, F. Van. III. Series.
QA251.3.C6522 1999
512'.24—dc21 99-17653
CIP

This book is printed on acid-free paper.

Headquarters
Marcel Dekker, Inc.
270 Madison Avenue, New York, NY 10016
tel: 212-696-9000; fax: 212-685-4540

Eastern Hemisphere Distribution
Marcel Dekker AG
Hutgasse 4, Postfach 812, CH-4001 Basel, Switzerland
tel: 41-61-261-8482; fax: 41-61-261-8896

World Wide Web
http://www.dekker.com

The publisher offers discounts on this book when ordered in bulk quantities. For more information, write to Special Sales/Professional Marketing at the headquarters address above.

Copyright © 1999 by Marcel Dekker, Inc. All Rights Reserved.

Neither this book nor any part may be reproduced or transmitted in any form or by any means, electronic or mechanical, including photocopying, microfilming, and recording, or by any information storage and retrieval system, without permission in writing from the publisher.

Current printing (last digit)
10 9 8 7 6 5 4 3 2

PRINTED IN THE UNITED STATES OF AMERICA

To Mario We know you love books, in particular old books and art books that are like treasures to you. We hope you will treasure this new book in remembrance of all your old and new friends.

Preface

For this book we have collected papers in the areas of commutative algebra and algebraic geometry, dedicated to Mario Fiorentini on the occasion of his retirement. Most of these papers represent talks given at the meeting at the University of Ferrara.

This meeting was organized by two friends of Mario, P. Ellia and A. Lascu, and it fitted the personality of Mario perfectly; there was good mathematics, many cultural activities, good food, and more friendship than I ever observed at any meeting. In the name of all participants, I thank P. Ellia and A. Lascu for organizing the perfect meeting, and thanks to all speakers and authors for their contribution to the success of the meeting and these proceedings.

Freddy Van Oystaeyen

Contents

Preface v
Contributors ix
A Simple Article E. Sernesi xi
Mario Fiorentini: Bibliography xv

1. The Projective Normality of Smooth Degree Nine Varieties 1
A. Alzati and G. M. Besana

2. Minimal Generating Sets for a Family of Monomial Curves in A4 5
Henrik Bresinsky and Lê Tuân Hoa

3. Cohomology of Surfaces $X \subseteq \mathbf{P}^4$ with Degree < 2r – 2 15
M. Brodmann

4. On Certain Spaces Associated to Tetragonal Curves of Genus 7 and 8 35
G. Casnati and A. Del Centina

5. Gorenstein Algebras with Pure Resolution 47
Mihai Cipu

6. Smooth Specializations of Space Curves: Questions and Examples 53
Ph. Ellia and R. Hartshorne

7. On Codimension Two *k*-Buchsbaum Subvarieties of P^n 81
Ph. Ellia and A. Sarti

8. Curves Contractable in General Surfaces 93
D. Franco and A. T. Lascu

9. On Nagata's Theorem for the Class Group, II 117
Stefania Gabelli

10. Catalecticant Varieties 143
Anthony V. Geramita

11. Bounds for the Betti Numbers of Shellable Simplicial Complexes and Polytopes 157
Jürgen Herzog and Enzo Maria Li Marzi

12. Generators for the Generic Rational Space Curve: Low Degree Cases 169
Monica Idà

13. Equisingularity, Multiplicity, and Dependence 211
Steven L. Kleiman

14. Picard Bundles and Syzygies of Canonical Curves 227
Giuseppe Pareschi

15. Variations on Green's Theorem Concerning the Hilbert Functions 237
Dorin Popescu

16. On the Dimension Filtration and Cohen-Macaulay Filtered Modules 245
Peter Schenzel

17. Monomial Conjecture and Auslander's δ-Invariant 265
Ann-Marie Simon and Jan R. Strooker

18. Hilbert Function and Numerical Invariants of Closed Subschemes of P^n 275
Mario Valenzano

19. A Deformation of Projective Schemes 295
Freddy Van Oystaeyen

Contributors

A. Alzati University of Milan, Milan, Italy

G. M. Besana Eastern Michigan University, Ypsilanti, Michigan

Henrik Bresinsky University of Main, Orono, Maine

M. Brodmann University of Zürich, Zürich, Switzerland

G. Casnati University of Padua, Padua, Italy

A. Del Centina University of Ferrara, Ferrara, Italy

Mihai Cipu Institute of Mathematics of the Romanian Academy, Bucharest, Romania

Ph. Ellia University of Ferrara, Ferrara, Italy

D. Franco University of Ferrara, Ferrara, Italy

Stefania Gabelli Università di Roma "La Sapienza," Rome, Italy

Anthony V. Geramita Queen's University, Kingston, Ontario, Canada, and University of Genoa, Genoa, Italy

R. Hartshorne University of California, Berkeley, California

Jürgen Herzog University of Essen, Essen, Germany

Lê Tuân Hoa Institute of Mathematics, Hanoi, Vietnam

Monica Idà University of Bologna, Bologna, Italy

Steven L. Kleiman Massachusetts Institute of Technology, Cambridge, Massachusetts

A. T. Lascu University of Ferrara, Ferrara, Italy

Enzo Maria Li Marzi University of Messina, Messina, Italy

Giuseppe Pareschi Università di Roma "La Sapienza," Rome, Italy

Dorin Popescu University of Bucharest, Bucharest, Romania

A. Sarti Göttingen University, Göttingen, Germany

Peter Schenzel Martin Luther University, Halle, Germany

Anne-Marie Simon University of Brussels, Brussels, Belgium

Jan R. Strooker University of Utrecht, Utrecht, The Netherlands

Mario Valenzano University of Turin, Turin, Italy

Freddy Van Oystaeyen University of Antwerp, Antwerp, Belgium

Mario Fiorentini

A Simple Article

Mario Fiorentini has spent almost all his scientific life in Ferrara, where he was appointed to the chair of Geometria Superiore in 1971, and where he remained until his retirement in 1996.

Mario came from Roma where, while a secondary school teacher, he had studied mathematics as an autodidact. His starting point was classical projective algebraic geometry, whose tradition was still alive in Roma, thanks to B. Segre and his school; but his interests concentrated mostly on homological methods in commutative algebra and algebraic geometry, in connection with the most advanced ideas of Grothendieck and his school. Mario was especially interested in Cohen-Macaulay and Gorenstein rings, flatness, factoriality and semifactoriality, liaison, determinantal varieties. These topics have always remained at the center of his interests.

In a paper published in 1971 in J. of Algebra he generalized regular sequences, introducing the notion of "relative regular sequence", which is more natural in several homological and geometrical situations, especially in connection with determinantal ideas. That paper remained unnoticed for many years; only in the 1980's several authors introduced and used notions which turned out to be equivalent to relative regular sequences.

Not long after his settling in Ferrara, Mario succeeded in bringing there the algebraists G. Baccella and S. Gabelli, and the geometers F. Ghione, M. Letizia and myself, whom he had met in Roma in 1970 while attending the lectures of O. Zariski during a short visit of his. G. Tomassini, another newly appointed professor of geometry, was also in Ferrara, and later on V. Ancona joined the group of geometers.

At that time Ferrara was partly a commuter's university : several mathematicians, like L. Cattabriga, A. Orsatti and later on F. Menegazzo and C. Parenti commuted every day from Bologna and Padova. A few years later more mathematicians moved to Ferrara, notably A. Ambrosetti and M. Giaquinta. This had a positive influence on the atmosphere.

Mario always considered research as the first priority. He was concerned about creating and maintaining good conditions for doing research : he was constantly active in inviting mathematicians from everywhere and in organizing short meetings; often this was made possible by the presence of visiting professors elsewhere in Italy. First to come were D. Buchsbaum, F. Gaeta, S. Kleiman, A. Micali, P. Ribenboim. Later on C. Peskine and L. Szpiro started visiting Ferrara : they came quite often and sometimes for long periods. P. Roberts spent the entire academic year 1974-1975 in Ferrara, giving a course (in Italian) on intersection theory. H. Kurke spent the fall term of 1974 giving seminars on deformation theory. There was always someone from Romania : I remember visits of L. Badescu, A. Brezuleanu, I. Bucur, M. Iurchescu, M. Morianu, C. Nastasescu, and many others. It is difficult to make a complete list of the mathematicians who came to Ferrara in those years. They included M. André, L. Avramov, A. Bialinycki-Birula, M. Boratynski, J. Carlson, A. Douady, M. Hermann, J. Herzog, T. Jozefiuak, H. Matsumura, H. Popp, P. Pragatz, D. B. Scott, C. S. Seshadri, D. Simson, R. Treger, W. Vogel.

Mario had frequent contacts with Genova, where an active group was working under the direction of P. Salmon and with Bologna, Padova and Roma. The seminars and conferences organized by Mario were opportunities for mathematicians coming from those cities to get together in Ferrara.

Thanks to this situation, in the seventies in Ferrara the atmosphere was exceptionally stimulating. At that time the mathematical institute was located in via Savonarola n. 9 : its premises were displayed around a beautiful renaissance cloister. A quite rich library and the offices of a few professors were there. Everybody else had office space in an apartment located in the same street, a few steps away : here we spent long winter days, studying, working and discussing. In the good season we worked outside as well. I remember many afternoons spent in Parco Massari or Piazza Ariostea, where we took our children to play, talking of mathematics with M. Bester, and where sometimes Mario joined us; or long hours spent discussing with L. Szpiro, who liked to do it sitting in one of the squares of Ferrara, sipping a beer.

The city life was well organized but slow, so that plenty of time was left for extra-mathematical activities. Mario has always been very interested in figurative arts. It was common to go with him to the art exhibitions of Palazzo dei Diamanti, sometimes together with a visiting professor; or to buy engravings directly in the artists' workshops. He had always ideas on how to spend a Sunday morning, and quite often we went out together to visit some new place; and spending a day with him meant discussing all the time politics, art, and mathematics of course.

In 1980 the situation in Ferrara changed, because a new "concorso" moved many people around in Italy. I also moved from Ferrara, not without sorrow : I was leaving a place where I spent seven productive and intense years among friends. The memory of those years identifies with Mario and with his extraordinary energy and enthusiasm.

E. Sernesi
Terza Università di Roma

Mario Fiorentini: Bibliography

1. Fiorentini, M., *Dall'analisi matematica classica all'analisi funzionale all'analisi generale*, Ed. Cultura e Scuola, Roma, 1964.

2. Fiorentini, M., *Geometria delle matrici rettangolari*, Ist. Mat. G. Castelnuovo, Roma, 1965.

3. Fiorentini, M., *Generalizzazione di un teorema di Pascal*, Ist. Mat. G. Castelnuovo, Roma, 1969.

4. Fiorentini, M., Marruccelli, A., *Mutamenti nell'oggetto della Matematica*, Ed. Cultura e Scuola, Roma, 1969.

5. Fiorentini, M., *La varietà di Veronese e due indici*, Ist. Mat. G. Castelnuovo, Roma, 1969.

6. Fiorentini, M., *Integrità dell'algebra simmetrica di una famiglia di ideali generati da una successione regolare generalizzata*, Ist. Mat. G. Castelnuovo, Roma, 1969.

7. Fiorentini, M., *Sopra una speciale famiglia di anelli locali regolari*, Ist. Mat. G. Castelnuovo, Roma, 1969.

8. Fiorentini, M., *Una speciale famiglia di ideali di classe principale ganeralizzata*, Rend. di Matematica, Vol. 3, Roma, 1970.

9. Fiorentini, M., *Una rappresentazione primbasis per gli ideali di classe principale generalizzata*, Ist. Mat. G. Castelnuovo, Roma, 1969.

10. Fiorentini, M., Lorenzani, M., *Ideali sizigietici e successioni regolari relative*, Ist. Mat. G. Castelnuovo, Roma, 1970.

11. Fiorentini, M., Marruccelli, A., *Complementi di Matematiche Moderne*, Ed. Cedam, Padova, 1970.

12. Fiorentini, M., *Proprietà aritmetiche delle curve razionali normali*, Ist. Mat. G. Castelnuovo, Roma, 1970.

13. Fiorentini, M., Marruccelli, A., *Il problema dei fondamenti nella Matematica*, Ed. Cultura e Scuola, Roma, 1970.

14. Fiorentini, M., *Una rappresentazione primbasis per gli ideali di classe principale generalizzata*, Ist. Mat. G. Castelnuovo, Roma, 1970.

15. Fiorentini, M., *Esempi di anelli di Cohen-Macaulay che non sono di Gorenstein*, Acc. Naz. Lincei, Roma, febbraio 1971.

16. Fiorentini, M., *On relative regular sequences*, Journal of Algebra, May 1971.

17. Fiorentini, M., Tomassini, G., *Un ampliamento del linguaggio degli schemi. Minischemi di mojezon e spazi algebrici di M. Artin*, Ist. G. Castelnuovo, Roma, 1971.

18. Fiorentini, M., Sernesi, E., *Sull'algebra esterna sopra un modulo*, Ist. Mat. G. Castelnuovo, Roma, 1971.

19. Fiorentini, M., Sernesi, E., *Proprietà di stabilità di certe algebre associative*, Ist. Mat. G. Castelnuovo, Roma, 1971.

20. Fiorentini, M., Marruccelli, A., *The Objects and Foundations of Mathematics*, Ed. Scientia, Milano, 1971.

21. Fiorentini, M., *Esempi di anelli di Cohen-Macaulay semifattoriali che non sono di Gorenstein*, Acc. Naz. Lincei, Roma, Maggio 1971.

22. Fiorentini, M., Lascu, A. T., *Un Teorema sulle trasformazioni monoidali di spazi algebrici*, Annali Scuola Normale Superiore di Pisa, 1972.

23. Fiorentini, M., Badescu, L., *Criteri di semifattorialità e di fattorialità per gli anelli locali con applicazioni geometiche*, Ann. di Mat. Pura e Applicata, Bologna, 1975.

24. Fiorentini, M., *Trois conférences à la XIXme session du Séminaire de Mathématiques Supérieures de l'Université de Montréal*, Juillet, 1979.

25. Fiorentini, M., *Successioni regolari relative e ideali sizigietici*, Ann. Univ. Ferrara, Vol. XXVI, 1980.

26. Fiorentini, M., *Intersections résiduelles dans les anneaux de Cohen-Macaulay et un critère de régularité rélative*, Teubner Text für Mathematik, Band 40(1980), pp. 50-62.

27. Fiorentini, M., *Esempi di curvi di Buchsbaum che non sono di Macaulay*, Sem. di Variabili Complesse, Univ. Bologna, 1981.

28. Fiorentini, M., Lascu, A. T., *Una formula di geometria numerativa*, Ann. Univ. Ferrara, Vol. XXVII, 1982.

29. Fiorentini, M., Lascu, A. T., *A Criterion for Quasi-Complete Intersections and Related Embedded Questions*, Ann. Univ. Ferrara, Vol. XXVIII, 1982.

30. Fiorentini, M., Lascu, A. T., *Two Theorems of Giuseppe Gherardelli*, Lect. Notes in Math. Springer-Verlag, 1982.

31. Fiorentini, M., Lascu, A. T., *On the Homogeneous Ideal of a Quasi-complete Intersections in P^N*, Ann. Univ. Ferrara, Vol. XXIX, 1983.

32. Fiorentini, M., Ellia, Ph., *Défaut de postulation et singularité du schéma de Hilbert*, Ann. Univ. Ferrara, Vol. XXX, 1984.

33. Fiorentini, M., Lascu, A. T., *Varietà proiettive quasi-complete intersezioni*, Sem. di Geometria, Università di Bologna, 1984.

34. Fiorentini, M., *Curve di Buchsbaum quasi-complete intersezioni*, Seminari di geometria, Bologna, 1985.

35. Fiorentini, M., Portelli, D. *La varietà di Veronese e le sue proiezioni*, Seminari di Geometria, Univ. Bologna, 1986.

36. Fiorentini, M., Ellia, Ph., *Quelques remarques sur les courbes arithmétiquement Buchsbaum de l'espace projectif*, Ann. Ferrara, Vol. XXXIII, 1987.

37. Fiorentini, M., Lascu, A. T., *Subbundles of Maximal Degree of the Normal Bundle of Algebraic Space Curves almost Complete Intersection of Special Type*, Ann. Univ. Ferrara, Vol, XXXIII, 1987.

38. Fiorentini, M., Lascu, A. T., *Projective Embedding and Linkage*, Rend. Sem. Mat. Fis. di Milano, Vol. LVII, 1987.

39. Fiorentini, M., Bouchard, P., *A Commutative Algebra and Algebraic Geometry Laboratory with the Use of Computer*, Convegno di Algebra, Dip. Mat. Univ. Roma, 1989.

40. Fiorentini, M., Spangher, W., *Embedded Rank and Analytic Spread of Projective Varieties*, Quaderni Univ. Trieste, Dip. Mat., 1989.

41. Fiorentini, M., Hoa, L. T., *On Monomial k-Buchsbaum Curves in $P^r(k)$*, Annali Univ. Ferrara, Vol. XXXVI, 1990.

42. Fiorentini, M., Van Oystaeyen, F., *Some General Approaches to Regularity Conditions on Sequences*, Dep. Wiskunde en Informatica, UIA, Antwerpen, Belgium, March 1991.

43. Fiorentini, M., Lascu, A. T., *Linkage among Subcanonical and Quasi-complete-intersection Projective Schemes*, Dip. Mat. Univ. Ferrara, 1991.

44. Fiorentini, M., Vogel, W., *Old and New Results and Problems om Buchsbaum Modules*, Sem. di Geometria, Univ. di Bologna, 1991.

45. Fiorentini, M., *Semi-normalité et semi-factorialité en géométrie et algèbre*, Dep. Wiskunde en Informatica, UIA, Antwerpen, Belgium, February, 1993.

46. Fiorentini, M., Lascu, A.T., *Deux theorèmes d'acyclicité pour les schémas sous-canoniques*, Dip. Mat. Univ. Ferrara, 1993.

47. Fiorentini, M., *Propriété Gorenstein, Cohen Macaulay, Buchsbaum de l'anneau de Rees $R_A(I)$ et du gradué $G_I(A)$ associé à un ideal IA de classe principale faible*, Dep. de Alg. Univ. Almeria, Spain, 1993.

48. Fiorentini, M., Cipu, M.., *Ubiquity of Relative Regular Sequences and Proper Sequences*, Proceedings of Conference on Algebraic Geometry and Ring Theory in Honour of Michael Artin, Kluwer Academic Publishers, Vol. 8, no. 1, January, 1994.

49. Fiorentini, M., Simon, A.M., *Residual Intersections and Linkage for Ideals of the Weak Principal Class*, Comm. Alg. World Scientific, 14-25, September, 1992, ICTP, Trieste, 1994.

50. Fiorentini, M., M-Hoa, L.T., *Some Remarks on Generalized Cohen-Macaulay Rings*, Bull. Belg. Mat. Soc, September, 1994.

51. Fiorentini, M., Curtis, F., Bresinsky, H., Hoa, L.T., *On the Structure of Local Cohomology Modules for Curves in P^3*, Nagoya Math. Journal, Vol. 136, 1994.

The Projective Normality of Smooth Degree Nine Varieties

A. Alzati (*)
Dipartimento di Matematica, Università di Milano via C.Saldini 50 20133-Milano (Italy)
E-mail: ALZATI@VMIMAT.MAT.UNIMI.IT

G. M. Besana
Department of Mathematics, Eastern Michigan University Ypsilanti MI 48197 (U.S.A.)
E-mail: GBESANA@EMUNIX.EMICH.EDU

Abstract: the projective normality of smooth, complex, projective, linearly normal, degree 9, n-dimensional varieties, $n \geq 3$, is proved.
Mathematics subject classification: 14J40.
Key words: algebraic varieties, projective normality.

§1. **INTRODUCTION:** by using adjunction theory, thank to the work of many authors, (see: [A], [A-D-S], [I1], [I2], [I3], [O1], [O2], [L]), during the last few years the complete classification of smooth, complex, linearly normal, projective varieties of degree up to 8 was completed. If the degree is bigger than or equal to 9 the classification is more difficult: the surface case is so intricated that it has to be considered separately, moreover we have only maximal lists, i.e. lists containing some members for which the existence is not sure. Up to now we have the classification for degree 9 surfaces, (by combining [A-R] and [L]), for n-dimensional varieties with $n \geq 3$ ([F-L;1]) and for degree 10 varieties with $n \geq 3$ ([F-L;2]). The classification of degree 11 varieties with $n \geq 3$ is in preparation by G. M. Besana and A. Biancofiore.

It is very natural to ask whether these linearly normal varieties are projectively normal (p.n.) or not. If the variety X is embedded in $\mathbb{P}^N$ its being p.n. means that the hypersurfaces in $\mathbb{P}^N$ of any degree $t \geq 1$ cut complete linear systems on X. Equivalently the maps: $H^0(\mathbb{P}^N, \mathcal{O}_{\mathbb{P}^N}(t)) \to H^0(X, \mathcal{O}_X(t))$ are surjective for every $t \geq 1$, where $\mathcal{O}_X$ is the structural sheaf of X.

In [A-B-B] the projective normality of varieties of degree up to 8 is studied and the problem is completely solved. The projective normality of degree 9 surfaces is studied in [B-D]. The solution of the problem for them has an important consequence:

Theorem: every smooth, complex, n-dimensional projective variety of degree 9 which is linearly normal is projectively normal when $n \geq 3$.

In this short note we give the proof of this theorem.

(*) The author is member of G.N.S.A.G.A. of the Italian C.N.R. This research has been done within the framework of the M.U.R.S.T. national project "Geometria Algebrica".

§2. PROOF OF THE THEOREM: varieties of degree nine were classified in [F-L;1]. A handy list of them appears in [F-L;2]. We consider this list and we examine the varieties case by case. Note that in spite of the fact that the existence of 3 types of varieties listed in [F-L;2] is not sure, our theorem is not affected because we prove that these varieties are p.n. if they exist.

First of all we recall the following results of Fujita which we now translate and reduce for our aim:

Lemma: (see [F], th. 3.5 and 2.5) let X be a smooth, n-dimensional, complex projective variety, linearly normal, embedded in $\mathbb{P}^N$. Let d be the degree of X, let Δ = d-1-N+n be the delta-genus of X and let g be the sectional genus of X. Assume that $g \geq \Delta$, then:

1) if $d \geq 2\Delta+1$, X is p.n. ;

2) if $d \geq 2\Delta-1$ and the generic hyperplane section is p.n. in $\mathbb{P}^{N-1}$, X is p.n.

The assumption $d \geq 2\Delta-1$ yelds that the generic hyperplane section is linearly normal in $\mathbb{P}^{N-1}$, this fact explains the assumption on the section.

By recalling that every linearly normal elliptic scroll is p.n. (see [A-B-B] th. 2.8) and by using the previous lemma, part 1), from the list in [F-L;2] we get that we have to consider only the varieties for which $\Delta \geq 5$.

There are no varieties for which $\Delta \geq 8$ and if $\Delta = 7$ X is a hypersurface, obviously p.n. If $\Delta = 6$ we have complete intersections, which are p.n.; or scrolls over a K_3 surface, which are p.n. (see the resolution of their ideal sheaves for instance in [B-S-S]) ; or 3-varieties in $\mathbb{P}^5$, which are directly linked with $\mathbb{P}^3$ or with the cubic scroll in $\mathbb{P}^5$.

For any r = 0,1,..., N the r-th deficiency module of X in $\mathbb{P}^N$ is additively defined as the direct sum, over $j \in \mathbb{Z}$, of $H^r(\mathbb{P}^N, \mathcal{J}_X(j))$, where $\mathcal{J}_X$ is the ideal sheaf of X in $\mathbb{P}^N$. If X is directly linked to another n-variety Y the first deficiency module of X coincides, up to duality and a suitable twist, with the n-th deficiency module of Y. In our cases it is very easy to see that the third deficiency module of $\mathbb{P}^3$ and of the cubic scroll (both considered embedded in $\mathbb{P}^5$) vanishes, so that $H^1(\mathbb{P}^N, \mathcal{J}_X(t)) = 0$ for any $t \geq 1$. This implies that X is p.n.

If $\Delta = 5$ we have 5 types of varieties: if g = 7 or g = 6 we can use the previous lemma, part 2): the generical hyperplane section is p.n. by [B-D], so we can conclude that X is p.n. If g = 6 we have a type of variety which is also considered by Okonek in [O3] p.435, there it is proved that these varieties are directly linked in $\mathbb{P}^6$ with the Segre embedding of $\mathbb{P}^1 \times \mathbb{P}^2$ in $\mathbb{P}^5$. So we can also argue as for the previous liaisons.

If $\Delta = 5$ and g = 5 we have 3 types of varieties whose existence is not sure, in any case they are 3-folds in $\mathbb{P}^6$ with $d \geq 2\Delta-1$; so to prove that they are p.n. it suffices to show that their generic hyperplane section S is p.n. in $\mathbb{P}^5$. By [B-D] we get that the only possibility for such a surface to be not p.n.

is that it is a scroll over a trigonal genus 5 curve, or the blowing up Σ in 15 generic points of a $\mathbb{F}_e$, with $5 \geq e \geq 0$, embedded in $\mathbb{P}^5$ by the very ample divisor: $p^*[2C_0 + (6+e)F] - E_1 ..... - E_{15}$ where p is the blowing up, C_0 and F are the generators of $Pic(\mathbb{F}_e)$ and the E_j are the exceptional divisors of Σ. In any case $(K_\Sigma)^2 = -7$.

Let us consider the 3 cases according to the list in [F-L;2].

1) X is a scroll over $\mathbb{F}_1$; from Prop. (2,2) of [F-L;1] we get that the generic section S of X is the blowing up of $\mathbb{F}_1$ in 12 generic points, embedded in $\mathbb{P}^5$ by the very ample divisor: $p^*[3C_0 + 5F] - E_1 ..... - E_{12}$. It is easy to see that S is not a scroll over a genus five curve by looking at the irregularity and that it can not coincide with Σ as $(K_S)^2 = -4$.

2) X is a scroll over the blowing up Y of $\mathbb{P}^2$ in 5 generic points; let L be the generator of $Pic(\mathbb{P}^2)$, $q: Y \to \mathbb{P}^2$ the blowing up, $E_1, ... ,E_5$ the exceptional divisor of Y. From Prop. (2,2) of [F-L;1] we get that the generic section S of X is the blowing up of Y in 7 generic points, embedded in $\mathbb{P}^5$ by the very ample divisor: $p^*[q^*(6L) - 2E_1 ..... - 2E_5] - E_1 ..... - E_7$. It is easy to see that S is not a scroll over a genus five curve by looking at the irregularity and that it can not coincide with Σ as $(K_S)^2 = -3$.

3) Let (X,L) be the last polarized 3-fold to be considered. In this case we know only that the first reduction of (X,L) is (Q,2H), where Q is a smooth hyperquadric in $\mathbb{P}^4$ and H is the class of hyperplane divisors of Q.

The notion of first reduction is fully treated in [B-S]. For our present purposes it is enough to recall that it implies the existence of a blow up $f: X \to Q$ at a finite set of distinct points such that: $K_X + 2L = f^*(K_Q + 4H)$, where K is the canonical divisor.

The generic hyperplane section S' of (Q,2H) is a Del Pezzo surface with $(K_{S'})^2 = 4$, in fact, by adjunction theory: $K_{S'} = (K_Q+2H)_{|S'} = -H_{|S'}$. Hence S' is isomorphic to the blowing up of $\mathbb{P}^2$ in 5 generic points (we use the same notation as before). S' is polarized with $-2K_{S'} = 2H_{|S'} = q^*(6L) - 2E_1 .... - 2E_5$. Now we recall that f induces another blow up $\varphi: S \to S'$ at a finite set of m distinct points, where S is the generic hyperplane section of X, as usual. Let $E_1, E_2, ..., E_m$ be the exceptional divisors of S. A priori we do not know m, but, by adjunction theory, there exists a bijection between the elements of the linear system $|L|$ and the elements of $|2H|$ passing through the points at which Q is blown up. This implies that $|L_{|S}| = |\varphi^*(2H_{|S'}) - E_1 - E_2 - .... - E_m|$. Hence $m = [q^*(6L) - 2E_1 ..... - 2E_5]^2 - 9 = 7$. Now it is easy to see that S can not coincide with Σ as $(K_S)^2 = -3$, moreover S is not a scroll over a genus five curve by looking at the irregularity.

The proof of the theorem is completed. □

REFERENCES

[A] J.Alexander: "Surfaces rationelles non speciales dans $\mathbb{P}^4$". *Math. Zeit.* **200** (1) (1988) pp. 87-110.

[A-B-B] A.Alzati-M.Bertolini-G.M.Besana: "Projective normality of varieties of small degree". *Comm. in Algebra.* **25** (12) (1997) pp. 3761-3771.

[A-D-S] H.Abo-W.Decker-N.Sasakura: "An elliptic conic bundle in $\mathbb{P}^4$ arising from a stable rank-3 vector bundle". Preprint (1997).

[A-R] A.B.Aure-K.Ranestad: "The smooth surfaces of degree 9 in $\mathbb{P}^4$". In *Complex Projective Geometry*, vol. **179** of *London Math. Soc. Lecture Notes*, pp. 32-46.

[B-D] G.M.Besana-S.Di Rocco: "On the projective normality of smooth surfaces of degree nine". To appear in *Geometriae Dedicata* (1997).

[B-S] M.Beltrametti-A.J.Sommese: "The Adjunction Theory of Projective Varieties". De Gruyter Exposition in Mathematics 16 (1995).

[B-S-S] M.Beltrametti-M.Schneider-A.J.Sommese: "Threefolds of degree 9 in $\mathbb{P}^5$". *Math. Ann.* **288** (1990) pp. 613-644.

[F] T.Fujita: "Classification Theories of Polarized Varieties". *London Math. Soc. Lecture Note* n. **155**. Cambridge University Press, (1990).

[F-L;1] M.L.Fania-E.L.Livorni: "Degree nine manifolds of dimension n $\geq$ 3". *Math. Nachr.* **169** (1994) pp. 117-134.

[F-L;2] M.L.Fania-E.L.Livorni: "Degree ten manifolds of dimension n $\geq$ 3". To appear in: *Math. Nachr.* (1997) .

[I1] P.Ionescu: "Embedded projective varieties of small invariants". In *Proceedings of the Week of Algebraic Geometry*, Bucharest 1982. Springer L.N.M. **1056** (1984) pp. 142-186.

[I2] P.Ionescu: "Embedded projective varieties of small invariants II". *Rev. Roumaine Math. Pures Appl.* **31** (1986) pp. 539-544.

[I3] P.Ionescu: "Embedded projective varieties of small invariants III". In *Algebraic Geometry*, L'Aquila 1988. Springer L.N.M. **1417** (1990) pp. 138-154.

[L] E.L.Livorni: "On the existence of some surfaces". In *Algebraic Geometry*, L'Aquila 1988. Springer L.N.M. **1417** (1990) pp. 155-179.

[O1] C.Okonek: "Uber 2-codimensionale untermannigfaltigkeiten vom grad 7 im $\mathbb{P}^4$ und $\mathbb{P}^5$". *Math. Zeit.* **187** (1984) pp. 209-219.

[O2] C.Okonek: "Flachen vom grad 8 im $\mathbb{P}^4$". *Math. Zeit.* **191** (1986) pp. 207-223.

[O3] C.Okonek: "Notes on Varieties of Codimension 3 in $\mathbb{P}^N$". *Manuscr. Math.* **84** (1994) pp. 421-442.

Minimal Generating Sets for a Family of Monomial Curves in A^4

HENRIK BRESINSKY AND LÊ TUÂN HOA

Department of Mathematics, University of Maine
Orono, Maine 04469-5752, USA
E-mail: Henrik@maine.maine.edu

Institute of Mathematics
Box 631, Bò Hô, Hanoi, Vietnam

Dedicated to Mario Fiorentini

ABSTRACT. The paper constructs minimal generating sets for a family of monomial curves in $\mathbf{A}^4$ such that, with increasing integers, the number of relations needed to obtain the generating binomials in all four variables becomes arbitrarily large.

INTRODUCTION

In [B] minimal binomial generating sets were described for prime ideals $\mathfrak{p}(n_1, n_2, n_3, n_4) = \mathfrak{p}$ subject to:

(i) n_1, n_2, n_3, n_4 are distinct positive integers with g.c.d.$(n_1, n_2, n_3, n_4) = 1$ (here $n_1, n_2,$ n_3, n_4 are not necessarily listed by increasing size),

(ii) $\mathfrak{p} = \ker\phi : K[x_1, ..., x_4] \to K[t]$, K a field,

$$x_1 \to t^{n_1},\ x_2 \to t^{n_2},\ x_3 \to t^{n_3},\ x_4 \to t^{n_4}.$$

To explain our purpose here we first introduce some terminology and notation. By a binomial we always will mean a binomial with coefficients ± 1. The set of all these binomials in $\mathfrak{p}$ will be denoted by $B(\mathfrak{p})$. A binomial such that each monomial term has two variables (with positive exponents) will be said to be of 2-type (or type 2) and of subtype for instance (i, j), if it becomes necessary to identify the particular variables of a monomial term. A binomial with a pure power term is of 1-type and of subtype say (i), if it is necessary to identify the variable of the pure power term. In [B] the elements of type 2 and subtype, say (i, j), in a minimal generating set were denoted by $\mathcal{A}_{ij}$. If one arranges the elements of $\mathcal{A}_{ij}$ by, for instance, increasing x_j-exponent (and therefore decreasing x_i-exponent), it follows that the successor of an element in $\mathcal{A}_{ij}$ is determined by another unique element of type 2 in the minimal generating set. We will show that this uniqueness property may change for different binomials in $\mathcal{A}_{ij}$. That is we will construct a family of prime ideals $\mathfrak{p}(n_1, n_2, n_3, n_4)$ such that with increasing integers the number of relations to determine different successive members in $\mathcal{A}_{ij}$ of a minimal generating set becomes arbitrarily large. This corrects a statement of Theorem 4 in [B] and shows that an explicit algorithmic determination of a minimal binomial generating set for $\mathfrak{p}(n_1, n_2, n_3, n_4) \subseteq k[x_1, ..., x_4]$ is much more intricate than the analogous problem in $k[x_1, x_2, x_3]$ (see [H]).

Some concepts and definitions introduced in [B] and needed here are:

(1) Given two binomials $b_1 = m_{11} - m_{12}$ and $b_2 = m_{21} - m_{22}$ in $B(\mathfrak{p})$ assume that one of the monomial terms, say m_{21}, of b_2 divides one of the monomial terms, say m_{11}, of b_1. Then if $m_{11} = qm_{21}, b_3 = qm_{22} - m_{12} \in B(\mathfrak{p})$.

(2) We say an element $b_1 \in B(\mathfrak{p})$ is not reducible mod $\mathfrak{m}$, $\mathfrak{m} \subseteq B(\mathfrak{p})$, if the process in (1), with some $b_2 \in \mathfrak{m} \setminus \{b_1\}$, applied finitely many times, does not result in a binomial b_3 divisible by one of the variables. This of course includes the possibility, that the process in (1) cannot be applied at all.

(3) It was shown in [B] that $\mathfrak{m} \subseteq B(\mathfrak{p})$ is a generating set of $\mathfrak{p}$ iff for $b \in B(\mathfrak{p}) \setminus \mathfrak{m}$, b is reducible mod $\mathfrak{m}$.

Some preliminary definitions and elementary properties

Definition 1. Let $b_i = m_{i1} - m_{i2} \in B(\mathfrak{p})$, $i = 1, 2$, and $d = \text{g.c.d.}(m_{11}m_{22}, m_{12}m_{21})$. Then $b(b_1, b_2) = (m_{11}m_{22} - m_{12}m_{21})/d \in B(\mathfrak{p})$ is called the cross-product of b_1 and b_2.

We define next the following sequences of integers

(1) $A_0 = 1,\ B_0 = 0;\ A_i = A_{i-1} + B_{i-1},\ B_i = A_i + B_{i-1} = A_{i-1} + 2B_{i-1},\ i \geq 1.$
(2) $C_{i-1} = B_{i-1},\ D_{i-1} = A_i,\ i \geq 1.$

We then have

$$C_i = B_i = A_i + B_{i-1} = C_{i-1} + D_{i-1},\ i \geq 1,$$

and

$$D_i = A_{i+1} = A_i + B_i = C_i + D_{i-1} = C_{i-1} + 2D_{i-1},\ i \geq 1.$$

Let

$$v_0 = (3, 3),\ w_0 = (2, 4),\ v_i = A_i v_0 + B_i w_0,\ w_i = C_i v_0 + D_i w_0,\ i \geq 0.$$

Example 1. For $0 \leq i \leq 3,\ 0 \leq j \leq 2$, we have:

$v_0 = (3, 3) = v_0,$ $\qquad\qquad$ $w_0 = (2, 4) = w_0,$

$$v_1 = (5,7) = v_0 + w_0, \qquad w_1 = (7,11) = v_0 + 2w_0,$$
$$v_2 = (12,18) = 2v_0 + 3w_0, \qquad w_2 = (19,29) = 3v_0 + 5w_0,$$
$$v_3 = (v_{31}, v_{32}) = (31,47) = 5v_0 + 8w_0.$$

Here $A_3 = 5,\ B_3 = 8,\ v_{31} + v_{32} = 13 \cdot 6 = (A_3 + B_3)6$.

Properties of $\mathbf{v_i}$, $\mathbf{w_i}$, $\mathbf{A_i}$, $\mathbf{B_i}$, $\mathbf{C_i}$, $\mathbf{D_i}$.

1. $w_{i-1} = w_i - v_i,\ v_{i-1} = v_i - w_{i-1},\ i \geq 1$.

Proof. $C_i - A_i = B_i - A_i = B_{i-1} = C_{i-1}$ and $D_i - B_i = A_i = D_{i-1}$ proves the first equation and $A_i - C_{i-1} = A_i - B_{i-1} = A_{i-1},\ B_i - D_{i-1} = B_i - A_i = B_{i-1}$ the second.

2. g.c.d.$(A_i, B_i) =$ g.c.d.$(C_i, D_i) = 1,\ i \geq 1$ (we omit g.c.d. in the proof).

Proof. For $i = 1,\ A_1 = B_1 = C_1 = 1,\ D_1 = 2$, which proves the claim for $i = 1$. Assume $(A_j, B_j) = 1,\ j \geq 1$. Then $(A_{j+1}, B_{j+1}) = (A_j + B_j, A_j + 2B_j) = (A_j + B_j, B_j) = (A_j, B_j)$. The proof for $(C_i, D_i) = 1$ is analogous.

3. If $v_i = (v_{i1}, v_{i2})$, then g.c.d.$(v_{i1}, v_{i2}) = 1$ iff $2 \nmid A_i$ and $3 \nmid B_i$.

Proof. $v_i = A_i v_0 + B_i w_0 = (3A_i + 2B_i, 3A_i + 4B_i)$. Thus g.c.d.$(3A_i + 2B_i, 3A_i + 4B_i) = J$g.c.d.$(3A_i, 2B_i) = 1$ iff $2 \nmid A_i$ and $3 \nmid B_i$ since g.c.d.$(A_i, B_i) = 1$.

4. Let $i \geq 0$. We get A_i is odd iff $i \equiv 0, 1 \bmod 3$, B_i is odd iff $i \equiv 1, 2 \bmod 3$.

Proof. The statement is true for $i = 0$. Let o denote odd and e even. Then the matrix

$$\begin{Bmatrix} o & e \\ o & o \\ e & o \end{Bmatrix},$$

where the first column denotes successively the parity of A_i, the second of B_i, proves the statement by induction.

5. Let $i \geq 1$. Then $A_i \equiv 1 \bmod 3$ iff $i \equiv 0, 1 \bmod 4$, $A_i \equiv 2 \bmod 3$ iff $i \equiv 2, 3 \bmod 4$, $B_i \equiv 0 \bmod 3$ iff $i \equiv 0, 2 \bmod 4$, $B_i \equiv 1 \bmod 3$ iff $i \equiv 1 \bmod 4$, $B_i \equiv 2 \bmod 3$ iff $i \equiv 3 \bmod 4$.

Proof. The the statement is true for i = 1. Then the matrix

$$\begin{Bmatrix} 1 & 1 \\ 2 & 0 \\ 2 & 2 \\ 1 & 0 \end{Bmatrix},$$

where the columns denote successively congruence of A_i and B_1 mod 3, proves the remainder.

6. Let $i \geq 1$. $2 \nmid A_i$ and $3 \nmid B_i$ iff $i \equiv 1, 3, 7, 9 \bmod 12$.

Proof. The equivalence $2 \nmid A_i$ and $3 \nmid B_i$ iff $i \equiv 0, 1 \bmod 3$ and $i \equiv 1, 3 \bmod 4$ follows from 4. and 5.. The statement now follows by easy congruence calculations.

7. Let $i \geq 0$. $\begin{vmatrix} A_{i+1} & B_{i+1} \\ A_i & B_i \end{vmatrix} = -1$ and $\begin{vmatrix} C_{i+1} & D_{i+1} \\ C_i & D_i \end{vmatrix} = 1$.

Proof. $\begin{vmatrix} A_1 & B_1 \\ A_0 & B_0 \end{vmatrix} = \begin{vmatrix} 1 & 1 \\ 1 & 0 \end{vmatrix} = -1$. Since

$$\begin{aligned} \begin{vmatrix} A_i & B_i \\ A_{i-1} & B_{i-1} \end{vmatrix} &= \begin{vmatrix} A_i + B_i & B_i \\ A_{i-1} + B_{i-1} & B_{i-1} \end{vmatrix} = \begin{vmatrix} A_{i+1} & B_i \\ A_i & B_{i-1} \end{vmatrix} \\ &= \begin{vmatrix} A_{i+1} & A_{i+1} + B_i \\ A_i & A_i + B_{i-1} \end{vmatrix} = \begin{vmatrix} A_{i+1} & B_{i+1} \\ A_i & B_i \end{vmatrix}, \end{aligned}$$

the proof follows by induction. The proof for $\begin{vmatrix} C_{i+1} & D_{i+1} \\ C_i & D_i \end{vmatrix} = 1$ is analogous.

8. Let $i \geq 1$, $v_i = (v_{i,1}, v_{i,2})$, $v_{i-1} = (v_{i-1,1}, v_{i-1,2})$. Then $\begin{vmatrix} v_{i-1,1} & v_{i-1,2} \\ v_{i,1} & v_{i,2} \end{vmatrix} = 6$ and $v_{i,1} + v_{i,2} = 6(A_i + B_i)$.

Proof. We have

$v_{i-1} = A_{i-1}(3,3) + B_{i-1}(2,4) = (3A_{i-1} + 2B_{i-1}, 3A_{i-1} + 4B_{i-1})$,

$v_i = (A_{i-1}+B_{i-1})(3,3)+(A_{i-1}+2B_{i-1})(2,4) = (5A_{i-1}+7B_{i-1}, 7A_{i-1}+11B_{i-1})$.

Hence

$$\begin{aligned} \begin{vmatrix} v_{i-1,1} & v_{i-1,2} \\ v_{i,1} & v_{i,2} \end{vmatrix} &= \begin{vmatrix} 3A_{i-1} + 2B_{i-1} & 3A_{i-1} + 4B_{i-1} \\ 5A_{i-1} + 7B_{i-1} & 7A_{i-1} + 11B_{i-1} \end{vmatrix} \\ &= \begin{vmatrix} 3A_{i-1} + 2B_{i-1} & 2B_{i-1} \\ 5A_{i-1} + 7B_{i-1} & 2A_{i-1} + 4B_{i-1} \end{vmatrix} \\ &= \begin{vmatrix} 3A_{i-1} & 2B_{i-1} \\ 3A_{i-1} + 3B_{i-1} & 2A_{i-1} + 4B_{i-1} \end{vmatrix} = 6 \begin{vmatrix} A_{i-1} & B_{i-1} \\ A_i & B_i \end{vmatrix} = 6 \quad \text{(by 7.)} \end{aligned}$$

and $v_{i,1} + v_{i,2} = 12A_{i-1} + 18B_{i-1} = 6(2A_{i-1} + 3B_{i-1}) = 6(A_{i-1} + B_{i-1} + A_{i-1} + 2B_{i-1}) = 6(A_i + B_i)$.

Definition of n_1, n_2, n_3, n_4

Let $v_\ell = (v_{\ell,1}, v_{\ell,2})$ be such that g.c.d. $(v_{\ell,1}, v_{\ell,2}) = 1$, $\ell \geq 1$ (see 3. and 6.). Let α_3 and α_4 be positive integers such that:

(i) α_3, α_4, $v_{\ell,1}, v_{\ell,2}$ are pairwise relatively prime.

(ii) $v_{\ell,2} < \min\{\alpha_3, \alpha_4\}$.

We define n_1, n_2, n_3, n_4 as the signed 3×3 minors of the matrix

$$M = \begin{bmatrix} -v_{\ell,1} & 0 & 0 & v_{\ell,2} \\ v_{\ell-1,1} & 1 & -1 & -v_{\ell-1,2} \\ -\alpha_4 & \alpha_3 & 0 & 0 \end{bmatrix}.$$

That is

$$n_1 = -\begin{vmatrix} -v_{\ell,1} & 0 & 0 \\ v_{\ell-1,1} & 1 & -1 \\ -\alpha_4 & \alpha_3 & 0 \end{vmatrix} = v_{\ell,1}\alpha_3,$$

$$n_2 = \begin{vmatrix} -v_{\ell,1} & 0 & v_{\ell,2} \\ v_{\ell-1,1} & 1 & -v_{\ell-1,2} \\ -\alpha_4 & \alpha_3 & 0 \end{vmatrix} = -\begin{vmatrix} -v_{\ell,1} & v_{\ell,2} \\ v_{\ell-1,1} & -v_{\ell-1,2} \end{vmatrix}\alpha_3 + v_{\ell,2}\alpha_4 = 6\alpha_3 + v_{\ell,2}\alpha_4 \text{ (by 8.)},$$

$$n_3 = \begin{vmatrix} -v_{\ell,1} & 0 & v_{\ell,2} \\ v_{\ell-1,1} & -1 & -v_{\ell-1,2} \\ -\alpha_4 & 0 & 0 \end{vmatrix} = v_{\ell,2}\alpha_4,$$

$$n_4 = \begin{vmatrix} 0 & 0 & v_{\ell,2} \\ 1 & -1 & -v_{\ell-1,2} \\ \alpha_3 & 0 & 0 \end{vmatrix} = v_{\ell,2}\alpha_3.$$

Since g.c.d. $(n_1, n_3, n_4) = 1$, g.c.d. $(n_1, n_2, n_3, n_4) = 1$ and we let $\mathfrak{p} = \mathfrak{p}(n_1, n_2, n_3, n_4)$ be as defined in the introduction.

POLYNOMIAL RELATIONS IN $\mathfrak{p}$

From M we have immediately

$$\{x_4^{v_{\ell,1}} - x_1^{v_{\ell,2}} = h_0,\ x_3^{\alpha_3} - x_4^{\alpha_4},\ x_4^{v_{\ell-1,1}}x_3^{B_1} - x_2^{B_1}x_1^{v_{\ell-1,2}} = g_1,\ x_3^{\alpha_3} - x_4^{\alpha_4 - v_{\ell,1}}x_1^{v_{\ell,2}}\} \subseteq \mathfrak{p}.$$

1-type binomials in $B(\mathfrak{p})$

Of particular interest are 1-type binomials in $B(\mathfrak{p})$ with the exponent of the pure power minimal. We denote this exponent by $\text{minmult}(n_r)$, $r = 1, 2, 3, 4$.

1-type of subtype (4). Assume $y_4(v_{\ell,2}\alpha_3) = y_{42}(6\alpha_3 + v_{\ell,2}\alpha_4) + y_{43}(v_{\ell,2}\alpha_4) + y_{41}(v_{\ell,1}\alpha_3)$ (with nonnegative integer coefficients) $\Rightarrow \alpha_3 | y_{42} + y_{43}$.

(i) $y_{42} \neq 0$ or $y_{43} \neq 0 \Rightarrow y_{42} + y_{43} = m\alpha_3$, $m \geq 1 \Rightarrow y_4 v_{\ell,2}\alpha_3 \geq \alpha_3 v_{\ell,2}\alpha_4 \Rightarrow y_4 \geq \alpha_4$. Denote the minimal positive integer for such a relation by $\text{minmult}(n_4, y_{42} \neq 0$ or $y_{43} \neq 0)$. Thus $\text{minmult}(n_4, y_{42} \neq 0$ or $y_{43} \neq 0) \geq \alpha_4$.

(ii) $y_{42} = 0 \Rightarrow \alpha_3 | y_{43}$ and $v_{\ell,2} | y_{41} \Rightarrow y_4 = z_1\alpha_4 + z_2 v_{\ell,1} \Rightarrow y_4 \geq \min\{\alpha_4, v_{\ell,1}\} = v_{\ell,1}$.

From (i) and (ii) $\text{minmult}(n_4) = v_{\ell,1}$ and $\text{minmult}(n_4, y_{42} \neq 0$ or $y_{43} \neq o) = \alpha_4$.

1-type of subtype (1). Let $y_1(v_{\ell,1}\alpha_3) = y_{12}(6\alpha_3 + v_{\ell,2}\alpha_4) + y_{13}(v_{\ell,2}\alpha_4) + y_{14}(v_{\ell,2}\alpha_3)$.

(i) $y_{12} \neq 0 \Rightarrow y_{12} + y_{13} = m\alpha_3$, $m \geq 1 \Rightarrow y_1 v_{\ell,1} = 6y_{12} + (m\alpha_4 + y_{14})v_{\ell,2} \geq v_{\ell,1}v_{\ell,2} \Rightarrow y_1 \geq v_{\ell,2}$.

(ii) $y_{12} = 0 \Rightarrow v_{\ell,2} | y_1 \Rightarrow y_1 \geq v_{\ell,2}$.

Thus $\text{minmult}(n_1) = v_{\ell,2}$.

1-type of subtype (3). Let $y_3(v_{\ell,2}\alpha_4) = y_{32}(6\alpha_3 + v_{\ell,2}\alpha_4) + y_{34}(v_{\ell,2}\alpha_3) + y_{31}(v_{\ell,1}\alpha_3)$.

(i) $y_{32} \neq 0 \Rightarrow \alpha_3 | (y_3 - y_{32})$.

1. $y_3 - y_{32} > 0 \Rightarrow y_3 = m\alpha_3 + y_{32},\ m \geq 1 \Rightarrow y_3 \geq \alpha_3$.
2. $y_3 - y_{32} \leq 0 \Rightarrow y_3(v_{\ell,2}\alpha_4) < y_{32}(6\alpha_3 + v_{\ell,2}\alpha_4)$, which contradicts the initial equation.

(ii) $y_{32} = 0 \Rightarrow \alpha_3 | y_3 \Rightarrow y_3 \geq \alpha_3$.

Thus $\mathrm{minmult}(n_3) = \alpha_3$.

1-type of subtype (2). Since g.c.d.$(v_{\ell,1}, v_{\ell,2}) = 1$ we have by 3., that $6, v_{\ell,1}, v_{\ell,2}$ are pairwise relatively prime. Consider the numerical semigroup $S = \langle 6, v_{\ell,1}, v_{\ell,2} \rangle = \{z;\ z = z_1 6 + z_2 v_{\ell,1} + z_3 v_{\ell,2};\ z_i \text{ nonnegative integers},\ 1 \leq i \leq 3\}$. Since $6, v_{\ell,1}, v_{\ell,2}$ are pairwise relatively prime, by [H] S is not symmetric.

By 8., $(2A_{\ell-1} + 3B_{\ell-1})6 = (A_\ell + B_\ell)6 = v_{\ell,1} + v_{\ell,2}$, thus by [H] $A_\ell + B_\ell$ is minimal such that $(A_\ell + B_\ell)6 \in \langle v_{\ell,1}, v_{\ell,2} \rangle$. Now let $y_2(6\alpha_3 + v_{\ell,2}\alpha_4) = y_{24}(v_{\ell,2}\alpha_3) + y_{23}(v_{\ell,2}\alpha_4) + y_{21}(v_{\ell,1}\alpha_3) \Rightarrow (y_2 6 - y_{21}v_{\ell,1}J - y_{24}v_{\ell,2})\alpha_3 = (y_{23} - y_2)v_{\ell,2}\alpha_4$.

(i) $y_{23} - y_2 \geq 0 \Rightarrow y_2 6 - y_{21}v_{\ell,1} - y_{24}v_{\ell,2} = m v_{\ell,2},\ m \geq 0 \Rightarrow y_2 6 \in \langle v_{\ell,1}, v_{\ell,2} \rangle \Rightarrow y_2 \geq A_\ell + B_\ell$.

(ii) $y_2 > y_{23} \Rightarrow y_2 = m\alpha_3 + y_{23},\ m \geq 1 \Rightarrow y_2 \geq \alpha_3 > v_{\ell,2} = 3A_\ell + 4B_\ell > A_\ell + B_\ell$

.

Thus $\mathrm{minmult}(n_2) = A_\ell + B_\ell$ (with $y_{23} = A_\ell + B_\ell,\ y_{24} = y_{21} = 1$).

We will collect the preceding as follows.

Lemma 1. *With* n_1, n_2, n_3, n_4 *as defined,* $\mathrm{minmult}(n_1) = v_{\ell,2}$, $\mathrm{milmult}(n_2) = A_\ell + B_\ell$, $\mathrm{minmult}(n_3) = \alpha_3$ *and* $\mathrm{minmult}(n_4) = v_{\ell,1}$. *Furthermore* $\mathrm{minmult}(n_4,\ y_{42} \neq 0$ *or* $y_{43} \neq 0) = \alpha_4$.

Binomials of 2-type in $\mathfrak{p}$

2-type and subtype (2,4) or (3,4). The polynomials

$$h_0 = x_4^{v_{\ell,1}} - x_1^{v_{\ell,2}},\ g_1 = x_4^{v_{\ell-1,1}} x_3^{B_1} - x_2^{B_1} x_1^{v_{\ell-1,2}} = x_4^{v_{\ell-1,1}} x_3 - x_2 x_1^{v_{\ell-1,2}},$$

and the equations $w_{i-1} = w_i - v_i,\ v_{i-1} = v_i - w_{i-1}$ define successively the sets

$$\begin{aligned}
\mathcal{A}_{43} =& \{g_1, \ldots, g_r = x_4^{v_{\ell-r,1}} x_3^{B_r} - x_2^{B_r} x_1^{v_{\ell-r,2}}, \ldots, g_\ell = x_4^3 x_3^{B_\ell} - x_2^{B_\ell} x_1^3 = x_4^{v_{0,1}} x_3^{B_\ell} - x_2^{B_\ell} x_1^{v_{0,2}}\},\\
\mathcal{A}_{42} =& \{h_1 = x_4^{w_{\ell-1,1}} x_2 - x_3 x_1^{w_{\ell-1,2}} = x_4^{w_{\ell-1,1}} x_2^{A_1} - x_3^{A_1} x_1^{w_{\ell-1,2}}, \ldots,\\
& h_r = x_4^{w_{\ell-r,1}} x_2^{A_r} - x_3^{A_r} x_1^{w_{\ell-r,2}}, \ldots, h_\ell = x_4^2 x_2^{A_\ell} - x_3^{A_\ell} x_1^4 = x_4^{w_{0,1}} x_2^{A_\ell} - x_3^{A_\ell} x_1^{w_{0,2}}\}
\end{aligned}$$

in $B(\mathfrak{p})$.

Example 2. For $\ell = 3$, $\mathcal{A}_{43}$ and $\mathcal{A}_{42}$ are obtained successively as in Definition 1:

$$\begin{array}{ll}
 & h_0 = x_4^{31} - x_1^{47} \\
g_1 = x_4^{12} x_3 - x_2 x_1^{18} & h_1 = x_4^{19} x_2 - x_3 x_1^{29} \\
g_2 = x_4^5 x_3^3 - x_2^3 x_1^7 & h_2 = x_4^7 x_2^2 - x_3^2 x_1^{11} \\
g_3 = x_4^3 x_3^8 - x_2^8 x_1^3 & h_3 = x_4^2 x_2^5 - x_3^5 x_1^4.
\end{array}$$

Note that one could finish the left hand column with $x_4 x_3^{13} x_1 - x_2^{13} =: -b(\mathrm{minmult}(n_2))$.

Corollary 1. *For $g_r \in \mathcal{A}_{43}$, $h_r \in \mathcal{A}_{42}$, $1 \le r \le \ell$, neither g_r nor h_r is reducible by a 1-type binomial in $B(\mathfrak{p})$.*

Proof. This follows immediately by Lemma 1, since all exponents involved are smaller than the positive integers defined in Lemma 1.

2-type and subtype (2,3). The binomials $x_3^{\alpha_3} - x_4^{\alpha_4}$ and $h_1 = x_4^{w_{\ell-1,1}} x_2 - x_3 x_1^{w_{\ell-1,2}}$ in $B(\mathfrak{p})$ determine

$$f_1 = x_2 x_3^{\alpha_3 - 1} - x_4^{\alpha_4 - w_{\ell-1,1}} x_1^{w_{\ell-1,2}} \in B(\mathfrak{p}).$$

Note that $w_{\ell-1,1} < v_{\ell,1}$, $w_{\ell-1,2} < v_{\ell,2}$. If $\ell \ge 3$, then from f_1 and $g_1 = x_4^{v_{\ell-1,1}} x_3 - x_2 x_1^{v_{\ell-1,2}}$,

$$f_2 = x_2^2 x_3^{\alpha_3 - 2} - x_4^{\alpha_4 - (w_{\ell-1,1} - v_{\ell-1,1})} x_1^{w_{\ell-1,2} - v_{\ell-1,2}} \in B(\mathfrak{p}),$$

is obtained (again we have $w_{\ell-1,s} - v_{\ell-1,s} = w_{\ell-2,s} < v_{\ell-1,s} < v_{\ell,s}$, $s = 1, 2$).

Lemma 2. *Assume $f_t = x_2^t x_3^{\alpha_3 - t} - x_4^{\alpha_4 - z_{t4}} x_1^{z_{t1}}$, $1 \le t < \operatorname{minmult}(n_2) - 1$, $z_{t4} < v_{\ell,1}$, $z_{t1} < v_{\ell,2}$, has been defined. Then*
(i) $z_{t1} > v_{\ell-1,2}$ iff $z_{t4} > v_{\ell-1,1}$.
(ii) $z_{t1} = v_{\ell-1,2}$ or $z_{t4} = v_{\ell-1,1}$ are not possible.

Proof. f_t and g_1 imply the equation $(t+1)n_2 + (\alpha_3 - (t+1))n_3 = [\alpha_4 - (z_{t4} - v_{\ell-1,1})]n_4 + (z_{t1} - v_{\ell-1,2})n_1$. $z_{t1} > v_{\ell-1,2}$ but $z_{t4} \le v_{\ell-1,1}$ imply $t+1 \ge \operatorname{minmult}(n_2)$ (since $\alpha_4 n_4 = \alpha_3 n_3$), contrary to assumption. $z_{t1} \le v_{\ell-1,2}$ but $z_{t4} > v_{\ell-1,1}$ imply $\operatorname{minmult}(n_4, y_{42} \ne 0 \text{ or } y_{43} \ne 0) < \alpha_4$, contradicting Lemma 1. $z_{t1} = v_{\ell-1,2}$ implies by Lemma 1 $z_{t4} \le v_{\ell-1,1}$ which implies $t+1 > \operatorname{minmult}(n_2)$, a contradiction. $z_{t4} = v_{\ell-1,1}$, implies, if $z_{t1} - v_{\ell-1,2} \le 0$, then $\operatorname{minmult}(n_3) < \alpha_3$ and if $z_{t1} - v_{\ell-1,2} > 0$, then $t+1 > \operatorname{minmult}(n_2)$, in either case a contradiction.

We now define f_{t+1} and distinguish two cases.

(i) $v_{\ell,2} > z_{t1} > v_{\ell-1,2}$, $v_{\ell,1} > z_{t4} > v_{\ell-1,1}$. Let

$$f_{t+1} = x_2^{t+1} x_3^{\alpha_3 - (t+1)} - x_4^{\alpha_4 - (z_{t4} - v_{\ell-1,1})} x_1^{z_{t1} - v_{\ell-1,2}}.$$

Thus $z_{(t+1)4} = z_{t4} - v_{\ell-1,1} < v_{\ell,1}$, $z_{(t+1)1} = z_{t1} - v_{\ell-1,2} < v_{\ell,2}$.

(ii) $z_{t1} < v_{\ell-1,2}$, $z_{t4} < v_{\ell-1,1}$. Let

$$f_{t+1} = x_2^{t+1} x_3^{\alpha_3 - (t+1)} - x_4^{\alpha_4 - (z_{t4} + w_{\ell-1,1})} x_1^{z_{t1} + w_{\ell-1,2}}.$$

Thus $z_{(t+1)4} = z_{t4} + w_{\ell-1,1} < v_{\ell-1,1} + w_{\ell-1,1} = v_{\ell,1}$, $z_{(t+1)1} = z_{t1} + w_{\ell-1,2} < v_{\ell-1,2} + w_{\ell-1,2} = v_{\ell,2}$.

Let

$$\mathcal{A}_{23} = \{f_1, \ldots, f_t, \ldots, f_{\operatorname{minmult}(n_2)-1}\}.$$

Note that $\alpha_4 - z_{t4} > v_{\ell,2} - v_{\ell,1} = 2B_\ell \ge 2A_\ell$ for $\ell \ge 1$. For the sequel let

$$b(\operatorname{minmult}(n_2)) = x_2^{A_\ell + B_\ell} - x_3^{A_\ell + B_\ell} x_1 x_4$$

and $b(i,j) \in B(\mathfrak{p})$ is of 2-type and subtype (i,j).

Lemma 3. *(i) For* $f_t = x_2^t x_3^{\alpha_3 - t} - x_4^{\alpha_4 - z_{t4}} x_1^{z_{t1}} \in \mathcal{A}_{23}$, $\alpha_3 - t > 2A_\ell + 3B_\ell$.

(ii) $b(2,3) \notin \mathcal{A}_{23}$ *is reducible mod*$\{\mathcal{A}_{23},\ b(\mathrm{minmult}(n_2)),\ x_3^{\alpha_3} - x_4^{\alpha_4 - v_{\ell,1}} x_1^{v_{\ell,2}}\}$.

Proof. (i) Since $\alpha_3 > v_{\ell,2} = 3A_\ell + 4B_\ell$, $\alpha_3 - (\mathrm{minmult}(n_2) - 1) > 3A_\ell + 4B_\ell - A_\ell - B_\ell + 1 > 2A_\ell + 3B_\ell$.

(ii) W. l.o.g. assume $b(2,3) = x_2^{\gamma_2} x_3^{\gamma_3} - x_4^{\gamma_4} x_1^{\gamma_1}$, $1 \le \gamma_2 = t \le \mathrm{minmult}(n_2) - 1$, $1 \le \gamma_3 \le \alpha_3 - 1$. Suppose $\gamma_3 < \alpha_3 - t$. Then the binomial $b^* = x_4^{\gamma_4} x_1^{\gamma_1} x_3^{\alpha_3 - t - \gamma_3} - x_4^{\alpha_4 - z_{t,4}} x_1^{z_{t,1}} \in B(\mathfrak{p})$ implies a contradiction to Lemma 1. Thus $\gamma_3 \ge \alpha_3 - t$ and reducibility as required follows.

Definition 2. Assume $\mathfrak{m} \subseteq B(\mathfrak{p})$, $b \in B(\mathfrak{p})$. We say b is not divisible mod $\mathfrak{m}$ if no monomial term of any $b' \in \mathfrak{m} \setminus \{b\}$ divides a monomial term of b.

Definition 3. Let $B(\mathfrak{p}, 1) \subseteq B(\mathfrak{p})$ be the binomials of 1-type. Let $\{r, s, q, t\} = \{1, 2, 3, 4\}$. Assume $b_1(r, s) = b_1$ and $b_2(r, s) = b_2$ are of 2-type and subtype $(r, s) \in \{(2,3), (4,2), (4,3)\}$ in $B(\mathfrak{p})$. Assume also that b_1 and b_2 are not divisible mod $B(\mathfrak{p}, 1)$ and have respective monomial terms $x_r^{\gamma_{1r}} x_s^{\gamma_{1s}}$, $x_r^{\gamma_{2r}} x_s^{\gamma_{2s}}$, $\gamma_{1r} > \gamma_{2r}$, $\gamma_{1s} < \gamma_{2s}$. If there does not exist
$b(r, s) \in B(\mathfrak{p})$ (of 2-type and subtype (r, s)) with monomial term $x_r^{\gamma_r} x_s^{\gamma_s}$ such that $\gamma_{1s} < \gamma_s < \gamma_{2s}$, $\gamma_{2r} < \gamma_r < \gamma_{1r}$, then b_1 and b_2 are said to be successive (or b_2 is a successor of b_1, or b_1 is a predecessor of b_2). If b_1 is as indicated, but does not have a predecessor (resp. successor), b_1 is said to be first (resp. last).

Lemma 4. *Assume* $b_1(r, s) = b_1$ *and* $b_2(r, s) = b_2$ *are successive elements. Then the element* $b(b_1, b_2) \in B(\mathfrak{p})$ *determined by* b_1 *and* b_2 *has the following properties:*
(i) $b(b_1, b_2)$ *is of 2-type and either subtype* (r, q) *or* (r, t).
(ii) $b(b_1, b_2)$ *is not divisible mod* $B(\mathfrak{p})$.

Proof. Since neither binomial $b_1 = x_r^{\gamma_{1r}} x_s^{\gamma_{1s}} - x_q^{\gamma_{1q}} x_t^{\gamma_{1t}}$ nor $b_2 = x_r^{\gamma_{2r}} x_s^{\gamma_{2s}} - x_q^{\gamma_{2q}} x_t^{\gamma_{2t}}$ is divisible mod $B(\mathfrak{p}, 1)$ we have either
(a) $\gamma_{1q} > \gamma_{2q}$, $\gamma_{1t} < \gamma_{2t}$ or
(b) $\gamma_{1q} < \gamma_{2q}$, $\gamma_{1t} > \gamma_{2t}$.
From this for (a)

$$b(b_1, b_2) = x_r^{\gamma_{1r} - \gamma_{2r}} x_t^{\gamma_{2t} - \gamma_{1t}} - x_s^{\gamma_{2s} - \gamma_{1s}} x_q^{\gamma_{1q} - \gamma_{2q}},$$

or for (b)

$$b(b_1, b_2) = x_r^{\gamma_{1r} - \gamma_{2r}} x_q^{\gamma_{2q} - \gamma_{1q}} - x_s^{\gamma_{2s} - \gamma_{1s}} x_t^{\gamma_{1t} - \gamma_{2t}},$$

which proves (i). Since all exponents in $b(b_1, b_2)$ are less than the corresponding exponents in either b_1 or b_2 , no monomial term of $b(b_1, b_2)$ is divisible by the pure power term of some $b \in B(\mathfrak{p}, 1)$. From this $b(b_1, b_2)$ not divisible mod $B(\mathfrak{p}, 1)$ is immediate. Next suppose $b(b_1, b_2)$ is divisible mod $B(\mathfrak{p}) \setminus B(\mathfrak{p}, 1)$. This would imply existence of a binomial $b' \in B(\mathfrak{p})$ of the same type as $b(b_1, b_2)$ with smaller exponents throughout, which would imply b_1 and b_2 are not successive. (Note that b_2 is obtained from b_1 by cross multiplying with the monomial terms of $b(b_1, b_2)$.)

Remark 1. Starting with Definition 3, all may be repeated if one considers the negative monomial term in each binomial.

Lemma 5.

(i) $y_4n_4 + y_3n_3 = y_2n_2 + y_1n_1$ *and* $0 < y_3 < \alpha_3,\ 0 < y_4 < v_{\ell,1} < \alpha_4,\ 0 < y_2,\ 0 < y_1$ *imply*
$y_4 \geq 3$.

(ii) $y_4n_4 + y_2n_2 = y_3n_3 + y_1n_1$ *and* $0 < y_3 < \alpha_3,\ 0 < y_1 < v_{\ell,2} < \alpha_4,\ 0 < y_4,\ 0 < y_2$ *imply* $y_1 \geq 4$.

Proof. We consider $\langle 6, v_{\ell,1}, v_{\ell,2}\rangle$ and $\mathfrak{p}(6, v_{\ell,1}, v_{\ell,2}) = (x_1^{A_\ell+B_\ell} - x_2x_3, x_2^4 - x_1^{A_\ell}x_3^2, x_3^3 - x_1^{B_\ell}x_2^3)$ (see [H] and the previous calculations for 1-type of subtype (2)). The relation in (i) implies $y_4(v_{\ell,2}\alpha_3) = y_2(6\alpha_3) + (y_2 - y_3)(v_{\ell,2}\alpha_4) + y_1(v_{\ell,2}\alpha_3)$, thus $\{y_4 + [(y_3 - y_2)/\alpha_3]\alpha_4\}v_{\ell,2} = y_2(6) + y_1(v_{\ell,1})$. Therefore $y_4 + [(y_3 - y_2)/\alpha_3]\alpha_4 > 0$. Since $y_4 < v_{\ell,1} < \alpha_4$, $(y_3 - y_2)/\alpha_3 \geq 0$. Since $y_3 < \alpha_3$ and $(y_3 - y_2)/\alpha_3$ is an integer, this implies $y_3 - y_2 = 0$. Now the generating set for $\mathfrak{p}(6, v_{\ell,1}, v_{\ell,2})$ implies $y_4 \geq 3$.

The proof for (ii) is similar.

For the next lemma we order the binomials of 2-type as follows:

Subtype $(2,3)$ by increasing x_2-exponent, decreasing x_3-exponent of the positive term.

Subtype $(4,3)$ by increasing x_3-exponent, decreasing x_4-exponent of the positive term.

Subtype $(4,2)$ by increasing x_3-exponent and decreasing x_1-exponent of the negative term.

Lemma 6. *The elements in* $\mathcal{A}_{23} = \{f_1, ..., f_{\text{minmult}(n_2)-1}\}$, $\mathcal{A}_{42} = \{h_1, ..., h_\ell\}$, $\mathcal{A}_{43} = \{g_1, ..., g_\ell\}$ *satisfy Definition 3 with* f_1, g_1, h_1 *first and* $f_{\text{minmult}(n_2)-1}$, g_ℓ, h_ℓ *last.*

Proof. By Lemma 1 f_t, $1 \leq t \leq \text{minmult}(n_2) - 1$, h_s, g_s, $1 \leq s \leq \ell$ are not divisible by the pure power term for any $b \in B(\mathfrak{p}, 1)$. From this it follows that they are not divisible mod $B(\mathfrak{p}, 1)$. Clearly the elements in $\mathcal{A}_{23}$ are successive with f_1 first and $f_{\text{minmult}(n_2)-1}$ last. Also g_1 and h_1 are first and if the elements in $\mathcal{A}_{42}$ and $\mathcal{A}_{43}$ are successive, then g_ℓ and h_ℓ are last by Lemma 5. Assume $g_1, ..., g_r = x_4^{v_{\ell-r,1}}x_3^{B_r} - x_2^{B_r}x_1^{v_{\ell-r,2}}$ and $h_1, ..., h_r = x_4^{w_{\ell-r,1}}x_2^{A_r} - x_3^{A_r}x_1^{w_{\ell-r,2}}$, $r < \ell$, are successive. Assume that the successor h (of 2-type and subtype $(4,2)$) of h_r is different from h_{r+1}. Let $b = b(h_r, h)$. Since the x_3-exponent in h_{r+1} is $A_r + B_r = A_{r+1}$, the x_3-exponent in h is $< A_{r+1}$. By Lemma 3, (i) and (ii), and by Lemma 4 (ii), the binomial b (of 2-type) cannot be of subtype $(2,3)$. Thus b if of subtype $(4,3)$. Since $v_{\ell-1,2} > \cdots > v_{\ell-r+1,2} > w_{\ell-r,2}$ and by assumption $g_1, ..., g_r$ are successive, all binomials of 2-type and subtype $(4,3)$ not in this list have x_3-exponents $> B_r$. Note that $h = b(b, h_r)$ (see the end of the proof of Lemma 4) and $h_{r+1} = b(g_r, h_r)$. From that it follows that the x_3-exponent in h is $> A_{r+1}$, a contradiction. Thus h_{r+1} is the successor of h_r.

Similarly $b(h_{r+1}, g_r) = g_{r+1}$ is the successor of g_r. Thus inductively the claim.

Let $\mathcal{A} = \{x_4^{v_{\ell,1}} - x_1^{v_{\ell,2}},\ x_3^{\alpha_3} - x_4^{\alpha_4 - v_{\ell,1}}x_1^{v_{\ell,2}},\ b(\text{minmult}(n_2)) = x_2^{A_\ell+B_\ell} - x_3^{A_\ell+B_\ell}x_1x_4\}$. Then:

Theorem 1. $\mathfrak{m} = \mathcal{A} \cup \mathcal{A}_{23} \cup \mathcal{A}_{43} \cup \mathcal{A}_{42}$ *is a minimal generating set for* $\mathfrak{p}(n_1, n_2, n_3, n_4)$.

Proof. We show first that $\mathfrak{m}$ is a generating set, i.e. each $b \in B(\mathfrak{p})$ is reducible mod $\mathfrak{m}$. For b of 2-type and subtype $(2,3)$, this follows from Lemma 3, (ii). Assume $b \in B(\mathfrak{p}, 1)$. If b is of subtype (2), then clearly b has a reduction mod $\mathcal{A}$.

For $b = x_3^{\beta_3} - x_2^{\beta_{32}} x_1^{\beta_{31}} x_4^{\beta_{34}}$, we also have a reduction mod $\mathcal{A}$. For $b = x_1^{\beta_1} - x_2^{\beta_{12}} x_3^{\beta_{13}} x_4^{\beta_{14}}$, either a reduction mod $\mathcal{A}$ is obtained or b is reduced to be of 2-type and subtype $(2,3)$, thus a reduction by the previous. The proof for b of subtype (4) is similar. Next assume $b = x_4^{\gamma_4} x_3^{\gamma_3} - x_2^{\gamma_2} x_1^{\gamma_1}$ of 2-type and subtype $(4,3)$, $b \notin \mathcal{A}_{43}$ and b not divisible mod $\mathcal{A}_{43}$. If $\gamma_4 < 3$, then by Lemma 5, b is reducible mod $\mathcal{A}$. Assume $\gamma_3 \geq 3$. If $3 \leq \gamma_4 < v_{\ell,1}$, then b is divisible mod $\mathcal{A}_{43}$, since the elements in $\mathcal{A}_{43}$ are successive. Thus in this case by hypothesis $\gamma_4 \geq v_{\ell,1}$ and b is reducible mod $\mathcal{A}$. The case of b of 2-type and subtype $(4,2)$ is similar. Thus $\mathfrak{m}$ is a generating set. It is easily checked that every $b \in \mathfrak{m}$ has a monomial term m not divisible by any monomial term m' of $b' \in \mathfrak{m} \setminus \{b\}$. Since $\mathfrak{p}(n_1, n_2, n_3, n_4)$ is homogeneous with respect to the weighted grading $\deg(x_s) = n_s$, $s = 1, 2, 3, 4,$, this proves $\mathfrak{m}$ to be minimal.

Conclusion. Although $\mathfrak{m}$ in Theorem 1 exhibits with increasing integers an arbitrary large number of "switching between $\mathcal{A}_{42}$ and $\mathcal{A}_{13}$", the structure of $\mathfrak{m}$ is simple, since $\mathfrak{m}$ is basically determined by $\mathcal{A}$ and the first elements g_1, h_1, f_1. We have the following questions.

Question 1. Is it possible to characterize n_1, n_2, n_3, n_4 such that a minimal generating set for $\mathfrak{p}(n_1, n_2, n_3, n_4)$ is determined by at most four binomials in $B(\mathfrak{p}, 1)$ and the first elements of 2-type?

Question 2. More generally, is it possible to characterize and classify the ideals $\mathfrak{p}(n_1, n_2, n_3, n_4)$ by the interaction (as in Lemma 4) between the binomials of 2-type in a minimal generating set?

Acknowledgment: This paper was put into its final form while the authors were visiting Massey University, Palmerston North, New Zealand. Thanks are due to the Department of Mathematics, in particular for the technical support in preparation of this paper and the pleasant, cordial atmosphere in general.

References

[B] Bresinsky, H.: Binomial generating sets for monomial curves, with applications in $\mathbf{A}^4$. Rend. Sem. Mat. Univers. Politecn. Torino **46** (1988), 353-370.

[H] Herzog, J.: Generators and relations of Abelian semigroups and semigroup rings. manuscripta mathematica **3** (1970), 175-193.

Cohomology of Surfaces $X \subseteq P^r$ with Degree $\leq 2r-2$

M. Brodmann

Institute of Mathematics

University of Zurich

Winterthurerstrasse 190

8057 Zürich, Switzerland

(e-mail: brodmann@math.unizh.ch)

Abstract: *Let $X \subseteq \mathbb{P}^r_K$ be a non-degenerate surface defined over an arbitrary algebraically closed field K. Assume that X has at most finitely many non-normal points and that $\deg(X) \leq 2r-2$. We show that $h^1(X, \mathcal{O}_X(n))$ takes the same value $e^1(X)$ for all $n < 0$, that $h^1(X, \mathcal{O}_X) \geq h^1(X, \mathcal{O}_X(1))$ and that $h^1(X, \mathcal{O}_X(n+1)) \leq \max\{0, h^1(X, \mathcal{O}_X(n)) - 1\}$ for all $n > 0$. We distinguish three classes of surfaces X which may occur, one of them being the well studied class of "quasi-$K3$-surfaces".*

1. Introduction

Let K be an algebraically closed field and let $X \subseteq \mathbb{P}^r_K$ be a non-degenerate projective variety of dimension > 1. It is known that, at least if X is of dimension ≥ 6, it may occur that $h^1(X, \mathcal{O}_X(-1)) \neq 0$, even if X is smooth, (s.[L-Ra]). So, the vanishing theorems of Kodaira [K] and Mumford [Mu_2] (which both hold if $\mathrm{Char}(K) = 0$) may be hurt in a bad way if K is of positive characteristic. On the other hand it is still open, whether

$h^1(X, \mathcal{O}_X(n)) = 0$ for all $n < 0$ if X is a normal surface. Observe that the counterexamples found in [Mu$_2$] and [R] are both concerned with ample sheaves which are not very ample.

In the present paper we show that the latter question has a positive answer if the degree of X is not too large, e.g. if $d := \deg(X) \leq 2r - 2$. We attack the problem in a more general way. First of all, we assume that $d \leq r' + r'' - 1$, where $r' = h^0(X, \mathcal{O}_X(1)) - 1$ and $r'' = h^0(\mathcal{O}_Y, \mathcal{O}_Y(1)) - 1$, Y being a generic hyperplane section of X. In addition, we assume that $\#(X \backslash \mathrm{Nor}(X)) < \infty$, where $\mathrm{Nor}(X)$ denotes the normal locus of X. Moreover, we set $e^1(X) := \sum_{x \in X \text{ closed}} \mathrm{length}_{\mathcal{O}_{X,x}} \ (H^1_{\mathfrak{m}_{X,x}}(\mathcal{O}_{X,x}))$. This invariant counts in a "weighted" way the number of non Cohen-Macaulay points and is known to be the ultimate value of $h^1(X, \mathcal{O}_X(n))$ for $n << 0$, (cf.[Br$_1$]). Then we give ourselves the task to establish the equalities $h^1(X, \mathcal{O}_X(n)) = e^1(X)$ for all $n < 0$ and the estimates given in the abstract on the numbers $h^1(X, \mathcal{O}_X(n))$ for $n \geq 0$.

It turns out, that the sectional genus σ of X is always $\leq r'$ under our assumptions. We say that X is *subcritical* if $\sigma < r'$. In this case, the requested equalities and estimates may be deduced from results in [A-Br] (if $\mathrm{Char}(K) > 0$) or by means of Mumfords vanishing theorem in conjunction with methods that are used in [Br$_1$], [Br-Vo], [Br-N], [Br-Sc], [Br$_2$] and [Mat]. We are lead in a natural way to distinguish 3 classes of subcritical surfaces which may be characterized as follows

$$SC_1: \ d < r' + r'' - 1 \ ,$$
$$SC_2: \ d = r' + r'' - 1 \quad \text{and} \quad r'' \geq r' \ ,$$
$$SC_3: \ d = 2r'', \ r'' = r' - 1 \quad \text{and} \quad \sigma \leq r'' \ .$$

If on the other hand, if $\sigma = r'$, we say that X is *critical*. In this case, the results of [A-Br] do not furnish the requested equalities and estimates. But here, by the use of Cliffords theorem we see that the hyperplane section Y of X is a canonical curve. From this we conclude that X is a *quasi-K3-surface*, e.g. a normal surface with $H^1(X, \mathcal{O}_X) = 0$ and dualizing sheaf $\omega_X \cong \mathcal{O}_X$. This latter conclusion also may be found in a paper of Epema [Ep], where it is presented without a proof. So, we decided to present a proof which also will give a more detailed statement and which only relies on the fact that a canonically embedded non-hyperelliptic curve is arithmetically normal – a classical result due to M. Noether. So, in the critical case, the methods and ideas which have to be applied are completely different from those which are successful in the subcritical case. Luckily, in the case where these latter methods do not apply, we drop to a class of well-studied surfaces (cf.[Ep], [Mé], [U], [W]).

Finally, we apply our results in the special case where $d = \deg(X) \leq 2r - 2$. In this situation, the case SC_2 can not occur. Moreover, in the case SC_3, both X and Y have to be linearly normal, whereas the critical surfaces are precisely the quasi-$K3$-surfaces of minimal degree and thus are arithmetically normal and arithmetically Gorenstein. Altogether, this refines a result of Chern and Griffiths on smooth nondegenerate surfaces of degree $\leq 2r-2$ in complex r-space, a result which occurs as an exercise in [H]. We express our special thanks to N.V. Trung, for his valuable hint concerning canonical curves and the reference [W].

2. Preliminary Results

Let K be an algebraically closed field, let $r \geq 3$ and let $\mathbb{P}^r = \text{Proj}\big(K[\mathbf{x}_0, \ldots, \mathbf{x}_r]\big)$ denote the projective r-space over K. Let $X \subseteq \mathbb{P}^r$ be a non-degenerate projective surface of degree d, reduced and irreducible. Let $Y = X \cap \mathbb{P}^{r-1}$ be a generic hyperplane section of X. We know by Bertini that Y is again reduced and irreducible. Moreover, as a curve in the hyperplane $\mathbb{P}^{r-1}$, Y is again non-degenerate and of degree d. By σ, we denote the sectional genus of X, thus the arithmetic genus $p_a(Y)$ of the curve Y.

For a coherent sheaf $\mathcal{F}$ over a closed subscheme $Z \subseteq \mathbb{P}^r$ and for $i \in \mathbb{N}_0$ we write $h^i(Z, \mathcal{F})$ for the K-dimension of the i-th cohomology group $H^i(Z, \mathcal{F})$ of Z with coefficients in $\mathcal{F}$. The characteristic of $\mathcal{F}$ shall be denoted by $\chi(Z, \mathcal{F})$.

Finally, let us introduce the notation

$$\textbf{(2.1)} \qquad r' := h^0(X, \mathcal{O}_X(1)) - 1; \ r'' := h^0(Y, \mathcal{O}_Y(1)) - 1 \ .$$

In what follows, we repeatedly shall make use of the natural exact sequences

$$\textbf{(2.2)} \qquad \begin{aligned} 0 &\to H^0\big(X, \mathcal{O}_X(n)\big) \to H^0\big(X, \mathcal{O}_X(n+1)\big) \to H^0\big(Y, \mathcal{O}_Y(n+1)\big) \\ &\to H^1\big(X, \mathcal{O}_X(n)\big) \to H^1\big(X, \mathcal{O}_X(n+1)\big) \to H^1\big(Y, \mathcal{O}_Y(n+1)\big) \\ &\to H^2\big(X, \mathcal{O}_X(n)\big) \to H^2\big(X, \mathcal{O}_X(n+1)\big) \to 0; \quad (n \in \mathbb{Z}) \ . \end{aligned}$$

2.3. Remark: A) We keep in mind the obvious relations $d \geq r'' \geq r' - 1 \geq r - 1$ and that X is linearly normal if and only if $r' = r$ and that Y is linearly normal if and only $r'' = r - 1$.

B) If we apply the sequence (2.2) with $n = -1$, we see that

$$\sigma = h^1(Y, \mathcal{O}_Y) = \chi(X, \mathcal{O}_X(-1)) - \chi(X, \mathcal{O}_X) + 1 \ .$$

As $\sigma = 1 - \chi(Y, \mathcal{O}_Y)$ and $\chi(Y, \mathcal{O}_Y(1)) - \chi(Y, \mathcal{O}_Y) = d$, we also have

$$\sigma = d + h^1(Y, \mathcal{O}_Y(1)) - r'' \ .$$

Finally, keep in mind that for all $n \in \mathbb{Z}$ we have

$$\chi(X, \mathcal{O}_X(n)) = \frac{d}{2}n^2 + (\frac{d}{2} + 1 - \sigma)n + \chi(X, \mathcal{O}_X) \ . \qquad \bullet$$

For a set $\mathbb{M}$, let $\#\mathbb{M} \in \mathbb{N}_0 \cup \{\infty\}$ denote the cardinality of $\mathbb{M}$. Moreover, let $\mathrm{Nor}(X)$ denote the normal locus of X. Using this notation, we have

2.4. Lemma: *$\#(X \backslash Nor(X)) < \infty \Leftrightarrow Y$ is smooth.*

Proof: "$\Rightarrow$": If X has only finitely many non-normal points it is regular in codimension one and thus has only finitely many singularities, say $x_1, \ldots, x_m$. Therefore a generic hyperplane $\mathbb{P}^{r-1} \subseteq \mathbb{P}^r$ avoids $x_1, \ldots, x_m$. If we apply Bertini, in the form found in [J, (6.10; 11)] to the inclusion map $X \backslash \{x_1, \ldots, x_m\} \hookrightarrow \mathbb{P}^r$, we see that $Y = X \cap \mathbb{P}^{r-1}$ is smooth for a generic hyperplane $\mathbb{P}^{r-1} \subseteq \mathbb{P}^r$.

"$\Leftarrow$": If $Y = X \cap \mathbb{P}^{r-1}$ is smooth, $\mathbb{P}^{r-1}$ avoids the singular locus of X. So, this locus is finite and hence so is $X \backslash \mathrm{Nor}(X)$. ■

Next, let us introduce the invariant (s.[Br$_1$, (5.7)])

$$\text{(2.5)} \qquad e^1(X) := \sum_{x \in X \text{closed}} \mathrm{length}_{\mathcal{O}_{X,x}} \left(H^1_{\mathfrak{m}_{X,x}}(\mathcal{O}_{X,x})\right) \ ,$$

where $H^1_{\mathfrak{m}_{X,x}}$ denotes the first local cohomology functor with respect to the maximal ideal $\mathfrak{m}_{X,x} \subseteq \mathcal{O}_{X,x}$.

2.6. Remark: A) Since X is a surface, it has only finitely many non-Cohen-Macaulay points and these are precisely the closed points $x \in X$ where $H^1_{\mathfrak{m}_{X,x}}(\mathcal{O}_{X,x}) \neq 0$. But nevertheless this module is of finite length by Grothendieck's finiteness theorem (s.[Br-Sh, (9.5.2]). So, $e^1(X)$ counts in a weighted way the number of non-Cohen-Macaulay points of X.

B) Assume that $\#(X \backslash \text{Nor}(X)) < \infty$, so that X is non-singular in codimension one. In this situation, X is normal if and only if it is Cohen-Macaulay, hence if and only if $e^1(X) = 0$. •

If our surface X is normal and if $\mathcal{L}$ is an ample invertible sheaf of $\mathcal{O}_X$-modules, the cohomology groups $H^1(X, \mathcal{L}^{\otimes n})$ behave rather satisfactory for $n < 0$ (s.[Mu$_2$, Thm.2], [A, (5.9)]). The next result will allow us to make profit of this nice behaviour in a situation where we only know that $X \backslash$ Nor (X) is finite, (cf.[A-Br, (5.5)]).

2.7. Proposition: *Let* $\#(X \backslash \text{Nor}(X)) < \infty$ *and let* $\tilde{\nu} : \tilde{X} \to X$ *be a normalization of* X. *Then* $\nu^*(\mathcal{O}_X(1)) =: \mathcal{L}$ *is an ample invertible sheaf of* $\mathcal{O}_{\tilde{X}}$*-modules. Moreover*

a) $h^1(X, \mathcal{O}_X(n)) = h^1(\tilde{X}, \mathcal{L}^{\otimes n}) + e^1(X)$ *for all* $n \leq 0$

b) $h^1(X, \mathcal{O}_X(n)) \leq h^1(\tilde{X}, \mathcal{L}^{\otimes n}) + e^1(X)$ *for all* $n > 0$.

c) $h^2(X, \mathcal{O}_X(n)) = h^2(\tilde{X}, \mathcal{L}^{\otimes n})$ *for all* $n \in \mathbb{Z}$.

Proof: As $\nu : \tilde{X} \to X$ is a finite and surjective morphism and as $\mathcal{O}_X(1)$ is a (very) ample sheaf of $\mathcal{O}_X$-modules, the inverse image sheaf $\nu^*(\mathcal{O}_X(1))$ is indeed an ample invertible sheaf of $\mathcal{O}_{\tilde{X}}$-modules (s.[H, III Ex. 5.7 (d)]).

In order to prove the remaining statements, consider the natural exact sequence of coherent sheaves of $\mathcal{O}_X$-modules

$$0 \to \mathcal{O}_X \xrightarrow{\nu^\#} \nu_* \, \mathcal{O}_{\tilde{X}} \to \mathcal{F} \to 0 \, .$$

Then $\mathcal{F}$ has support $X \backslash$ Nor (X) and thus is a sheaf of finite length. Let $x \in X$ be a closed point. Then, the stalk $\mathcal{F}_x \cong (\nu_* \mathcal{O}_{\tilde{X}})_x / \nu_x^\# \mathcal{O}_{X,x}$ is an $\mathfrak{m}_{X,x}$ -torsion module, so that $\mathcal{F}_x \cong H^0_{\mathfrak{m}_{X,x}}(\mathcal{F}_x)$. Moreover, $\nu_* \mathcal{O}_{\tilde{X}}$ satisfis the Serre condition S_2 so that $H^i_{\mathfrak{m}_{X,x}}((\nu_* \mathcal{O}_{\tilde{X}})_x) = 0$ for $i = 0, 1$. If we localize the above exact sequence at x and apply local cohomology we thus get $\mathcal{F}_x \cong H^1_{\mathfrak{m}}(\mathcal{O}_{X,x})$. From this, we may conclude that length$(\mathcal{F}) = e^1(X)$.

If we twist the above exact sequence and apply cohomology, we now get an exact sequence

$$(*) \qquad \begin{array}{l} 0 \to H^0(X, \mathcal{O}_X(n)) \to H^0(X, (\nu_* \mathcal{O}_{\tilde{X}})(n)) \to K^{e^1(X)} \\ \to H^1(X, \mathcal{O}_X(n)) \to H^1(X, (\nu_* \mathcal{O}_{\tilde{X}})(n)) \to 0 \end{array}$$

and an isomorphism

$$(**) \qquad H^2(X, \mathcal{O}_X(n)) \cong H^2(X, (\nu_* \mathcal{O}_{\tilde{X}})(n))$$

for each $n \in \mathbb{Z}$.

By the projection formula (s.[H, II Ex. 5.1 (d)]) we also have

$$H^i(X, (\nu_* \mathcal{O}_{\tilde{X}})(n)) = H^i(X, \nu_* \mathcal{O}_{\tilde{X}} \otimes_{\mathcal{O}_X} \mathcal{O}_X(n)) \cong$$
$$H^i(\tilde{X}, \mathcal{O}_{\tilde{X}} \otimes_{\mathcal{O}_{\tilde{X}}} \nu^*(\mathcal{O}_X(n))) = H^i(\tilde{X}, \nu^*(\mathcal{O}_X(1)^{\otimes n})) \ .$$

As $\nu^*(\mathcal{O}_X(1)^{\otimes n}) \cong (\nu^* \mathcal{O}_X(1))^{\otimes n} = \mathcal{L}^{\otimes n}$ we thus get $H^i(X, (\nu_* \mathcal{O}_{\tilde{X}})(n)) \cong H^i(\tilde{X}, \mathcal{L}^{\otimes n})$ for all $n \in \mathbb{Z}$ and for all $i \in \mathbb{N}_0$. This allows us to rewrite $(*)$ and $(**)$ respectively in the form

$$(*') \qquad 0 \to H^0(X, \mathcal{O}_X(n)) \to H^0(\tilde{X}, \mathcal{L}^{\otimes n}) \to K^{e^1(X)}$$
$$\to H^1(X, \mathcal{O}_X(n)) \to H^1(\tilde{X}, \mathcal{L}^{\otimes n}) \to 0$$

and

$$(**') \qquad H^2(X, \mathcal{O}_X(n)) \cong H^2(\tilde{X}, \mathcal{L}^{\otimes n}) \ .$$

As $H^0(X, \mathcal{O}_X(n)) = H^0(\tilde{X}, \mathcal{L}^{\otimes n}) = 0$ for all $n < 0$ and as $H^0(X, \mathcal{O}_X(0)) = H^0(X, \mathcal{O}_X) = K = H^0(\tilde{X}, \mathcal{O}_{\tilde{X}}) = H^0(\tilde{X}, \mathcal{L}^{\otimes 0})$, statement a) is immediate by $(*')$. Statement b) is obvious by $(*')$, whereas statement c) follows from $(**')$. ■

3. Subcritical Surfaces

We keep the notations and hypotheses of the previous section. We now shall study the function $n \mapsto h^1(X, \mathcal{O}_X(n))$ if X is subject to some additional assumptions. Our first result gives a refinement of the bounds found in [Br$_1$, (5.9)] under the additional assumption that X has only finitely many non-normal points. For the definition of the occuring invariants r' and $e^1(X)$ see (2.1) resp. (2.5). Here σ is again used to denote the sectional genus of X.

3.1. Proposition: *Assume that* $\#(X \setminus \mathrm{Nor}(X)) < \infty$.

a) *If* $\mathrm{Char}(K) = 0$ *or if* $\sigma < r'$, *then* $h^1(X, \mathcal{O}_X(n)) = e^1(X)$ *for all* $n < 0$.

b) *If* $\mathrm{Char}(K) > 0$ *and if* σ *is arbitrary, then*

$$e^1(X) \le h^1(X, \mathcal{O}_X(n)) \le \max\{e^1(X), h^1(X, \mathcal{O}_X(n+1)) - r'\} \quad \text{for all} \quad n < 0 \ .$$

Proof: Statement b) follows from [A-Br, (5.2)a)], applied to the ample invertible sheaf $\mathcal{O}_X(1)$. If $\mathrm{Char}(K) > 0$ and if $\sigma < r'$, the equalities of statement a) follow by [A-Br, (5.5)a)].

So, let $\mathrm{Char}(K) = 0$ and let $\nu : \tilde{X} \to X$ be a normalization of X. Let $\mathcal{L} := \nu^*\big(\mathcal{O}_X(1)\big)$. By (2.7), $\mathcal{L}$ is an ample invertible sheaf of $\mathcal{O}_{\tilde{X}}$-modules. So $h^1(\tilde{X}, \mathcal{L}^{\otimes n}) = 0$ for all $n < 0$ by [Mu$_2$, Thm.2]. But now, (2.7) shows that $h^1\big(X, \mathcal{O}_X(n)\big) = e^1(X)$ for all $n < 0$. ■

3.2. Remark: By [Br$_1$, (5.9)], we always have

$$e^1(X) \le h^1(X, \mathcal{O}_X(n)) \le \max\Big\{e^1(X), h^1\big(X, \mathcal{O}_X(n+1)\big) - 1\Big\} \quad \text{for all } n < 0 \ ,$$

even if X has infinite non-normal local locus. Obviously, this bound is intrinsic, e.g. does not depend on the embedding of X into its ambient space $\mathbb{P}^r$. If $r \le 3$, then X is arithmetically Cohen-Macaulay, so that $h^1(X, \mathcal{O}_X(n)) = 0$ for all $n \in \mathbb{Z}$ in this case. Therefore (3.1)b) gives

$$e^1(X) \le h^1\big(X, \mathcal{O}_X(n)\big) \le \max\{e^1(X),\ h^1\big(X, \mathcal{O}_X(n+1)\big) - 4\} \quad \text{for all } n < 0 \ ,$$

whenever $\#(X \backslash \mathrm{Nor}(X)) < \infty$. •

By (3.1)a), the values of $h^1(X, \mathcal{O}_X(n))$ behave nicely for $n < 0$ whenever $\sigma < r'$. This will be an important ingredient of our investigations. For the sake of completeness, we also shall give some bounds on the values of $h^1\big(X, \mathcal{O}_X(n)\big)$ for $n > 0$. We begin with two auxiliary results.

3.3. Lemma: *For all $n \in \mathbb{Z}$ we have*

a) $h^2\big(X, \mathcal{O}_X(n+1)\big) \le \max\{0, h^2(X, \mathcal{O}_X(n)) - r'\}$,

b) $h^1\big(Y, \mathcal{O}_Y(n+1)\big) \le \max\{0, h^1(Y, \mathcal{O}_Y(n)) - r''\}$.

Proof: Let $Z \subseteq \mathbb{P}^r$ be a closed and integral subscheme with $h^0(Z, \mathcal{O}_Z(1)) - 1 = s$ and $\dim(Z) = t$. Then, for each $\gamma \in H^0(Z, \mathcal{O}_Z(1)))\backslash\{0\}$ and each $n \in \mathbb{Z}$ we have an epimorphism

$$H^t(Z, \mathcal{O}_Z(n)) \xrightarrow{\gamma} H^t(Z, \mathcal{O}_Z(n+1)) \to 0 \ .$$

Thus [Br$_1$, (3.2)] gives $h^t(Z, \mathcal{O}_Z(n+1)) \le \max\{0, h^t(Z, \mathcal{O}_Z(n)) - s\}$ and this proves our claim. ■

3.4. Lemma: *Let $s \in \mathbb{Z}$ such that $h^1(Y, \mathcal{O}_Y(s)) = 0$. Then*

a) $h^1(X, \mathcal{O}_X(s)) \leq h^1(X, \mathcal{O}_X(s-1))$

b) $h^1(X, \mathcal{O}_X(n)) \leq \max\{0, h^1(X, \mathcal{O}_X(n-1)) - 1\}$ *for all* $n > s$.

c) $h^2(X, \mathcal{O}_X(n)) = 0$ *for all* $n \geq s - 1$.

Proof: By our hypothesis, the coherent sheaf of $\mathcal{O}_X$-modules $\mathcal{O}_Y$ is $(s+1)$-regular in the sense of Castelnuovo and Mumford (cf.[Mu_1]) and so, statements a) and b) are a consequence of a result sometimes called the Lemma of Mumford - Le Potier (cf.[Br-N, (4.6)]). Statement c) follows immediate on application of the sequences (2.2) if we keep in mind (3.3). ■

Now, we obtain (cf.[Br_2, (3.2)]):

3.5. Proposition: *Assume that $\sigma < r'$. Then*

a) $h^1(X, \mathcal{O}_X) - h^1(X, \mathcal{O}_X(1)) = r'' - r' + 1 (\geq 0)$.

b) $h^1(X, \mathcal{O}_X(n)) \leq \max\{0, h^1(X, \mathcal{O}_X(n-1)) - 1\}$ *for all* $n > 1$.

c) $h^2(X, \mathcal{O}_X(n)) = 0$ *for all* $n \geq 0$.

Proof: As $h^1(Y, \mathcal{O}_Y) = \sigma < r'$, (2.3)A) gives $h^1(Y, \mathcal{O}_Y) \leq r''$. So, (3.3)b) shows that $h^1(Y, \mathcal{O}_Y(n)) = 0$ for all $n > 0$. If we apply (3.4)b) and c) with $s = 1$, we thus get statements b) and c). If we apply the sequence (2.2) with $n = 0$, we obtain statement a). ■

Now, we are in the position to study the surfaces X which we call subcritical. As a first step in this direction we prove.

3.6. Lemma: a) $d < 2r'' \Rightarrow h^1(Y, \mathcal{O}_Y(1)) = 0$.

b) *If $d \leq r' + r'' - 1$ then $h^1(Y, \mathcal{O}_Y(1))$ vanishes if and only if $\sigma < r'$.*

c) *If $d \leq r' + r'' - 1$ and $\sigma \geq r'$ then $r'' = r' - 1$ and $d = 2r''$.*

Proof: a): By (2.3)B) we have $\sigma - h^1(Y, \mathcal{O}_Y(1)) + r'' = d < 2r''$ thus $h^1(Y, \mathcal{O}_Y(0)) - h^1(Y, \mathcal{O}_Y(1)) < r''$. So, (3.3)b) gives $h^1(Y, \mathcal{O}_Y(1)) = 0$.

b): Let $d \leq r' + r'' - 1$. Assume first that $h^1(Y, \mathcal{O}_Y(1)) = 0$. Then (2.3)B) gives $\sigma = d - r'' \leq r' - 1 < r'$. If conversely $\sigma < r'$ we have $h^1(Y, \mathcal{O}_Y) = \sigma \leq r''$ (s.(2.3)A)) and so (3.3)b) gives $h^1(Y, \mathcal{O}_Y(1)) = 0$.

c): Let $d \leq r' + r'' - 1$ and $\sigma \geq r'$. By statement b) we obtain $h^1(Y, \mathcal{O}_Y(1)) \neq 0$. By statement a) we thus get $d \geq 2r''$ hence $2r'' \leq d \leq r' + r'' - 1$. Now (2.3)A) gives our claim. ■

3.7. *Definition:* We call the surface X *subcritical* if $d \leq r' + r'' - 1$ and if $\sigma < r'$. By (3.6)b) the second of these two conditions may be replaced by the requirement that $h^1(Y, \mathcal{O}_Y(1)) = 0$. •

3.8. *Proposition:* *Let X be subcritical. Then*

a) *Statements a), b) and c) of (3.5) hold.*

b) *If $\#(X \setminus Nor(X)) < \infty$ we have in addition $h^1(X, \mathcal{O}_X(n)) = e^1(X)$ for all $n < 0$.*

Proof: By (3.7) we have $\sigma < r'$. So, we may conclude by (3.5) and by (3.1)a). ■

3.9. *Remark:* A) If $d < r' + r'' - 1$, (2.3)A) shows that $d < 2r''$. So, (3.6)a) and b) show that X is subcritical. If $d = r' + r'' - 1$ and $r'' \geq r'$, we have again $d < 2r''$ and hence get by (3.6)a) and b) that X is subcritical. In view of (2.3)A) this means that X is subcritical precisely in the following three cases

$SC_1 : d < r' + r'' - 1$.

$SC_2 : d = r' + r'' - 1$ and $r'' \geq r'$.

$SC_3 : d = 2r''$, $r'' = r' - 1$ and $\sigma \leq r''$.

B) Let X be subcritical. Then (3.6)b) gives that $h^1\big(Y, \mathcal{O}_Y(1)\big) = 0$ and so by (2.3)B) we get $\sigma = d - r''$. In particular we see by (2.3)A) that $\sigma \leq r''$ with equality sign exactly in the case SC_3. Moreover, by (2.3)B) and (3.5)C) we obtain for all $n \in \mathbb{Z}$.

$$\chi\big(X, \mathcal{O}_X(n)\big) = \frac{d}{2}n^2 + \left(r'' - \frac{d}{2} + 1\right) n - h^1(X, \mathcal{O}_X) + 1 \ .$$

C) Assume that $d \leq 2r - 2$. Then (2.3)A) gives $d \leq r' + r'' - 1$. If the first inequality is strict, then so is the second. So, we are in the case SC_1 if $d < 2r - 2$.

Observe that the case SC_2 cannot occur if $d \leq 2r - 2$.

Assume in addition that X and Y are not both linearly normal so that $r < r'$ or $r - 1 < r'' - 1$. Then again $d < r' + r'' - 1$ and we are in the case SC_1. •

4. Critical Surfaces

The aim of this section is to study the so called critical surfaces. As already said in the introduction, we shall see that these belong to a class of well understood surfaces. We keep the previous hypotheses and notations.

4.1. Definition: We call the surface X *critical* if $d \leq r' + r'' - 1$ and $\sigma \geq r'$. •

4.2. Remark: If X is critical, we see by (3.9)A) that $d = 2r''$ and $r'' = r' - 1$ and hence that $\sigma = r'' + h^1(Y, \mathcal{O}_Y(1))$ (cf.(2.3)B)), with $h^1(Y, \mathcal{O}_Y(1)) \neq 0$ (see also (3.6)b)). •

The previous remark makes it obvious to ask what happens in the case $d = 2r''$ and $h^1(Y, \mathcal{O}_Y(1)) \neq 0$. Our next result is devoted to this question.

4.3. Proposition: *Assume that* $\#(X \backslash Nor(X)) < \infty$, *that* $d = 2r''$ *and that* $h^1(Y, \mathcal{O}_Y(1)) \neq 0$. *Then* Y *is a non-hyperelliptic curve of genus* $\sigma = r'' + 1$ *and with canonical sheaf* $\omega_Y \cong \mathcal{O}_Y(1)$. *In particular we have* $h^1(Y, \mathcal{O}_Y(1)) = 1$.

Proof: By (2.4) we know that Y is a smooth curve of genus σ. Using Serre duality we get

$$h^0(Y, \omega_Y \otimes_{\mathcal{O}_Y} \mathcal{O}_Y(1)^\vee) = h^1(Y, \mathcal{O}_Y(1)) \neq 0 ,$$

so that the very ample line bundle $\mathcal{O}_Y(1)$ is special (cf.[H, IV 1.3.4]). Moreover, $\mathcal{O}_Y(1)$ is of degree $d = 2r'' = 2h^0(Y, \mathcal{O}_Y(1))$. Therefore, as $\mathcal{O}_Y(1) \not\cong \mathcal{O}_Y$, Cliffords theorem (s.[H, IV 5.4]) leaves us with the following two possibilities:

(α) $\mathcal{O}_Y(1) \cong \omega_Y$.

(β) Y is hyperelliptic and $\mathcal{O}_Y(1) \cong \mathcal{L}^{\otimes t}$ for some $t \in \mathbb{Z}$ where $\mathcal{L}$ is the unique line bundle on Y which is generated by its global sections and defines a finite morphism $Y \to \mathbb{P}^1_K$ of degree 2 (s.[H, IV 5.3]).

In the case (α) we have $\sigma = h^0(Y, \omega_Y) = r'' + 1 > 2$. But then [H, IV 5.2] shows that Y is not hyperelliptic. Moreover by duality $h^1(Y, \mathcal{O}_Y(1)) = h^1(Y, \omega_Y) = h^0(Y, \mathcal{O}_Y) = 1$. This proves our claim in the case (α).

So, let us assume that we are in the case (β). As $\mathcal{L}$ is of degree 2 we see that $\mathcal{L}^{\otimes t}$ is of degree $2t$ so that $t = r''$. According to [H, IV, 5.3] we may write $\omega_Y \cong \mathcal{L}^{\otimes(\sigma-1)}$. Let $s := h^1(Y, \mathcal{O}_Y(1)) - 1$. Then (2.3)B) gives $\omega_Y \cong \mathcal{L}^{\otimes(r''+s)} = \mathcal{L}^{\otimes r''} \otimes_{\mathcal{O}_Y} \mathcal{L}^{\otimes s} \cong \mathcal{O}_Y(1) \otimes_{\mathcal{O}_Y} \mathcal{L}^{\otimes s}$.

As $s \geq 0$ and as $\mathcal{L}$ is generated by its global sections, $\mathcal{L}^{\otimes s}$ is generated by its global sections. Therefore, ω_Y is very ample (s.[H, II Ex. 7.5 (d)], as $\sigma = r'' + s \geq 2$, [H, IV 5.2] gives us that Y is not hyperelliptic. This contradiction shows us that the case (β) does not occur at all. ■

4.4. *Remark:* Assume that X is critical and that $\#(X \backslash \mathrm{Nor}(X)) < \infty$. Then, the hypotheses of (4.3) are satisfied. Therefore Y is a non-hyperelliptic curve with canonical sheaf $\omega_Y \cong \mathcal{O}_Y(1)$ and $\sigma = r'' + 1 = r'$. Moreover $h^1(Y, \mathcal{O}_Y(1)) = 1$. •

The previous observation $\omega_Y \cong \mathcal{O}_Y(1)$ brings us close to the situation of a surface $X \subseteq \mathbb{P}^r$ with a canonically embedded hyperplane section Y. This situation has been considered from different point of views by other authors (s.[Ep], [W]). So, after embedding X to $\mathbb{P}^{r'}$ by means of the very ample sheaf $\mathcal{O}_X(1)$, we could use results of Epema [Ep] to conclude directly that X is a normal Gorenstein surface with trivial canonical divisor – at least if K is of characteristic $\neq 2, 3$. But we prefer to prove these two properties of X by elementary cohomological arguments from a classical result of Max Noether, which says that a canonically embedded non-hyperelliptic curve is arithmetically normal (s.[No]), for a modern treatment in the case $k = \mathbb{C}$ see [Ar-C-G-Ha, pg 117], for the general case see the last equality in [So, Thm.2.1])

4.5. *Corollary:* *Assume that X is critical with $\#(X \backslash \mathrm{Nor}(X)) < \infty$ and let $i : X \hookrightarrow \mathbb{P}^{r'}$ be the non-degenerate closed immersion induced by the very ample sheaf $\mathcal{O}_X(1)$. Let $Y' = \mathbb{P}^{r'-1} \cap i(X)$ be a generic hyperplane section of $i(X) \subseteq \mathbb{P}^{r'}$. Then*

a) *$Y' \subseteq \mathbb{P}^{r'-1}$ is a smooth, non-degenerate and non-hyperelliptic canonical curve of genus r'.*

b) *The homogeneous coordinate ring $B = K \oplus B_1 \oplus B_2 \oplus \cdots$ of $Y' \subseteq \mathbb{P}^{r'}$ is Gorenstein with $*$canonical module $B(1)$.*

Proof: a): Let $A = K \oplus A_1 \oplus A_2 \oplus \cdots$ be the homogeneous coordinate ring of X in $\mathbb{P}^r$. Then, A is a graded subring of the graded domain $\Gamma := \oplus_{n \in \mathbb{N}_0} \Gamma(X, \mathcal{O}_X(n))$ and the homogeneous coordinate ring R of $i(X) \subseteq \mathbb{P}^{r'}$ may be identified with the graded subdomain $K[\Gamma(X, \mathcal{O}_X(1))]$ of the ring Γ. As A and Γ coincide in all large degrees, so do A and R. So, the inclusion homomorphism $A \to R$ induces an isomorphism $\alpha : i(X) \xrightarrow{\cong} X$. Now, let $n \in \mathbb{Z}$. Then, we have $\mathcal{O}_{i(X)}(n) = R(n)^{\sim}$ and hence $\alpha_* \mathcal{O}_{i(X)}(n) = \alpha_*(R(n)^{\sim}) \cong$

$(R(n)\restriction_A)^\sim$, where $\cdot\restriction_A$ denotes scalar restriction to A. As the graded A-modules $R(n)\restriction_A$ and $A(n)$ coincide in large degrees, we thus get $\alpha_*\mathcal{O}_{i(X)}(n) \cong A(n)^\sim = \mathcal{O}_X(n)$. But this shows, that $h^0(i(X), \mathcal{O}_{i(X)}(n)) = h^0(X, \mathcal{O}_X(n))$ for all $n \in \mathbb{Z}$. In particular we have $h^0(i(X), \mathcal{O}_{i(X)}(1)) - 1 = r'$. Moreover $X \subseteq \mathbb{P}^r$ and $i(X) \subseteq \mathbb{P}^{r'}$ have the same Hilbert polynomal. This means in particular, that $i(X)$ is of degree d and of sectional genus σ (s.(2.3)B)). Let $r''' := h^0(Y', \mathcal{O}_{Y'}(1)) - 1$, so that $r''' \geq r' - 1$. As X is critical, we have $\sigma \geq r'$ and $d = 2r'' - 2$ (s.(4.2)) so that $d \leq r' + r''' - 1$. So $i(X)$ is critical by (4.1). But now, (4.4) gives $\omega_{Y'} \cong \mathcal{O}_{Y'}(1)$ and this proves (claim a).

b): By Max Noethers theorem on canonically embedded curves, we have

$$B = \oplus_{n\in\mathbb{Z}}\Gamma(Y', \mathcal{O}_{Y'}(n))$$

, so that B is CM. As $Y' \subseteq \mathbb{P}^{r'-1}$ is a canonical curve, its canonical sheaf $\omega_{Y'}$ is isomorphic to $\mathcal{O}_{Y'}(1)$. Let Ω be a $*$canonical module of B. Then, as Ω is a CM-module, we have

$$\Omega = \oplus_{n\in\mathbb{Z}}\Gamma(Y', \tilde{\Omega}(n)) = \oplus_{n\in\mathbb{Z}}\Gamma(Y', \omega_{Y'}(n)) =$$
$$\oplus_{n\in\mathbb{Z}}\Gamma(Y', \mathcal{O}_{Y'}(1)(n)) = \oplus_{n\in\mathbb{Z}}\Gamma(Y', \mathcal{O}_{Y'}(n))(1) = B(1)\ .$$

This proves statement b), [Bru-He, 3.6.1]). ■

4.6. Proposition: *Assume that X is critical with $\#(X\setminus \mathrm{Nor}(X)) < \infty$ and let $i : X \hookrightarrow \mathbb{P}^{r'}$ be the non-degenerate closed immersion induced by the very ample sheaf $\mathcal{O}_X(1)$. Then, the homogeneous coordinate ring $R = K \oplus R_1 \oplus \cdots$ of $i(X) \subseteq \mathbb{P}^{r'}$ is a normal Gorenstein ring with $*$-canonical module R.*

Proof: Let $f \in R_1\setminus\{0\}$ be a generic linear form. For $i \in \mathbb{N}_0$, let $H^i_{R_+}$ denote the i-th local cohomology functor with respect to the irrelevant ideal $R_+ = R_1 \oplus R_2 \oplus \cdots$ of R, considered as a functor from the category of graded R-modules to itself (s.[Br-Sh, (12.3.3)]. Then $B := (R/fR)/H^0_{R_+}(R/fR)$ is the homogeneous coordinate ring of a generic hyperplane section $Y' = \mathbb{P}^{r'-1} \cap i(X)$ of $i(X) \subseteq \mathbb{P}^{r'}$. By (4.5)b), B is a Gorenstein ring of dimension 2, thus in particular a CM-ring. Therefore $H^1_{R_+}(B) = 0$ and hence $H^1_{R_+}(R/fR) = 0$. If we apply local cohomology to the exact sequence $0 \to R(-1) \xrightarrow{f} R \to R/fR \to 0$ we see that the multiplication map $f : H^1_{R_+}(R)(-1) \to H^1_{R_+}(R)$ is surjective. As $H^1_{R_+}(R)$ vanishes in all negative degrees (s.[Br-Sh, (14.1.6) (ii)]) we thus get $H^1_{R_+}(R) = 0$. As R is a domain and $R_+ \neq 0$, we have $H^0_{R_+}(R) = 0$. Applying once more cohomology to the above sequence we see now, that $H^0_{R+}(R/fR) = 0$ so that $R/fR \cong B$ is a Gorenstein ring. As $f \in R_+$ is a non-zero divisor, R becomes a Gorenstein ring. Therefore, R has a $*$canonical module of the form $\Omega = R(n)$, with some $n \in \mathbb{Z}$ (s.[Bru-He, 3.6.11]). But then $\big(R(n)/fR(n)\big)(1) \cong \Omega(1)/f\Omega(1)$ is $*$canonical module of $R/fR \cong B$ (s.[Bru-He, 3.6.14]) so that $\big(R(n)/fR(n)\big)(1) \cong (R/fR)(1)$, hence $n = 0$, thus $\Omega = R$.

Finally $i(X) \cong X$ has only finitely many non-normal points. By what we just have shown, X moreover is Gorenstein, hence in particular CM. So X is normal (s.(2.6)B)). This means that $i(X)$ is normal, so that Spec(R) is regular in codimension one. As R is CM, it follows that R is normal. ∎

Now, we are ready to prove the main result of this section, which should be compared with a similar result of Epema (s.[Ep, Prop.2]).

4.7. Theorem: *Let X be critical and such that $\#\big(X \backslash Nor(X)\big) < \infty$. Let $i : X \hookrightarrow \mathbb{P}^{r'}$ be the closed immersion induced by the very ample sheaf $\mathcal{O}_X(1)$. Then:*

a) *The surface $i(X) \subseteq \mathbb{P}^{r'}$ is arithmetically normal and arithmetically Gorenstein.*

b) *X is a normal Gorenstein surface whose dualizing sheaf $\dot{\omega}_X$ is isomorphic to $\mathcal{O}_X$.*

c) *$h^1\big(X, \mathcal{O}_X(n)\big) = 0$ for all $n \in \mathbb{Z}$.*

d) *$h^2(X, \mathcal{O}_X) = 1$.*

Proof: a): Follows from (4.6).

b): By (4.6) we have $\dot{\omega}_{i(X)} \cong \mathcal{O}_{i(X)}$. So we may conclude by the natural isomorphism $i(X) \cong X$.

c): Using duality we get $h^2(X, \mathcal{O}_X) = h^0(X, \dot{\omega}_X) = h^0(X, \mathcal{O}_X) = 1$.

d): By (4.2) we have $h^0(X, \mathcal{O}_X(1)) = r' + 1 = (r'' + 1) + 1 = h^0\big(Y, \mathcal{O}_Y(1)\big) + h^0(X, \mathcal{O}_X)$. By (4.3) we have $h^1\big(Y, \mathcal{O}_Y(1)\big) = 1$. So, if we apply the sequence (2.2) with $n = 0$, we see that $h^1\big(X, \mathcal{O}_X(1)\big) \geq h^1(X, \mathcal{O}_X)$. Using duality we thus have $h^1\big(X, \mathcal{O}_X(-1)\big) = h^1\big(X, \dot{\omega}_X(1)\big) = h^1\big(X, \mathcal{O}_X(1)\big) \geq h^1(X, \mathcal{O}_X)$. By (2.6)B), we have $e^1(X) = 0$. So (3.1), applied with $n = -1$, gives $h^1(X, \mathcal{O}_X) = 0$. A repeated use of (3.1) now gives us $h^1\big(X, \mathcal{O}_X(n)\big) = 0$ for all $n \leq 0$. Finally, by duality we obtain the equalities $h^1\big(X, \mathcal{O}_X(n)\big) = h^1\big(X, \dot{\omega}_X(-n)\big) = h^1\big(X, \mathcal{O}_X(-n)\big) = 0$ for all $n \geq 0$. Altogether, this proves our claim. ∎

4.8. Remark: A) According to [Ep, Cor 5] more could be said on the nature of critical surfaces – at least for $\mathrm{Char}(K) \neq 2, 3$: the singularities of X are all rational double points.

B) An obvious consequence of (3.8) and of (4.7) is that for any non-degenerate surface $X \subseteq \mathbb{P}^r$ of degree $d \leq r' + r'' - 1$ and with $\#\big(X\backslash \mathrm{Nor}(X)\big) < \infty$ we have $h^1\big(X, \mathcal{O}_X(n)\big) = e^1(X)$ for all $n < 0$, and moreover statement a) and b) of (3.5) hold. •

5. THE CASE $d \leq 2r - 2$

In this last section we summarize what we have shown in the previous sections and apply it to non-degenerate surfaces $X \subseteq \mathbb{P}^r$ of degree $d \leq 2r - 2$. We first give a definition which is appropriate to what we have shown in section 4.

5.1. Definition: A projective surface X is called a *quasi-$K3$-surface* if it is normal with $h^1(X, \mathcal{O}_X) = 0$ and if its dualizing sheaf ω_X is isomorphic to $\mathcal{O}_X$. •

5.2. Remark: A) As a quasi-$K3$-surface satisfies $\omega_X \cong \mathcal{O}_X$, it is Gorenstein. Moreover a quasi-$K3$-surface is a $K3$-surface if and only if it is smooth (cf.[H, V, Thm. 6.3 (1)]).

B) Let $X \subseteq \mathbb{P}^r$ be a non-degenerate quasi-$K3$-surface of degree d. Then, by (2.6)B) we have $e^1(X) = 0$ and so (3.1) shows that $h^1\big(X, \mathcal{O}_X(n)\big) = 0$ for all $n \leq 0$. By duality it follows $h^1\big(X, \mathcal{O}_X(n)\big) = 0$ for all $n \in \mathbb{Z}$ and $h^2(X, \mathcal{O}_X) = 1$. Moreover the sequence (2.2), applied with $n = 0$, gives $r'' = r' - 1$ and $h^1(Y, \mathcal{O}_Y(1)) = 1$. So, in view of (2.3)B) we have $\sigma = d + 2 - r'$ and $\chi(X, \mathcal{O}_X(n)) = \frac{d}{2}n^2 + \big(r'' - \frac{d}{2}\big)n + 2$ for all $n \in \mathbb{Z}$. In particular, $r'' + 2 = \chi\big(X, \mathcal{O}_X(1)\big) = h^0\big(X, \mathcal{O}_X(1)\big) = h^2\big(X, \omega_X(-1)\big) = h^2\big(X, \mathcal{O}_X(-1)\big) = \chi\big(X, \mathcal{O}_X(-1)\big) = d - r'' + 2$, so that $d = 2r'' = 2r' - 2$ and $\sigma = r'$. In particular, X is critical, (s.(4.1)). •

5.3. Corollary: *Let $X \subseteq \mathbb{P}^r$ be a non-degenerate surface with $\#\big(X\backslash \mathrm{Nor}(X)\big) < \infty$. Then, the following statements are equivalent:*

(i) *X is critical.*

(ii) *X is a quasi-$K3$-surface.*

(iii) $d \leq r' + r'' - 1$ *and* $h^2(X, \mathcal{O}_X) \neq 0$.

Proof: "(i) $\Rightarrow$ (ii)": Clear by (4.7)b,c).

"(ii) $\Rightarrow$ (iii)": This is verified in (5.2)B).

"(iii) $\Rightarrow$ (i)": As $d \leq r' + r'' - 1$, X is either subcritical or critical. As $h^2(X, \mathcal{O}_X) \neq 0$, the first possibility is excluded by (3.5)c). ■

5.4. Corollary: *Let $X \subseteq \mathbb{P}^r$ be a non-degenerate surface of degree $d \leq r' + r'' - 1$ with $\#(X \backslash Nor(X)) < \infty$. Then, the following statements are equivalent:*

(i) *X is a quasi-$K3$-surface.*

(ii) $h^2(X, \mathcal{O}_X) \neq 0$.

(iii) $\chi(X, \mathcal{O}_X(1)) = \chi(X, \mathcal{O}_X(-1))$.

Proof: "(i) $\Rightarrow$ (ii)" is clear from (5.3).

"(i) $\Rightarrow$ (iii)" is obvious by duality.

"(iii) $\Rightarrow$ (i)": Assume that X is not a quasi-$K3$-surface. Then by (5.3) the surface X is not critical and hence subcritical. By (3.9)A) this means that $d < 2r'' + 2$, hence that $r'' - \frac{d}{2} + 1 \neq 0$. By (3.9)B) we get $\chi\big(X, \mathcal{O}_X(-1)\big) \neq \chi\big(X, \mathcal{O}_X(1)\big)$. ■

5.5. Remark: Let $X \subseteq \mathbb{P}^r$ be a non-degenerate surface of degree $d \leq r' + r'' - 1$ and such that $\#(X \backslash \mathrm{Nor}(X)) < \infty$. We introduce the invariant $\triangle(X) := \chi(X, \mathcal{O}_X(1)) - \chi(X, \mathcal{O}_X(-1)) = d + 2 - 2\sigma$, (s.(2.3)B). Observe that in the cases SC_1 and SC_2 of (3.9)A) we have $d < 2r''$, whereas in the case SC_3 we have $d = 2r''$. Thus, by (3.9)B) and (5.4) we can say:

$\triangle(X) = 0 \Leftrightarrow X$ is a quasi-$K3$-surface.

$\triangle(X) = 2 \Leftrightarrow X$ belongs to the class SC_3.

$\triangle(X) > 2 \Leftrightarrow X$ belongs to the class SC_1 or SC_2. •

The next lemma shall pave the way to the consideration of the case where $d \leq 2r - 2$.

5.6. Lemma: *Let $X \subseteq \mathbb{P}^r$ be a non-degenerate quasi-$K3$-surface. Then X is of degree $d \geq 2r - 2$. Moreover, the following statements are equivalent:*

(i) $d = 2r - 2$.

(ii) *X is arithmetically Gorenstein.*

(iii) *X is arithmetically S_2.*

(iv) *X is arithmetically normal.*

(v) *X is linearly normal.*

(vi) *The generic hyperplane section $Y = \mathbb{P}^{r-1} \cap X$ is linearly normal.*

Proof: "(i) $\Rightarrow$ (ii)": Let $d = 2r - 2$. Then, by (2.3)A) and (5.3), X is critical. Moreover by (5.2)B), we have $d = 2r' - 2$ and hence $r = r'$. Now (4.7)a) shows that $X \subseteq \mathbb{P}^r$ is arithmetically Gorenstein.

"(ii) $\Rightarrow$ (iii)" is obvious, whereas "(iii) $\Rightarrow$ (iv)" follows easily as X is normal. Finally "(iv) $\Rightarrow$ (v)" is a general fact, too.

"(v) $\Rightarrow$ (vi)": Let $X \subseteq \mathbb{P}^r$ be linearly normal, so that $r' = r$. As X is quasi-$K3$, we have $h^1(X, \mathcal{O}_X(n)) = 0$ for all $n \in \mathbb{Z}$, (s.(5.2)B)). If we apply the sequence (2.2) with $n = 0$ it thus follows $r'' = r' - 1 = r - 1$ and therefore Y is linearly normal.

"(vi) $\Rightarrow$ (i)": Assume that Y is linearly normal, so that $r'' = r - 1$. By (5.2)B) we have $d = 2r'' = 2r - 2$. ∎

5.7. *Remark and Definition:* It is well known, that for each $r \geq 3$ there is non-degenerate $K3$-surface $X \subseteq \mathbb{P}^r$ of degree $d = 2r - 2$, (s.[B, (VIII. 5), + Appendice]). By (5.4), each non-degenerate quasi-$K3$-surface $X \subseteq \mathbb{P}^r$ has degree $d \geq 2r - 2$. So it makes sense to call such a surface $X \subseteq \mathbb{P}^r$ *a non-degenerate quasi-$K3$-surface of minimal degree,* if it satisfies the equivalent conditions (5.4) (i) - (vi). •

We now consider non-degenerate surfaces $X \subseteq \mathbb{P}^r$ of degree $d \leq 2r - 2$ which have only finitely many non-normal points. What we get is:

5.8. Theorem: *Let $X \subseteq \mathbb{P}^r$ be a non-degenerate surface of degree $d \leq 2r - 2$ and with $\#(X \setminus Nor(X)) < \infty$. Then*

a) *If $h^2(X, \mathcal{O}_X) = 0$, X must belong to one of the two classes SC_1 or SC_3 of (3.9).*

b) *If $h^2(X, \mathcal{O}_X) \neq 0$, then X is a quasi-$K3$-surface of minimal degree.*

Proof: "a)": If $h^2(X, \mathcal{O}_X) = 0$, (5.3) shows that X is not critical and thus must be subcritical. As $d \leq 2r - 2$, the case SC_2 may not occur (s.(3.9)C)).

"b)": By (5.4) we see that X is a quasi-$K3$-surface. As $d \leq 2r - 2$, we see from (5.6) that X is of minimal degree. ■

5.9. Corollary: *Let $X \subseteq \mathbb{P}^r$ be as in (5.8). Then, in the notation of (5.5) we have:*

$\triangle(X) = 0 \Leftrightarrow X$ *is a quasi-$K3$-surface of minimal degree.*

$\triangle(X) = 2 \Leftrightarrow X$ *belongs to the class SC_3.*

$\triangle(X) > 2 \Leftrightarrow X$ *belongs to the class SC_1.*

Proof: Clear from (5.8) and (5.5). ■

5.10. Remark: A) Theorem (5.8) extends a result of Chern and Griffiths, which says that a smooth non-degenerate surface $X \subseteq \mathbb{P}^r_{\mathbb{C}}$ of degree $d \leq 2r - 2$ is either of geometric genus 0 or a $K3$-surface of degree $d = 2r - 2$. In [H, V Ex.6.2.pg 423] the proof of this latter result is posed as an exercise. The hint which is given for the solution of this exercise proposes to use Cliffords theorem and Kodaira vanishing.

B) Let $X \subseteq \mathbb{P}^r$ be as in (5.8) and assume that X belongs to the class SC_3. Then:
X and Y are both linearly normal (s.(3.9)C)); $\sigma = r - 1$, (s.(3.9)B));
$h^1(X, \mathcal{O}_X) = h^1(X, \mathcal{O}_X(1))$, (s.(3.5)a)),
$\chi(X, \mathcal{O}_X(n)) = (r-1)n^2 + n - h^1(X, \mathcal{O}_X) + 1$ for all $n \in \mathbb{Z}$, (s.(3.9)B)). •

BIBLIOGRAPHIE

[A] Albertini C.: *Schranken für die Kohomologie ampler Divisoren über normalen projektiven Varietäten in positiver Charakteristik;* Dissertation, Universität Zürich, 1996

[A-Br] Albertini C., Brodmann M.: *A Bound on the cohomology for ample line bundles in positive characteristic;* preprint 1997

[Ar-C-G-Ha] *Arbarello E., Cornalba M., Griffiths P.A., Harris J.,*: *Geometry of algebraic curves* Vol I, Grundlehren der Math. Wissenschaften 267; Springer, New York 1985

[B] *Beauville A.*: *Surfaces algébriques complexes;* astérisque 54, SMF 1978

[Br_1] *Brodmann M.*: *Bounds on the cohomological Hilbert functions of a projective variety;* Journal of Algebra 109 (1987) 352-380

[Br_2] *Brodmann M.*: *Cohomology of certain projective surfaces with low sectional genus and degree;* to appear in the proceedings of the Hanoi Conference 1996 on commutative algebra, Volume dedicated to the memory of Wolfgang Vogel

[Br-Na] *Brodmann M., Nagel U.*: *Bounding cohomological Hilbert functions by hyperplane sections;* Journal of Algebra 174 (1995) 323-348

[Br-Sc] *Brodmann M., Schenzel P.*: *Cohomological, homological and geometric properties of curves in $\mathbb{P}^r$ having degree $r+2$;* in preparation

[Br-Sh] *Brodmann M., Sharp R.Y.*: *Local cohomology – an algebraic introduction with geometric applications;* to appear in "Cambridge studies in advanced mathematics", CUP, Cambridge

[Br-Vo] *Brodmann M., Vogel W.*: *Bounds for the cohomology and the Calstelnuovo regularity of certain surfaces;* Nagoya Math. Journal 131 (1993) 109-126

[Ep] *Epema D.*: *Surfaces with canonical hyperplane sections;* Proc.Kon.Nedelandse Akad. v. Wetenschappen 86 (1983) 173-184

[Har] *Hartshorne R.*: *Algebraic geometry, Graduate texts in mathematics 52;* Springer, Heidelberg 1977

[J] *Jouanolou J.-P.*: *Théorèmes de Bertini et applications;* Progress in Math., Birkhäuser 1983

[K] *Kodaira K.*: *On a differential geometric method in the theory of analytic stacks*; Proc.Nat.Acad.Sci, USA 39 (1953) 1268-1273

[L-Ra] *Lauritzen N., Rao A.P.*: *Elementary counterexamples to Kodaira vanishing in prime characteristic*; preprint

[Ma] *Matsumura H.*: *Commutative Ring Theory;* Cambridge Studies in advanced mathematics, Cambridge University Press 1989

[Mat] *Matteotti C.*: *A priori-Abschätzungen für die kohomologischen Defektfunktionen projektiver (resp. eigentlicher) Schemata;* Dissertation, University of Zürich 1993

[Mé] *Mérindol J.Y.*: *Surfaces normales, dont le faiseau dualisant est trivial;* C.R. Acad.Sc. Paris 293 (1981) 417-420

[Mu_1] *Mumford D.*: *Lectures on curves on an algebraic surface;* Annals of Mathematical Studies, No 59, Princeton University Press 1966

[Mu_2] *Mumford D.*: *Pathologies III;* Amer.J.Math. 89 (1967) 94-104

[No] *Noether M.*: *Über invariante Darstellung algebraischer Funktionen*; Mathematische Annalen 17 (1880) 263-284

[R] *Raynaud M.*: *Contre-example au "vanishing theorem" en characteristique $p > 0$;* in: C.P. Ramanujam - a tribute, TIFR Stud. in Math 8 (1978) 273-278

[So] *Šokurov V.V.*: *The Noether-Enriques theorem on canonical curves*; Math. USSR Sbornik. 15 (1971) No 3, 361- 403

[U] *Umezu Y.*: *On normal projective surfaces with trivial dualizing sheaf;* Tokoy J. Math. 4 (1981) 343-354

[W] *Wahl J.*: *On cohomology of the square of an ideal sheaf;* Journal of algebraic geometry 6 (1997) 481-511

On Certain Spaces Associated to Tetragonal Curves of Genus 7 and 8

G. CASNATI AND A. DEL CENTINA

Gianfranco Casnati, Dipartimento di Matematica Pura ed Applicata, Università degli Studi di Padova, via Belzoni 7, 35131 Padova, Italy
E-mail address: casnati@galileo.math.unipd.it

Andrea Del Centina, Dipartimento di Matematica, Università degli Studi di Ferrara, via Machiavelli 35, 44100 Ferrara, Italy
E-mail address: cen@ifeuniv.unife.it

0. INTRODUCTION AND NOTATIONS

Let $\mathfrak{M}_g$ be the coarse moduli space of smooth curves of genus g over the complex field $\mathbb{C}$. Let $\mathfrak{T}_g \subseteq \mathfrak{M}_g$ be the locus parametrizing isomorphism classes $[C]$ of tetragonal curves C, i.e. curves C carrying at least one g^1_4 but no g^1_d with $d \leq 3$.

It has been shown in [A–C] that $\mathfrak{T}_g$ is irreducible and unirational, its dimension is $2g+3$ if $g \geq 7$ and that the general tetragonal curve C has exactly one g^1_4. If $7 \leq g \leq 9$ there exist tetragonal curves C carrying more than one g^1_4 without being bielliptic, i.e. a double cover of an elliptic curve. If $g \geq 10$ each tetragonal curve C carries exactly one g^1_4 unless it is bielliptic, in which case it necessarily has infinitely many g^1_4's (see e.g. [DC–G], lemma 2.9).

Let $7 \leq g \leq 9$ and denote by $\mathfrak{T}^{(n)}_g$ and $\mathfrak{M}^{be}_g$ the locus in $\mathfrak{T}_g$ of isomorphism classes $[C] \in \mathfrak{T}_g$ of curves carrying exactly n linear series g^1_4's and that of bielliptic curves respectively. We have that $\mathfrak{T}_7 = \mathfrak{T}^{(1)}_7 \cup \mathfrak{T}^{(2)}_7 \cup \mathfrak{T}^{(3)}_7 \cup \mathfrak{M}^{be}_7$, $\mathfrak{T}_8 = \mathfrak{T}^{(1)}_8 \cup \mathfrak{T}^{(2)}_8 \cup \mathfrak{M}^{be}_8$ and $\mathfrak{T}_9 = \mathfrak{T}^{(1)}_9 \cup \mathfrak{T}^{(2)}_9 \cup \mathfrak{M}^{be}_9$. The main results proved in sections 2, 3 and 4 are summarized in the following

Theorem 0.1. *The loci* $\mathfrak{T}^{(3)}_7$, $\mathfrak{T}^{(2)}_7 \subseteq \mathfrak{M}_7$ *and* $\mathfrak{T}^{(2)}_8 \subseteq \mathfrak{M}_8$ *are rational, irreducible subvarieties of dimensions* 16, 15 *and* 17 *respectively.* □

For $g = 9$ the problem of the rationality of $\mathfrak{T}^{(2)}_9$ remains open.

In section 5 we focus our attention on particular subloci of $\mathfrak{T}_7$, namely the locus $\mathfrak{M}^3_7$ of points representing curves C carrying a theta–characteristic $\mathcal{L}$ satisfying $h^0(C, \mathcal{L}) = 3$, and its intersection $(\mathfrak{M}^3_7)' := \mathfrak{M}^3_7 \cap \mathfrak{T}^{(2)}_7 \subseteq \mathfrak{M}_7$. The locus $\mathfrak{M}^3_7$ has been carrefully described by

The first author has been supported by the framework of the SCIENCE contract n. SCI–0398–C(A). The second author has been partially supported by MURST 40% funds.

D.S. Nagaraj in [Na]. In particular it is proved in [Na] that $\mathfrak{M}_7^3$ is an irreducible subvariety of dimension 15 contained in $\mathfrak{T}_7^{(3)}$. The following theorem completes theorem 4 of [Na].

Theorem 0.2. *The locus* $\mathfrak{M}_7^3$ *is rational. Moreover* $(\mathfrak{M}_7^3)' \subseteq \mathfrak{M}_7$ *is irreducible and rational of dimension* 14. □

Notations. GL_n (resp. PGL_n) is the general (resp. projective) linear group on $\mathbb{C}$. $Bir(\mathbb{P}^2_{\mathbb{C}})$ is the group of birational automorphism of $\mathbb{P}^2_{\mathbb{C}}$. If $X \subseteq \mathbb{P}^2_{\mathbb{C}}$ is a subset we denote by $Bir_X(\mathbb{P}^2_{\mathbb{C}})$ its stabilizer inside $Bir(\mathbb{P}^2_{\mathbb{C}})$.

If g is an element of a certain group G then $\langle g \rangle$ denotes the subgroup of G generated by g. $\mathbb{I}$ is the trivial 1–dimensional representation and $\mathbb{S}_n$ is the standard permutation n–dimensional representation of the symmetric group $\mathfrak{S}_n$ of order n.

We denote by $diag(\alpha_1, \ldots, \alpha_n)$ the $n \times n$ diagonal matrix $A := (\alpha_{i,j})_{1 \le i,j \le n}$ with diagonal entries $\alpha_{i,i} = \alpha_i$, $i = 1, \ldots, n$.

We denote by $\cong$ isomorphisms and by $\approx$ birational equivalences.

For all the other notations and definition we always refer to [Ht].

1. Preliminaries

Let C be a tetragonal curve of genus $g \ge 7$. We recall that a g_4^1, say $|D|$, is said to be of type I (resp. of type II) if $\dim |2D| = 2$ (resp. $\dim |2D| = 3$).

Now assume that $g = 7, 8, 9$ and that C carries at least two distinct g_4^1's, say $|D_1|$ and $|D_2|$. These curves have been studied by M. Coppens in [Co1] (see also [Co2]). Here we summarize what we need in the sequel with a somewhat different approach.

We recall that a common pair to $|D_1|$ and $|D_2|$ is a divisor $P + Q$ on C contained in a divisor both of $|D_1|$ and of $|D_2|$. If $g = 7$ (resp $g = 8$) $|D_1|$ and $|D_2|$ have two (resp. one) common pairs by a classical formula (see [E–C], p. 74).

Suppose $g = 7$ and let $P' + Q'$ and $P'' + Q''$ be the two common pairs. Then both $|E'| := |D_1 + D_2 - P' - Q'|$ and $|E''| := |D_1 + D_2 - P'' - Q''|$ are two g_6^2 (Riemann–Roch) such that $|E'| = |K - E''|$ (K is a canonical divisor on C). In particular, if $P' + Q' + P'' + Q''$ belongs neither to $|D_1|$ nor to $|D_2|$, C is birational to a plane sextic $\widetilde{C}$ with three nodes which are non–collinear if $|E'| \ne |E''|$ (the general case) and collinear if $|E'| = |E''|$ (in this case $|E'| = |E''|$ is half–canonical). In any case $[C] \in \mathfrak{T}_7^{(3)}$. Conversely, by Appendix A, 1.20 of [A–C–G–H], the only g_4^1's on C are those cut out on $\widetilde{C}$ by the pencils of lines through the nodes and they are all of type I. Let us denote by $\mathfrak{T}_7^{(3)} \subseteq \mathfrak{M}_7$ the locus of isomorphism classes $[C]$ of curves C carrying exactly three g_4^1's (all of type I). The following proposition is a particular case of a more general result (see [Hr] or [Ra]).

Proposition 1.1. $\mathfrak{T}_7^{(3)} \subseteq \mathfrak{M}_7$ *is an irreducible subvariety.*

Proof. Fix two points $P_1, P_2 \in \mathbb{P}^2_{\mathbb{C}}$ and, for each $P \in \mathcal{U} := \mathbb{P}^2_{\mathbb{C}} \setminus \{P_1, P_2\}$, let $\mathcal{W}_P \subseteq \mathbb{C}[x, y, z]$ be the 19–dimensional subspace representing curves $\widetilde{C} \subseteq \mathbb{P}^2_{\mathbb{C}}$ having singularities at P_1, P_2, P. The variety $\mathcal{W} := \cup_{P \in \mathcal{U}} \mathcal{W}_P$ is irreducible by theorem I.6.8 of [Sh]. We have a family $\mathcal{X} \subseteq \mathcal{W} \times \mathbb{P}^2_{\mathbb{C}}$ over $\mathcal{W}$, whose fibre over w is the corresponding curve, thus it is flat (see [Ht], theorem III.9.9). Denoting by $\mathcal{W}^0 \subseteq \mathcal{W}$ the open subvariety corresponding to integral curves having at most nodes at P_1, P_2, P, we have a morphism $\vartheta: \mathcal{W}^0 \to \mathfrak{M}_7$ whose image is $\mathfrak{T}_7^{(3)}$, which is then an irreducible subvariety. □

Remark 1.2. Choose P not lying on the line joining P_1 and P_2. Then the above proof shows that $\vartheta_{|\mathcal{W}_P}$ is dominant. □

On the other hand, if $P'+Q'+P''+Q''$ belongs either to $|D_1|$ or to $|D_2|$, the curve C is birationally isomorphic to a plane sextic $\widetilde{C}$ with a node and a tacnode (i.e. a singular point analitically isomorphic to $x^2+y^4=0$). In this second case we have that the lines through the tacnode cut out on $\widetilde{C}$ a g^1_4 of type II and all the g^1_4 on C are cut out on $\widetilde{C}$ by the pencils of lines through the singular points (see [Co1]). Hence C carries exactly one g^1_4 of type II and one of type I. The tacnodal tangent line contains the node if and only if $|E'|=|E''|$ and in this case $|E'|=|E''|$ is half–canonical. Again, in any case, $[C]\in\mathfrak{T}^{(2)}_7$. Let $\mathfrak{T}^{(2)}_7\subseteq\mathfrak{M}_7$ be the locus corresponding to curves C carrying exactly two g^1_4's (one of type I and one of type II).

Proposition 1.3. *$\mathfrak{T}^{(2)}_7\subseteq\mathfrak{M}_7$ is an irreducible subvariety.*

Proof. Fix a point P_1 and a line r in $\mathbb{P}^2_{\mathbb{C}}$. Recall that if x, y and z are coordinates in $\mathbb{P}^2_{\mathbb{C}}$ such that $P_1=[1,0,0]$ and $r=\{z=0\}$, then P_1 is a tacnode on the curve C if and only if the corresponding form w lies in (z^2,zy^2,y^4). For each $P\in\mathcal{U}:=\mathbb{P}^2_{\mathbb{C}}\setminus\{P_1\}$ let $\mathcal{W}_P\subseteq\mathbb{C}[x,y,z]$ be the 19–dimensional subspace representing curves $\widetilde{C}\subseteq\mathbb{P}^2_{\mathbb{C}}$ having at least a tacnode at P_1 with tangent line r and another singularity at P. As in the proof of proposition 1.1, $\mathcal{W}:=\cup_{P\in\mathcal{U}}\mathcal{W}_P$ is irreducible, we can define an open subscheme $\mathcal{W}^0$ and a morphism $\vartheta\colon\mathcal{W}^0\to\mathfrak{M}_7$ whose image is $\mathfrak{T}^{(2)}_7$. □

Remark 1.4. Choosing $P\notin r$, $\vartheta_{|\mathcal{W}_P}$ is dominant. □

In both the cases, on the model $\widetilde{C}$, the g^2_6 is cut out either by the lines or (if there actually exists a second g^2_6) by the conics through the singularities (and which are tangent to the tacnodal tangent line). We will show in sections 2 and 4 that $\mathfrak{T}^{(3)}_7$, $\mathfrak{T}^{(2)}_7$ are both rational of dimension 16 and 15 respectively. Moreover, since each g^1_4 not of type I on a curve C of genus g is the specialization of two different g^1_4's in a suitable family of curves of the same genus (see [Co2] and [Co3]), then $\mathfrak{T}^{(2)}_7$ is contained in the closure of $\mathfrak{T}^{(3)}_7$ inside $\mathfrak{M}_7$.

Now assume $g=8$ and that $P+Q$ is the common pair. Then $|E|:=|D_1+D_2-P-Q|$ is a g^2_6 on C. In particular C is birationally isomorphic to a plane sextic $\widetilde{C}$ with two nodes. Again by Appendix A, 1.20 of [A–C–G–H], $|E|$ is the unique g^2_6 on C and $|D_1|$, $|D_2|$ are the unique g^1_4's (both of type I).

Let us denote by $\mathfrak{T}^{(2)}_8\subseteq\mathfrak{M}_8$ the locus of isomorphism classes of curves C carrying exactly two g^1_4's (of type I). We have the following trivial result.

Proposition 1.5. *$\mathfrak{T}^{(2)}_8\subseteq\mathfrak{M}_8$ is an irreducible subvariety.* □

We will show in section 3 that $\mathfrak{T}^{(2)}_8$ is rational of dimension 17.

Finally if $g=9$ then C is isomorphic to a curve of bidegree $(4,4)$ on a smooth quadric. Obviously we have the following

Proposition 1.6. *$\mathfrak{T}^{(2)}_9\subseteq\mathfrak{M}_9$ parametrizing the isomorphism classes of such curves is irreducible, of dimension* 18. □

The problem of the rationality of $\mathfrak{T}_9^{(2)}$ remains open and it seems, at least to us, rather difficult (see [C–DC] for another possible approach).

2. The rationality of $\mathfrak{T}_7^{(3)}$

Let C be a general tetragonal curve of genus $g = 7$ carrying three g_4^1's of type I. As explained above, we thus have a plane birational model given by a plane sextic $\widetilde{C} \subseteq \mathbb{P}^2_{\mathbb{C}}$ with three non–collinear nodes P_1, P_2, P_3. Such a model depends on the choice of the g_6^2. Since C is general, up to a projective transformation we can assume that $P_1 := [1,0,0]$, $P_2 := [0,1,0]$, $P_3 := [0,0,1]$. Let $X := \{P_1, P_2, P_3\}$.

If $\varphi: C \xrightarrow{\sim} C'$ is an isomorphism then it sends g_n^r's on C into g_n^r's on C', thus it induces a birational automorphism $\phi \in Bir_X(\mathbb{P}^2_{\mathbb{C}})$, defined on the whole of $\mathbb{P}^2_{\mathbb{C}} \setminus X$, leaving X fixed and sending $\widetilde{C}$ to $\widetilde{C'}$.

The group $Bir_X(\mathbb{P}^2_{\mathbb{C}})$ acts on $\mathbb{P}^2_{\mathbb{C}}$. It is generated by the torus $PT \subseteq PGL_3$ acting on $\mathbb{P}^2_{\mathbb{C}}$ in the natural way, by the standard quadratic transformation $\mu(x,y,z) = (yz, xz, xy)$ permuting the two g_6^2's on $\widetilde{C}$ and by the group of permutations of the P_i's, which is $\mathfrak{S}_3$, via the usual representation with 3×3 permutations matrices. More precisely it is not difficult to check that $Bir_X(\mathbb{P}^2_{\mathbb{C}}) \cong PT \rtimes w_{2,3}$, where $w_{2,3} := \langle \mu \rangle \times \mathfrak{S}_3$ is the Weyl group of the root system associated to the blow up of $\mathbb{P}^2_{\mathbb{C}}$ at X (see also [D–O], VII.2).

Let $W \subseteq \mathbb{C}[x,y,z]$ be the 19–dimensional subspace of forms of degree 6 representing plane curves having singularities at the points P_i's. The action of $Bir_X(\mathbb{P}^2_{\mathbb{C}})$ on $\mathbb{P}^2_{\mathbb{C}}$ induces an action on $|W|$, which is linear, i.e. $Bir_X(\mathbb{P}^2_{\mathbb{C}})$ can be realized as a subgroup of $PGL(W)$.

Consider the natural map $p: GL(W) \to PGL(W)$ and let $G := p^{-1}(Bir_X(\mathbb{P}^2_{\mathbb{C}})) \subseteq GL(W)$. For the sake of simplicity we consider the corresponding action of G on $W \cong \mathbb{C}^{19}$. Obviously $G \cong T \rtimes w_{2,3}$, where $T \subseteq GL(W)$ is the torus. Via the isomorphism $GL_{19} \cong GL(W)$ the elements of T are represented by diagonal matrices: both μ and the elements of $\mathfrak{S}_3$ are represented by permutation matrices in GL_{19}.

By remark 1.2, it follows the existence of a dominant rational map $W \dashrightarrow \mathfrak{T}_7^{(3)}$, whose fibres are the G–orbits of W. We can summarize the above remarks in the following proposition.

Proposition 2.1. *The variety $\mathfrak{T}_7^{(3)} \subseteq \mathfrak{M}_7$ has dimension* 16. *Moreover $\mathfrak{T}_7^{(3)} \approx W/T \rtimes (\langle \mu \rangle \times \mathfrak{S}_3)$.* □

Let $W_1 \subseteq W$ be the 7–dimensional subspace corresponding to the sextic curves splitting into a cubic through the P_i's and the three coordinate lines. Since G is reductive there exists a G–invariant complement W_2. In particular the representation W of G splits as $W_1 \oplus W_2$.

An element of W_1 is of the form

$$w(x,y,z) := xyz(a_1x^2y + a_2x^2z + a_3xy^2 + a_4y^2z + a_5xz^2 + a_6yz^2 + a_7xyz).$$

It is easy to check that the stabilizer of a general $w \in W_1$ is

$$G_w := \{(diag(\lambda,\lambda,\lambda), id, id) \in T \rtimes w_{2,3} \mid \lambda^6 = 1\}.$$

Notice that the action of G_w on W is trivial.

Proposition 2.2. *W_1/G is rational.*

Proof. Since $G \cong T \rtimes w_{2,3}$ then $G/T \cong w_{2,3}$. We will study

$$\left(\mathbb{C}(a_1,\ldots,a_7)^T\right)^{G/T} \cong \left(\left(\mathbb{C}(a_1,\ldots,a_7)^T\right)^{\langle\mu\rangle}\right)^{\mathfrak{S}_3}.$$

To this purpose we divide the proof into three steps.

Step 2.2.1. *Let $Y_1 := a_1a_6/a_7^2$, $Y_2 := a_2a_4/a_7^2$, $Y_3 := a_3a_5/a_7^2$, $Y_4 := a_2a_3a_6/a_7^3$. Then $\mathbb{C}(a_1,\ldots,a_7)^T = \mathbb{C}(Y_1,Y_2,Y_3,Y_4)$.*

Proof of step 2.2.1. Since T leaves invariant each monic monomial, the field of T–invariant rational functions is generated by T–invariant fractional monomials. If $\tau \in T$ the condition

$$\tau(a_1^{h_1}\cdot\ldots\cdot a_7^{h_7}) = a_1^{h_1}\cdot\ldots\cdot a_7^{h_7}$$

yields to the linear system $Ah = 0$, where $h := (h_1,\ldots,h_7)^t$ and

$$A := \begin{pmatrix} 3 & 3 & 2 & 1 & 2 & 1 & 2 \\ 2 & 1 & 3 & 3 & 1 & 2 & 2 \\ 1 & 2 & 1 & 2 & 3 & 3 & 2 \end{pmatrix}.$$

Let

$$U := \begin{pmatrix} 1 & 0 & 0 \\ 1 & 1 & 1 \\ 0 & 0 & 1 \end{pmatrix}, \qquad V := \begin{pmatrix} 1 & -1 & 0 & 1 & -2 & 0 & -2 \\ 0 & 1 & 0 & 0 & 0 & 0 & 0 \\ -1 & -1 & 1 & -1 & 2 & 0 & 2 \\ 0 & 2 & -1 & 1 & -1 & -1 & 2 \\ 0 & 0 & 0 & 0 & 0 & 1 & 0 \\ 0 & -2 & 1 & 0 & -1 & 1 & -2 \\ 0 & 1 & -1 & -1 & 1 & 0 & 1 \end{pmatrix}.$$

Then

$$UAV = \begin{pmatrix} 1 & 0 & 0 & 0 & 0 & 0 & 0 \\ 0 & 0 & 0 & 0 & 0 & 0 & 6 \\ 0 & 0 & 0 & 0 & 0 & 1 & 0 \end{pmatrix},$$

thus $\mathbb{C}(a_1,\ldots,a_7)^T = \mathbb{C}(X_1,X_2,X_3,X_4)$ where $X_1 = a_2a_4^2a_7/a_1a_3a_6^2$, $X_2 := a_3a_6/a_4a_7$, $X_3 := a_1a_4/a_3a_7$, $X_4 := a_3^2a_5a_7/a_1^2a_4a_6$. Since $Y_1 = X_2X_3$, $Y_2 = X_1X_2^2X_3$, $Y_3 = X_2X_3^2X_4$ and $Y_4 = X_1X_2^3X_3$, it follows that $\mathbb{C}(X_1,X_2,X_3,X_4) = \mathbb{C}(Y_1,Y_2,Y_3,Y_4)$. □

Step 2.2.2. *Let $Y_5 := Y_4 + \mu(Y_4)$. Then $\mathbb{C}(Y_1,Y_2,Y_3,Y_4)^{\langle\mu\rangle} = \mathbb{C}(Y_1,Y_2,Y_3,Y_5)$.*

Proof of step 2.2.2. The action of μ is given by $\mu(Y_1) = Y_1$, $\mu(Y_2) = Y_2$, $\mu(Y_3) = Y_3$ and $\mu(Y_4) = Y_1Y_2Y_3/Y_4$. Let $Y_5 := Y_4 + \mu(Y_4) \in \mathbb{C}(Y_1,Y_2,Y_3,Y_4)^{\langle\mu\rangle}$. Then

$$\mathbb{C}(Y_1,Y_2,Y_3,Y_5) \subseteq \mathbb{C}(Y_1,Y_2,Y_3,Y_4)^{\langle\mu\rangle} \subset \mathbb{C}(Y_1,Y_2,Y_3,Y_4).$$

Since $[\mathbb{C}(Y_1,Y_2,Y_3,Y_4){:}\,\mathbb{C}(Y_1,Y_2,Y_3,Y_5)] = 2$ it follows that

$$\mathbb{C}(Y_1,Y_2,Y_3,Y_5) = \mathbb{C}(Y_1,Y_2,Y_3,Y_4)^{\langle\mu\rangle}. \qquad \square$$

Step 2.2.3. $\mathbb{C}(Y_1, Y_2, Y_3, Y_5)^{\mathfrak{S}_3}$ *is rational.*

Proof of step 2.2.3. Let us consider $\overline{W} := \langle Y_1, Y_2, Y_3, Y_5 \rangle \subseteq \mathbb{C}(Y_1, Y_2, Y_3, Y_5)$.

$\mathfrak{S}_3$ is generated by its transposition $\sigma_1 := (y, x, z)$, $\sigma_2 := (z, y, x)$, $\sigma_3 := (x, z, y)$. Thus the action of $\mathfrak{S}_3$ on $\mathbb{C}(Y_1, Y_2, Y_3, Y_5)$ is given by $\sigma_1(Y_1, Y_2, Y_3, Y_5) = (Y_3, Y_2, Y_1, Y_5)$, $\sigma_2(Y_1, Y_2, Y_3, Y_5) = (Y_1, Y_3, Y_2, Y_5)$, $\sigma_3(Y_1, Y_2, Y_3, Y_5) = (Y_2, Y_1, Y_3, Y_5)$.

It follows that $\overline{W}$ is a linear representation of $\mathfrak{S}_3$ isomorphic to $\mathbb{I} \oplus \mathbb{S}_3$. In particular $\mathfrak{S}_3$ is generated by pseudoreflections with respect to such an action. Thus $\mathbb{C}[\overline{W}]^{\mathfrak{S}_3} = \mathbb{C}[Z_1, Z_2, Z_3, Z_4]$ for suitable G–invariant elements $Z_i \in \mathbb{C}[\overline{W}] \subseteq \mathbb{C}(a_1, \dots, a_7)$ (see [P–V], theorem 8.1).

On the other hand the group of characters of $\mathfrak{S}_3$ is finite, hence $\mathbb{C}(\overline{W})^{\mathfrak{S}_3}$ coincides with the fields of fractions of $\mathbb{C}[\overline{W}]^{\mathfrak{S}_3}$, whence we finally obtain $\mathbb{C}(Y_1, Y_2, Y_3, Y_5)^{\mathfrak{S}_3} = \mathbb{C}(Z_1, Z_2, Z_3, Z_4)$, which is rational. □This completes the proof of proposition 2.2. □

Taking into account prposition 2.2, by applying to the decomposition $W = W_1 \oplus W_2$ corollary 1 and the remark in section 4 of [Do], the following theorem follows immediately.

Theorem 2.3. $\mathfrak{T}_7^{(3)}$ *is rational.* □

Remark 2.4. Notice that among the curves C representing points in $\mathfrak{T}_7^{(3)}$ there are those whose plane model $\widetilde{C}$ is invariant with respect to the action of μ. Thus such an action induces an analogous action of $\nu := \mu_{|C}$ on C.

It has been proved in [DC–R] that for generic C the map $C \to C/\nu$ is an unramified cover of degree 2 of a smooth curve of genus 4, and all such covers arise in this way. In particular the moduli space $\mathfrak{R}_4$ of unramified covers of curves of genus 4 is birationally isomorphic to W_0/G_0 where $W_0 \subseteq W$ is the subspace corresponding to μ–invariant sextic curves and G_0 is the subgroup of G sending W_0 into itself.

It is not difficult to check that $G_0 \cong T_0 \rtimes \mathfrak{S}_3$ where $T_0 \subseteq T$ the subgroup generated by the matrices

$$\begin{pmatrix} \alpha & 0 & 0 \\ 0 & -\alpha & 0 \\ 0 & 0 & \alpha \end{pmatrix}, \qquad \begin{pmatrix} \beta & 0 & 0 \\ 0 & \beta & 0 \\ 0 & 0 & -\beta \end{pmatrix},$$

$\alpha, \beta \in \mathbb{C}$. An easy computation shows that $T_0 \rtimes \mathfrak{S}_3 = \mathbb{C} \times \mathfrak{S}_4$. Essentially using such a description of $\mathfrak{R}_4$, F. Catanese proved its rationality in [Ca]. □

3. The rationality of $\mathfrak{T}_8^{(2)}$

Let C be a tetragonal curve of genus $g = 8$ carrying two g_4^1's(of type I). Again recall that we have a plane birational model given by a plane sextic $\widetilde{C} \subseteq \mathbb{P}^2_{\mathbb{C}}$ with two nodes P_1, P_2. Up to a projective transformation we can assume that $P_1 := [1, 0, 0]$, $P_2 := [0, 1, 0]$. Let $X := \{P_1, P_2\}$.

If $\varphi: C \xrightarrow{\sim} C'$ is an isomorphism then it sends g_n^r's on C into g_n^r's on C', thus it induces a birational automorphism $\phi \in Bir_X(\mathbb{P}^2_{\mathbb{C}})$ leaving X fixed, sending $\widetilde{C}$ to $\widetilde{C'}$ and the linear system g_6^2 into itself. In particular ϕ is induced by an automorphism of $\mathbb{P}^2_{\mathbb{C}}$ fixing X.

Let $W \subseteq \mathbb{C}[x, y, z]$ be the 22–dimensional subspace of forms of degree 6 representing plane curves having singularities at the points P_i's. Let $G_0 \subseteq GL_3$ be the group generated by the matrices

$$\begin{pmatrix} a_{0,0} & 0 & a_{0,2} \\ 0 & a_{1,1} & a_{1,2} \\ 0 & 0 & a_{2,2} \end{pmatrix}$$

and

$$\mu := \begin{pmatrix} 0 & 1 & 0 \\ 1 & 0 & 0 \\ 0 & 0 & 1 \end{pmatrix} \in GL_3$$

Notice that $G := G_0 \rtimes \langle \mu \rangle$ acts linearly on W.

Let $W_0 \subseteq W$ be the open subset corresponding to irreducible curves with nodes at P_1 and P_2. As in the proofs of propositions 1.1 and 1.3, we can build a flat family $\mathcal{X} \subseteq W_0 \times \mathbb{P}^2_{\mathbb{C}}$ over W_0, thus a morphism $\vartheta \colon W_0 \to \mathfrak{M}_8$ whose image is $\mathfrak{T}_8^{(2)}$ and whose fibres are the G–orbits of W_0. Hence we obtain the following proposition.

Proposition 3.1. *The locus* $\mathfrak{T}_8^{(2)} \subseteq \mathfrak{M}_8$ *is an irreducible subvariety of dimension* 17. *Moreover* $\mathfrak{T}_8^{(2)} \approx W/G_0 \rtimes \langle \mu \rangle$. □

If $w \in W$ then

$$w(x, y, z) := \alpha_1 x^4 z^2 + \alpha_2 y^4 z^2 + \alpha_3 x^4 y^2 + \alpha_4 x^2 y^4 + b x^4 y z + c x y^4 z + \\ + \text{ terms of degree } \leq 3 \text{ both in } x \text{ and in } y \; .$$

We have the following technical lemma (see [P–V], section 2.8 for the definition of (G, H)–section).

Lemma 3.2. *Let* $\overline{W} := \{ w \in W \mid b = c = 0 \}$, $T \subseteq GL_3$ *be the torus of diagonal matrices and* $H := T \rtimes \langle \mu \rangle \subseteq G$. *We claim that* $\overline{W}$ *is a* (G, H)*–section of* W.

Proof. Fix a general $w \in W$: then $\alpha_3\alpha_4 \neq 0$. Choosing $g \in G_0$ such that $g(x, y, z) = (x - c/2\alpha_4 z, y - b/\alpha_3 z, z)$, then $g(w) \in \overline{W}$. It follows that $\overline{W}$ is G–dense inside W_1. Now let $g \in G_0$ be represented by

$$\begin{pmatrix} a_{0,0} & 0 & a_{0,2} \\ 0 & a_{1,1} & a_{1,2} \\ 0 & 0 & a_{2,2} \end{pmatrix}.$$

If $w \in \overline{W}$ is such that $\alpha_3\alpha_4 \neq 0$ then $g(w) \in W$ has $b = 2\alpha_3 a_{0,0}^4 a_{1,1} a_{1,2}$, $c = 2\alpha_4 a_{0,0} a_{0,2} a_{1,1}^4$. It follows that the condition $g(w) \in \overline{W}$ yields $g \in T$. □

Now consider the subspace $\overline{W}_1 \subseteq \overline{W}$ generated by the polynomials

$$w(x, y, z) := z^2(a_1 x^4 + a_2 x^2 z^2 + a_3 x z^3 + a_4 y^4 + a_5 y^2 z^2 + a_6 y z^3 + a_7 z^4).$$

$\overline{W}_1$ is G–invariant: since H is reductive then there exists a H–invariant complement $\overline{W}_2$. In particular the representation $\overline{W}$ of H splits as $\overline{W}_1 \oplus \overline{W}_2$.

The above lemma allow us to prove the following proposition.

Proposition 3.3. *$\overline{W}_1/H$ is rational.*

Proof. Since $H = T \rtimes \langle\mu\rangle$ we will compute $\left(\mathbb{C}(a_1,\dots,a_7)^T\right)^{\langle\mu\rangle}$.

As in the proof of step 2.2.1 we have to compute the T–invariant fractional monomials in the a_i's. Let $X_1 := a_1a_7^3/a_3^4$, $X_2 := a_2a_7/a_3^2$, $X_3 := a_4a_7^3/a_6^4$, $X_4 := a_5a_7/a_6^2$. Then $\mathbb{C}(a_1,\dots,a_7)^T = \mathbb{C}(X_1,X_2,X_3,X_4)$. Since $\mu(X_1) = X_3$, $\mu(X_2) = X_4$ then the abelian group $\langle\mu\rangle$ acts linearly on $\langle X_1,X_2,X_3,X_4\rangle$, thus $\mathbb{C}(X_1,X_2,X_3,X_4)^{\langle\mu\rangle}$ is rational by [Fi]. $\square$

It is not difficult to check that the stabilizer H_w of a general $w \in \overline{W}_1$ is

$$H_w = \left\{(diag(\lambda,\lambda,\lambda), id) \in T \rtimes \langle\mu\rangle \mid \lambda^6 = 1\right\}$$

which acts trivially on $\overline{W}_2$. As in the previous section we have thus completed the proof of the following theorem.

Theorem 3.4. *$\mathfrak{T}_8^{(2)}$ is rational.* $\square$

4. The rationality of $\mathfrak{T}_7^{(2)}$

Let C be a general tetragonal curve of genus $g = 7$ carrying two g_4^1. Then C carries exactly one g_4^1 of type II and another g_4^1 of type I. In this case we have a plane birational model given by a plane sextic $\widetilde{C} \subseteq \mathbb{P}^2_{\mathbb{C}}$ with two singular points, namely a tacnode P_1 and a node P_2. Since C is general, up to a projective transformation we can assume that $P_1 := [1,0,0]$ with tacnodal tangent line is $r := \{z = 0\}$ and $P_2 := [0,0,1] \notin r$. Let $X := \{P_1,P_2,r\}$.

If $\varphi\colon C \xrightarrow{\sim} C'$ is an isomorphism, then it induces a birational automorphism $\phi \in Bir_X(\mathbb{P}^2_{\mathbb{C}})$ leaving X fixed and sending $\widetilde{C}$ to $\widetilde{C}'$.

Notice that each $\phi \in Bir_X(\mathbb{P}^2_{\mathbb{C}})$ either fix the two g_6^2 or it permutes the two series. In the first case ϕ is induced by a projectivity of the plane represented by a matrix

$$\begin{pmatrix} a_{0,0} & a_{0,1} & 0 \\ 0 & a_{1,1} & 0 \\ 0 & 0 & a_{2,2} \end{pmatrix}. \tag{4.1}$$

In the second case ϕ sends the linear system of lines in $\mathbb{P}^2_{\mathbb{C}}$ in that of conics through the P_i's and tangent to r at the point P_1.

Let $W \subseteq \mathbb{C}[x,y,z]$ the 19–dimensional subspace of forms of degree 6 representing curves with a tacnode at P_1 with tangent r and a node at P_2. Recall that such forms lie in $(y^4,y^2z,z^2) \cap (x^2,xy,y^2)$. Let $G_0 \subseteq GL_3$ be the subgroup generated by the matrices of the form (4.1) and $\mu(x,y,z) := (xz,yz,y^2)$. Then $Bir_X(\mathbb{P}^2_{\mathbb{C}})$ is the image via the natural projection $GL_3 \to PGL_3$ of $G := G_0 \rtimes \langle\mu\rangle$. As usual we study the linear action of G on W. Again we get a rational map $W \dashrightarrow \mathfrak{M}_7$ whose image is $\mathfrak{T}_7^{(2)}$ and whose fibres are the G–orbits of W.

Proposition 4.2. *The locus* $\mathfrak{T}_7^{(2)} \subseteq \mathfrak{M}_7$ *is an irreducible subvariety of dimension* 15. *Moreover* $\mathfrak{T}_7^{(2)} \approx W/G_0 \rtimes \langle\mu\rangle$. $\square$

Since the statements and proofs are analogous to those of theorems 2.3 and 3.4, we will omit all the details. If $w \in W$ then

$$w(x,y,z) := \alpha_1 x^4 z^2 + \alpha_2 x^3 y^2 z + \alpha_3 x^3 z^3 + b x^3 y z^2 + \text{ terms of degree } \leq 3 \text{ in } x\ .$$

Lemma 4.3. *Let* $\overline{W} := \{w \in W \mid b = 0\ \}$, $T \subseteq GL_3$ *be the torus of diagonal matrices and* $H := T \rtimes \langle\mu\rangle \subseteq G$. *Then* $\overline{W}$ *is a* (G,H)*-section of* W. $\square$

We now consider the subspace $\overline{W}_1 \subseteq \overline{W}$ generated by the polynomials

$$w(x,y,z) := y^2(a_1 xyz^2 + a_2 y^4 + a_3 y^3 z + a_4 y z^3 + a_5 z^4 + a_6 x^2 z^2).$$

Again there exists a complement $\overline{W}_2 \subseteq \overline{W}$ such that $\overline{W} = \overline{W}_1 \oplus \overline{W}_2$ as a representation of H.

Theorem 4.3. $\mathfrak{T}_7^{(2)}$ *is rational.*

Proof. Imitating the proof of step 2.2.1, we obtain $\mathbb{C}(a_1,\dots,a_6)^T = \mathbb{C}(X_1,X_2,X_3)$ where $X_1 := a_2a_6/a_3a_4$, $X_2 := a_2a_5a_6^2/a_1^4$ and $X_3 := a_3^2a_6/a_2a_1^2$. Since $\mu(X_1) = X_1$, $\mu(X_2) = X_2$, $\mu(X_3) = X_2/X_1^2X_3$, as in the proof of step 2.2.1 we obtain

$$\mathbb{C}(a_1,\dots,a_6)^H = \left(\mathbb{C}(a_1,\dots,a_6)^T\right)^{\langle\mu\rangle} = \mathbb{C}(X_1,X_2,X_3)^{\langle\mu\rangle} = \mathbb{C}(X_1,X_2,X_4),$$

where $X_4 := X_3 + \mu(X_3)$.

On the other hand the stabilizer in H of a general $w \in \overline{W}_1$ is

$$H_w := \left\{(diag(\lambda,\lambda,\lambda), id) \in T \rtimes \langle\mu\rangle \mid \lambda^6 = 1\ \right\}.$$

In particular H_w acts trivially on $\overline{W}_2$ then $\overline{W}/H$ is rational, again by section 4 of [Do]. $\square$

5. Some particular subloci of $\overline{\mathfrak{T}_7^{(3)}}$

In this section we will study some subloci of $\overline{\mathfrak{T}_7^{(3)}}$ (other than $\mathfrak{T}_7^{(2)}$) whose points correspond to sextic curves having particular configurations of their singularities

Let $\mathfrak{M}_7^3$ be the set of $[C] \in \mathfrak{M}_7$ representing curves C carrying a theta–characteristic $\mathcal{L}$ satisfying $h^0(C,\mathcal{L}) = 3$. Each general point of $[C] \in \mathfrak{M}_7^3$ represents a curve C carrying three g_4^1 and one g_6^2. The corresponding plane model $\widetilde{C}$ is a sextic curve with three alligned nodes P_1, P_2, P_3. In particular $\mathfrak{M}_7^3 \subseteq \overline{\mathfrak{T}_7^{(3)}}$.

Let $(\mathfrak{M}_7^3)' := \mathfrak{M}_7^3 \cap \mathfrak{T}_7^{(2)}$ and choose a general $[C] \in (\mathfrak{M}_7^3)'$. Its plane model $\widetilde{C}$ is a sextic curve with a tacnode P_1 whose tangent line r contains a node P_2.

In this section we will prove the rationality of $\mathfrak{M}_7^3$ and $(\mathfrak{M}_7^3)'$. The proofs of the rationality of these two spaces is similar to the proofs of theorem 2.3 and 3.4, thus we will omit in the sequel the details.

We begin by studying $\mathfrak{M}_7^3$. To this purpose, let $\widetilde{C}$ have three alligned nodes P_1, P_2, P_3 ($X := \{P_1, P_2, P_3\}$). Up to a suitable transformation we can assume that $P_1 := [1,0,0]$, $P_2 := [0,1,0]$, $P_3 := [1,1,0]$.

Let $W \subseteq \mathbb{C}[x,y,z]$ be the 19–dimensional subspace of forms representing curves having singularities at the P_i's. We have a linear action of $Bir_X(\mathbb{P}^2_{\mathbb{C}})$ onto $|W|$. Let $G \subseteq GL(W)$ be the inverse image of $Bir_X(\mathbb{P}^2_{\mathbb{C}})$ via the natural map $p: GL(W) \to PGL(W)$.

It is not difficult to check that $G = G_0 \rtimes \mathfrak{S}_3$, where $G_0 \subseteq GL_3$ is the subgroup generated by the matrices

$$\begin{pmatrix} a_{0,0} & 0 & a_{0,2} \\ 0 & a_{0,0} & a_{1,2} \\ 0 & 0 & a_{2,2} \end{pmatrix}$$

and $\mathfrak{S}_3$ is the group of permutations of the elements of X. With our choice of coordinates $\mathfrak{S}_3 \subseteq GL(W)$ is the image via the natural map $GL_3 \to GL(W)$ of the group generated by the matrices

$$\begin{pmatrix} 0 & 1 & 0 \\ 1 & 0 & 0 \\ 0 & 0 & 1 \end{pmatrix}, \quad \begin{pmatrix} 1 & 0 & 0 \\ 1 & -1 & 0 \\ 0 & 0 & 1 \end{pmatrix}, \quad \begin{pmatrix} -1 & 1 & 0 \\ 0 & 1 & 0 \\ 0 & 0 & 1 \end{pmatrix}.$$

Theorem 5.1. *The locus $\mathfrak{M}_7^3 \subseteq \mathfrak{M}_7$ is an irreducible, rational subvariety of dimension* 15.

Proof. Let $T \subseteq G_0$ be the torus of diagonal matrices, $H := T \rtimes \mathfrak{S}_3$. If $w \in W$ then

$$\begin{aligned} w(x,y,z) :&= \alpha_1 x^4y^2 + \alpha_2 x^3y^3 + \alpha_3 x^3y^2z + \alpha_4 x^3yz^2 + \\ &+ bx^4yz + cx^3z^3 + \text{ terms of degree} \leq 2 \text{ in } x\ . \end{aligned}$$

As in the previous sections one verifies that $\overline{W} := \{ w \in W \mid b = c = 0 \}$ is a (G,H)–section of W.

Let $\overline{W}_1 \subseteq \overline{W}$ be the subspace of $w \in \overline{W}$ such that

$$w(x,y,z) := z^3(a_1z^3 + a_2x^2z + a_3xyz + a_4xz^2 + a_5y^2z + a_6yz^2).$$

$\overline{W} \cong \overline{W}_1 \oplus \overline{W}_2$ as a sum of representations of H. The stabilizer H_w of a general $w \in \overline{W}_1$ acts trivially on $\overline{W}_2$.

Now let $V_3 \subseteq \mathbb{C}[x,y]$ the subspace of forms of degree 3. We have a representation $U := V_3 \oplus \mathbb{C}z^3$ of G. The stabilizer of a general $u \in U$ acts trivially on $\overline{W}$ and $\dim(U) \leq \dim(\overline{W}_2)$, thus the statement follows by applying corollary 2 of section 4 of [Do]. □

Finally we describe $(\mathfrak{M}_7^3)'$. Let $\widetilde{C}$ have two singular points P_1, P_2. Assume that P_1 is a tacnode and that the tacnodal tangent line r contains P_2 ($X := \{P_1, P_2, r\}$. Up to a suitable transformation we can assume that $P_1 := [1,0,0]$, $P_2 := [0,1,0]$ and that $r := \{z = 0\}$.

Let $W \subseteq \mathbb{C}[x,y,z]$ be the 19–dimensional subspace of forms representing curves having singularities at the P_i's. Then we must consider the group of birational transformations of the plane fixing each element in X. One checks that such a group is generated by $g \in PGL_3$ represented by matrices of the form

$$\begin{pmatrix} a_{0,0} & 0 & a_{0,2} \\ 0 & a_{1,1} & a_{1,2} \\ 0 & 0 & a_{2,2} \end{pmatrix}.$$

Let $G \subseteq GL_3$ be the corresponding group. W is a representation of G and W/G is rational since G is triangular (see [P–V], theorem 2.11). On the other hand $(\mathfrak{M}_7^3)' \approx W/G$. Thus we have proved the following theorem.

Theorem 5.2. *The locus $(\mathfrak{M}_7^3)' \subseteq \mathfrak{M}_7$ is an irreducible, rational subvariety of dimension* 14. □

REFERENCES

[A–C] E. Arbarello, M. Cornalba, *Footnotes to a paper of Beniamino Segre*, Math. Ann. **256** (1981), pp. 341–362.

[A–C–G–H] E. Arbarello, M. Cornalba, P.A. Griffiths, J. Harris, *Geometry of algebraic curves*, vol. I, Springer, 1985.

[C–DC] G. Casnati, A. Del Centina, *The rationality of certain moduli spaces associated to half-canonical extremal curves*, preprint (1997).

[Ca] F. Catanese, *On the rationality of certain moduli spaces related to curves of genus* 4, Algebraic Geometry, Proceedings Ann Arbor 1981 (I. Dolgachev, ed.), L.N.M., vol. 1008, 1983.

[Co1] M. Coppens, *A study of* 4*–gonal curves of genus* ≥ 7, Preprint R.U. Utrecht (1981).

[Co2] M. Coppens, *One-dimensional linear systems of type II on smooth curves*, Ph–D thesis (1983).

[Co3] M. Coppens, *The existence of k–gonal curves possessing exactly two linear systems* g_k^1, Math. Ann. **307** (1997), pp. 291–297.

[DC–G] A. Del Centina, A. Gimigliano, *Bielliptic curves: special linear series and plane models*, The Curves Seminar at Queen's, Vol. IX (A. V. Geramita, ed.), Queen's papers in pure and applied mathematics, vol. 95, 1993.

[DC–R] A. Del Centina, S. Recillas, *Some projective geometry associated to unramified double covers of curves of genus* 4, Ann. Mat. Pura Appl. **XXXIII** (1983), pp. 125–140.

[Do] I.V. Dolgachev, *Rationality of the fields of invariants*, Algebraic Geometry, Bowdoin 1985 (Spencer J. Bloch, ed.), Proceedings of symposia in pure mathematics, vol. 46, 1987.

[D–O] I.V. Dolgachev, D. Ortland, *Points sets in projective spaces and theta functions*, Astérisque, vol. 165, 1988.

[E–C] F. Enriques, O. Chisini, *Teoria geometrica delle equazioni e delle funzioni algebriche*, vol. III, Zanichelli, 1924.

[Fi] E. Fischer, *Die Isomorphie der Invariantenkörpen der endlicher Abelschen Gruppen linearen Transformations*, Nachr. König. Ges. Wiss. Göttingen (1915), pp. 77–80.

[Hr] J. Harris, *On the Severi problem*, Invent. Math **84** (1986), pp. 445–461.

[Ht] R. Hartshorne, *Algebraic geometry*, Springer, 1977.

[Na] D.S. Nagaraj, *On the moduli of curves with theta–characteristics*, Comp. Math. **75** (1990), pp. 287–297.

[P–V] V.L. Popov, E.B. Vinberg, *Invariant theory*, Algebraic Geometry IV (A.N. Parshin and I.R. Shafarevich, eds.), Encyclopedia of Mathematical Sciences, vol. 55, 1991.

[Ra] Z. Ran, *On nodal plane curves*, Invent. Math. **86** (1986), pp. 529–534.

[Sh] I.R. Shafarevich, *Basic algebraic geometry*, Springer, 1977.

Gorenstein Algebras with Pure Resolution

Mihai Cipu
Institute of Mathematics of the Romanian Academy
P.O.Box 1–764, RO–70700 Bucharest, ROMANIA
mcipu@stoilow.imar.ro

Dedicated to Professor Mario Fiorentini on the occasion of the retirement

If a Gorenstein codimension four ideal in a polynomial ring over a field has a minimal graded resolution in which the non-zero entries in each transition matrix have the same degree, then this resolution is uniquely determined by the multiplicity of the ideal.

1 The Result

A homogeneous algebra S over a field K is a finitely generated K–algebra, generated over K by elements of degree one. Equivalently S is isomorphic to R/I, where $R := K[x_1, \ldots, x_n]$ is a polynomial ring and I is a homogeneous ideal contained in $(x_1, \ldots, x_n)$. Then S has a graded minimal free resolution of the form

$$0 \longrightarrow \bigoplus_{j=1}^{\beta_c} R(-d_{cj}) \longrightarrow \ldots \longrightarrow \bigoplus_{j=1}^{\beta_1} R(-d_{1j}) \longrightarrow R \longrightarrow S \longrightarrow 0\,, \qquad (1)$$

Partially supported by a grant from International Centre for Theoretical Physics Trieste, Italy

where the shifts d_{ij} and Betti numbers β_i are positive integers.

The interplay between the numerical information encoded in such a resolution and the algebraic properties of the given algebra is a question of long-standing interest. To deep the understanding of this relationship one had imposed various restrictions, either on the resolution itself or on the object of primary concern, namely S. A particularly satisfactory picture has been glained in the case of *pure resolutions*, when the shifts d_{ij} do not depend on i, *i.e.* $d_{ij} = d_i$ for all $j = 1, \ldots, \beta_i$, $i = 1, \ldots, c$ (see [2]). If it is pure, then the resolution (1) can be written in the form

$$0 \longrightarrow R(-d_c)^{\beta_c} \longrightarrow \ldots \longrightarrow R(-d_1)^{\beta_1} \longrightarrow R \longrightarrow S \longrightarrow 0 \quad . \tag{2}$$

The class of algebras having pure resolution is broad enough, encompassing certain coordinate rings of varieties well-known in algebraic geometry:

- a variety defined by maximal minors of a generic matrix, cf. Egon and Northcott, 1962 [4]
- a variety defined by the submaximal minors of a generic square matrix, cf. Gulliksen and Negárd, 1972 [5]
- a variety defined by generic maximal pfaffians, cf. Buchsbaum and Eisenbud, 1977 [3]
- the tangent cone of a minimally elliptic singularity of a rational surface singularity, cf. Wahl, 1977 [12].
- certain of the Segre–Veronese varieties, cf. Bărcănescu and Manolache, 1981 [1]

On the algebraic side, Herzog and Kühl [6] have expressed the Betti numbers of a graded Cohen-Macaulay algebra S with pure resolution in terms of the shifts appearing in (2):

$$\beta_i = (-1)^{i+1} \prod_{j \neq i} \frac{d_j}{d_j - d_i} \quad , \quad i = 1, \ldots, c \, . \tag{3}$$

From this formula it is apparent that the integers d_i are subject to stringent restrictions, as the right hand side must be an integer.

Huneke and Miller [8] have obtained a neat expression for the multiplicity of an algebra having the same properties:

$$e(S) = \left(\prod_{i=1}^{c} d_i \right) / c! \quad . \tag{4}$$

For a different proof and a conjectural extension of this formula to larger classes of homogeneous algebras the reader is referred to the paper [7].

Huneke and Miller put forward the question whether the relationship multiplicity–shifts is so strong that the whole resolution (2) can be written down knowing only the multiplicity of the ring S. Formally, they asked whether the values of the shifts d_i are uniquely determined by the value of the multiplicity e. In [8] they showed that the positive answer is granted for Gorenstein codimension four algebras of multiplicity at most a trillion.

Their proof of this result goes like this: first, from (3) and (4) it is found that, for a Gorenstein codimension four algebra with pure resolution, (2) has the form

$$0 \longrightarrow R(-2xy) \longrightarrow R^{y^2}(-xy-y) \longrightarrow R^{2y^2-2}(-xy) \longrightarrow R^{y^2}(-xy+y) \longrightarrow R.$$

From this it easily follows that the multiplicity for such an algebra is $x^4y^2(y^2-1)/12$. Therefore, the unicity of its resolution is equivalent to the fact that the Diophantine equation

$$x^4y^2(y^2-1) = u^4v^2(v^2-1) \tag{5}$$

has only the solutions $x = u$, $y = v$. Then the authors proceed to bound a certain concave function and check a table of prime factorization of integers up to 10^4 . This in turn implies that $xy = uv$, whence it follows that the equation (5) has only the obvious solutions, provided that each side is at most 12×10^{12} .

The aim of this paper is to remove the restriction on multiplicity in Huneke–Miller result. To this end one rephrases the unicity of the pure resolution still in number–theoretical terms, but relating it to another Diophantine equation. The next section is devoted to the fulfilment of this step. In Section 3 one shows that the equation we arrive at has at most one solution in positive integers. The main ingredient is a deep result of Ljunggren concerning biquadratic Diophantine equations. Modulo this result, our approach is straightforward and elementary.

The main result is the following:

Theorem 1.1 *Let R be a polynomial ring in finitely many indeterminates with coefficients in a field, and let S be a Gorenstein codimension four R–algebra with pure resolution. Then the graded minimal free resolution of S is uniquely determined by the multiplicity $e(S)$.*

2 A Diophantine Equation...

The multiplicity of an algebra satisfying the hypothesis of Theorem 1.1 is

$$e = \frac{x^4y^2(y^2-1)}{12} \quad .$$

The unicity asserted in the conclusion of Theorem 1.1 amounts to show that, apart from the trivial solutions in which $y = v = 1$, the Diophantine equation (5) has only the obvious solution $x = u$, $y = v$. The trivial solutions have no significance in our original algebraic setting and therefore we are not interested in them.

Let us consider (x, y) and (u, v) providing a solution for the equation (5) with $y > 1$, $v > 1$. Certainly one may remove common factors, so that we may supppose x and u are relatively prime. Then

$$\frac{y^2(y^2-1)}{u^4} = \frac{v^2(v^2-1)}{x^4} = t$$

for a certain natural number t . Let us note that $y^2(y^2-1)$ is always multiple of 4. Since u and x cannot be both even, it follows that 4 divides t . Therefore $t = 4d$, say, for a suitable positive integer d.

Upon multiplying by 4 one gets

$$(2y^2-1)^2 - d(2u)^4 = (2v^2-1)^2 - d(2x)^4 = 1 \quad .$$

Let us record the result obtained so far . It translates the problem from its original setting into a purely number–theoretic framework.

Proposition 2.1 *The conclusion of Theorem 1.1 holds if the Diophantine equation*

$$(2Y^2-1)^2 - DV^4 = 1 \tag{6}$$

has at most one solution in positive integers.

3 ...with at most One Positive Solution

The equation (6) is of the type

$$U^2 - DV^4 = 1 \quad , \quad D > 0 \text{ not a perfect square} \quad . \tag{7}$$

We are only interested in solutions (u, v) satisfying

$$u = 2y^2 - 1 \; > \; 1 \tag{8}$$

for a certain positive integer y .

According to a well-known result of Ljunggren [9] (*cf.* also Mordell [11], ch.28, th.9), the equation (7) has at most two solutions in positive integers. If there are two solutions, these are given by either $u + v^2\sqrt{D} = \varepsilon, \varepsilon^2$ or by $u + v^2\sqrt{D} = \varepsilon, \varepsilon^4$, where ε denotes the fundamental unit in the quadratic field $\mathbb{Q}(\sqrt{D})$.

Let us suppose that the Diophantine equation (6) has two positive solutions. Then one of them is given by $\varepsilon = u + v^2\sqrt{D}$, and the other is given by either

$$\varepsilon^2 = 2u^2 - 1 + 2uv^2\sqrt{D} \tag{9}$$

or

$$\varepsilon^4 = 8u^4 - 8u^2 + 1 + 4uv^2(2u^2 - 1)\sqrt{D} \ . \tag{10}$$

If ε and ε^2 are both solutions of (7) , then $2uv^2$ is a perfect square. Therefore u is twice a perfect square, so it can not be odd, as required by relation (8).

If ε and ε^4 give the solutions of (7) , then $u(2u^2 - 1)$ is a perfect square. Since u and $2u^2 - 1$ are coprime, each of them must be a perfect square. Hence $u = s^2$, $2u^2 - 1 = t^2$, for (s, t) a positive integer solution of the equation

$$2S^4 - T^2 = 1 \ . \tag{11}$$

It has been known for centuries that this equation is satisfied by $(s, t) = (1, 1)$ and $(13, 239)$. Ljunggren [10] was the first to prove that these are the only solutions of (11) . The pair $(s, t) = (1, 1)$ gives the trivial case $u = 1$, while $(s, t) = (13, 239)$ implies $u = 169$ and $D = 1785$. The (nontrivial) solutions of the Diophantine equation

$$U^2 - 1785V^4 = 1$$

are $(u, v) = (169, 2)$, $(6\ 525\ 617\ 281, 12\ 428)$. The former does not provide a solution for (6) , since u does not satisfy relation (8). Thus the equation (6) with $D = 1785$ has a unique positive solution $(x, y) = (57\ 121, 12\ 428)$.

This ends the proof of Theorem.

References

[1] S.BĂRCĂNESCU AND N.MANOLACHE, Betti numbers of Segre–Veronese singularities, *Rev.Roum.Math.Pures et Appl.*, **26**(1981), 549–565

[2] W.BRUNS AND J.HERZOG, *Cohen–Macaulay rings*, Cambridge University Press, 1993

[3] D.BUCHSBAUM AND D.EISENBUD, Algebra structures for some finite free resolutions and some structure theorems for ideals of codimension 3, *Amer.J.Math.*, **99**(1977), 447–485

[4] J.E.EGON AND D.G.NORTHCOTT, Ideals defined by matrices and a certain complex associated with them, *Proc.Roy.Soc.(Ser.A)*, **269**(1962), 188–205

[5] T.GULLIKSEN AND O.NEGÁRD, Un complexe résolvant pour certain idéaux déterminantiels, *CR Acad.Sci.Paris(Sér.A)*, **274**(1972), 16–18

[6] J.HERZOG AND M.KÜHL, On the betti numbers of finite pure and linear resolutions, *Commun.Algebra* , **12**(1984), 1627–1646

[7] J.HERZOG AND H.SRINIVASAN, Bounds for multiplicity, preprint, 1996

[8] C.HUNEKE AND M.MILLER, A note on the multiplicity of Cohen–Macaulay algebras with pure resolutions, *Canad. J. Math.*, **37**(1985), 1149–1162

[9] W.LJUNGGREN, Über die Gleichung $x^4 - Dy^2 = 1$, *Arch.Math.Natur.*, **45**(1942), nr.5

[10] W.LJUNGGREN, Zur Theorie der Gleichung $x^2 + 1 = Dy^4$, *Avh.Norske Vid.Akad.Oslo* , **1**(1942), nr.5

[11] L.J.MORDELL, *Diophantine Equations*, Academic Press, London, New York, 1969

[12] J.WAHL, Equations defining rational singularities, *Ann.Sci.École Norm.Sup.*, **10**(1977), 231–264

Smooth Specializations of Space Curves: Questions and Examples

PH. ELLIA AND R. HARTSHORNE

Dipartimento di Matematica
Universita di Ferrara
35, via Machiavelli
44100 Ferrara, Italy
phe@dns.unife.it

Department of Mathematics
University of California
Berkeley, California 94720-3840
robin@math.berkeley.edu

Introduction.

In this paper we consider several questions about specializations of smooth space curves. In each case it is a question whether one particular class of smooth curves can specialize to members of another particular class. While our knowledge of the classification of space curves has grown considerably in recent years, the questions we consider here serve rather to emphasize our ignorance. Nevertheless, some patterns emerge, and it seems timely to risk some conjectures even though we do not have enough evidence to feel sure of their truth. Our aim is to make some elementary observations, to give a few examples, and to identify the "smallest" particular cases which remain open.

We consider curves in the projective three-space $\mathbb{P}^3$ over an algebraically closed field k.

The first question (Conjecture A) is whether the smooth specialization of complete intersection curves is also a complete intersection (see problem list in [2]). Mohan Kumar [19] has given some counterexamples in positive characteristic, but his methods do not extend to characteristic zero, so the problem remains open there. We observe that a specialization of complete intersection curves of surfaces of degrees a and b must be *subcanonical of type* (a, b), meaning that the degree is $d = ab$ and the canonical sheaf is $\omega = \mathcal{O}(a + b - 4)$. Then we classify all subcanonical curves of type (a, b) on surfaces of degree ≤ 3. This enables us to show that any smooth specialization of complete intersection curves of type (a, b) with $a \leq 4$ is a complete intersection, except for the case $(a, b) = (4, 5)$. In this case, there are smooth subcanonical curves of type (4,5) which are not complete intersections, and we do not know if they can be specializations of complete intersection curves.

Our second question is under what conditions can the invariant s (the least degree of a surface containing the curve) drop under specialization of smooth curves? More precisely, Conjecture B states that if a family of curves C_t lying on normal surfaces X_t of degree $s = s(C_t)$ specializes to a curve C_0 with $s(C_0) < s$, then the degree of the curves is bounded by s^2. This conjecture is motivated by the detailed study of specializations of curves on nonsingular cubic surfaces, when the surface degenerates into a quadric plus a plane or three planes, used in the solution of Zeuthen's problem [16]. Unfortunately the techniques of that paper do not go far enough to settle this conjecture even for $s = 3$. For example, we do not know if there exist families of rational curves of degree $d \geq 8$ on smooth cubic surfaces which specialize to a smooth rational curve on a quadric surface.

We show that Conjecture B implies Conjecture A. Then we give some special cases and examples as evidence for Conjecture B.

Our third question (Conjecture C) is whether there exist families of curves in $\mathbb{P}^3$ in characteristic $p > 0$ which do not lift to characteristic 0. We suggest that this conjecture is implied by a Conjecture B′, closely related to Conjecture B, which concerns families of smooth curves on smooth surfaces specializing to an integral but non-normal surface. We give a few examples of existence and non-existence of such families as evidence for Conjecture B′.

§1. Complete Intersections

A *curve* is a one-dimensional closed subscheme of $\mathbb{P}^3_k$, over an algebraically closed field k, with no associated points, embedded or otherwise, hence locally Cohen-Macaulay. A curve C is a *complete intersection* if its homogeneous ideal I_C in $R = k[x, y, z, w]$ can be generated by two elements. If the degrees of those elements are a, b, with $a \leq b$, we say C is a complete intersection of *type* (a, b). A *family of curves* is a closed subscheme $C \subseteq \mathbb{P}^3_T$, flat over the parameter scheme T, such that all of the geometric fibers $C_t \subseteq \mathbb{P}^3_{\overline{k(t)}}$ for $t \in T$ are curves. We say a curve C_0 is a *specialization of complete intersection curves* if there is a family of curves $C \subseteq \mathbb{P}^3_T$ over an integral scheme T of dimension ≥ 1, with fiber C_0 over a point $0 \in T$, and such that for all $t \in T$, $t \neq 0$, the fiber C_t is a complete intersection.

Our first problem, which for definiteness we phrase in the form of a conjecture, is whether a specialization of complete intersection curves must be a complete intersection.

Conjecture A. *Let $C \subseteq \mathbb{P}^3_k$ be a smooth curve which is a specialization of complete intersection curves of type (a, b). Assume* char.$k = 0$. *Then C is a complete intersection.*

Remarks 1.1. a) The condition that C be smooth is necessary. For example, an elliptic quartic curve (which is a complete intersection of type (2,2)) can specialize to the union of a plane cubic curve plus a line meeting it at one point, which is not a complete intersection. However, we do not know any counterexample with C integral.

b) The restriction to $\mathbb{P}^3$ is necessary. For example, one can take a complete intersection of three quadric hypersurfaces in $\mathbb{P}^4$ and specialize it by projection to a smooth curve of degree 8 and genus 5 in $\mathbb{P}^3$ which is not a complete intersection. Or one can specialize the same curve to a trigonal canonical curve in $\mathbb{P}^4$, which is no longer a complete intersection.

c) Mohan Kumar [19] has shown that over a field of characteristic $p > 0$, for any a sufficiently large, there are smooth specializations of complete intersections of type (a, a), which are not complete intersections. He also has other examples with $a < b$. This accounts for the additional hypothesis char.$k = 0$.

Definitions. We denote the degree of a curve by d and its arithmetic genus by g. Let $s = s(C)$ denote the smallest degree of a surface containing the curve. Thus $s = \inf\{n \mid h^0(\mathcal{I}_C(n)) \neq 0\}$. Let ω_C denote the dualizing sheaf of C. We say C is *subcanonical* if there is an $\ell \in \mathbb{Z}$ such that $\omega_C \cong \mathcal{O}_C(\ell)$. We say C is *subcanonical of type* (a, b) if there exist integers $a \leq b$

such that $d = ab$ and $\omega_C \cong \mathcal{O}_C(a+b-4)$. Such integers a, b are uniquely determined, if they exist.

Proposition 1.2. *If $C \subseteq \mathbb{P}^3_k$ is an integral curve which is a specialization of complete intersection curves of type (a, b), then C is subcanonical of type (a, b).*

PROOF. It is well-known that a complete intersection curve of type (a, b) has degree $d = ab$ and $\omega = \mathcal{O}(a+b-4)$. The formation of ω is compatible with flat families, so by semicontinuity we see that $h^0(\omega_C(4-a-b)) > 0$. Since C is integral, by reason of degree, any non-zero section of $\omega_C(4-a-b)$ must generate this sheaf everywhere, so $\omega_C \cong \mathcal{O}_C(a+b-4)$ as required.

Because of this proposition, we are led to investigate subcanonical curves, and to ask in particular when they are complete intersections.

Lemma 1.3. *Let C be an integral subcanonical curve with $s(C) \leq 2$. Then C is a complete intersection.*

PROOF. If $s(C)=1$ there is nothing to prove. So we may assume $s(C)=2$. Since C is integral, is it contained in an integral surface Q of degree 2, namely a quadric cone or a nonsingular quadric surface. The divisor class of C on Q can be described by two integers s, t with $0 < s \leq t$, where $d = s + t$ and $g = (s-1)(t-1)$. (A curve on the quadric cone is determined up to linear equivalence by its degree d. If we write $s = t = \frac{1}{2}d$ for d even, and $s = \frac{1}{2}(d-1)$, $t = \frac{1}{2}(d+1)$ for d odd, then the genus is given by the same formula as on the smooth quadric.)

If C is subcanonical, then $\omega_C = \mathcal{O}_C(\ell)$ for some ℓ, and writing $\deg \omega_C = 2g-2$ we find

$$(s+t)\ell = 2st - 2s - 2t.$$

Of course if $s = t$, then C is a complete intersection of Q with another surface of degree s, so we will show that the case $s < t$ is impossible. Solving for ℓ, we find

$$\ell = \frac{2st}{s+t} - 2\,.$$

If $s < t$, then $\ell > s-2$, hence $\ell \geq s-1$, and so $H^1(\mathcal{O}_C(s-1)) \neq 0$.

On the other hand, we have the exact sequence on Q

$$0 \to \mathcal{I}_{C,Q} \to \mathcal{O}_Q \to \mathcal{O}_C \to 0.$$

Using $\mathcal{I}_{C,Q} \cong \mathcal{O}_Q(-s,-t)$ and twisting by $s-1$, we find

$$0 \to \mathcal{O}_Q(-1, s-t-1) \to \mathcal{O}_Q(s-1) \to \mathcal{O}_C(s-1) \to 0.$$

The first sheaf has $H^0 = H^1 = 0$, so we find

$$h^0(\mathcal{O}_C(s-1)) = h^0(\mathcal{O}_Q(s-1)) = s^2.$$

But by Riemann-Roch,

$$\begin{aligned} h^0(\mathcal{O}_C(s-1)) &= (s+t)(s-1) + 1 - (s-1)(t-1) + h^1(\mathcal{O}_C(s-1)) \\ &= s^2 + h^1(\mathcal{O}_C(s-1)). \end{aligned}$$

This contradicts $h^1(\mathcal{O}_C(s-1)) > 0$, so we have shown the case $s < t$ is impossible.

Remark 1.4. The hypothesis integral is necessary in this lemma. For example, the disjoint union of two lines is subcanonical with $s = 2$ but not complete intersection. Also, for any $g \leq 0$, there is a multiplicity two structure on a line which is subcanonical and has $s \leq 2$, but only the one with $g = 0$ is a complete intersection.

Lemma 1.5. *A subcanonical curve is a complete intersection if and only if it is arithmetically Cohen-Macaulay (ACM).*

PROOF. This is a result of Gherardelli [9]. By looking at the local ring of the vertex of the cone over the curve, it also follows from the fact that for local rings, "Gorenstein in codimension two implies complete intersection" [8, 21.20].

Proposition 1.6. *Let C be a subcanonical curve of type (a, b) with $0 < a \leq b$.*

a) $s(C) \leq a$
b) *If C is not a complete intersection, then $s(C) \leq a - 1$.*
c) *If $a = b$ and C is not a complete intersection, then $s(C) \leq a - 2$.*

PROOF. We use the method of associated vector bundles. There is a rank 2 vector bundle $\mathcal{E}$ on $\mathbb{P}^3$ and an exact sequence

$$0 \to \mathcal{O} \to \mathcal{E} \to \mathcal{I}_C(c_1) \to 0.$$

Here c_1 is the first Chern class of $\mathcal{E}$, and we have

$$d = c_2$$
$$\omega_C \cong \mathcal{O}_C(c_1 - 4),$$

where c_2 is the second Chern class [14, §1,2]. Thus in terms of a, b, we have

$$c_1 = a + b$$
$$c_2 = ab.$$

Since $c_1^2 - 4c_2 = (a-b)^2 \geq 0$, the bundle $\mathcal{E}$ cannot be stable [14, 8.4], and we conclude that $s(C) \leq \frac{1}{2}c_1$ [14, 3.1].

Let ℓ be the largest integer such that $h^0(\mathcal{I}_C(c_1 - \ell)) \neq 0$. Then $\ell \geq \frac{1}{2}c_1$ as we have seen. Also $\mathcal{E}(-\ell)$ is the smallest twist of $\mathcal{E}$ which has a global section, so it has a section corresponding to a curve (or possibly empty), and in particular, $c_2(\mathcal{E}(-\ell)) \geq 0$. Now

$$c_2(\mathcal{E}(-\ell)) = c_2 - c_1\ell + \ell^2.$$

Thus

$$\ell^2 - (a+b)\ell + ab \geq 0.$$

The zeros of this quadratic form are at $\ell = a$ and $\ell = b$. Since in any case $\ell \geq \frac{1}{2}c_1 = \frac{1}{2}(a+b)$, we conclude that $\ell \geq b$. Thus $c_1 - \ell \leq a$, so $s(C) \leq a$, which proves a).

If $s(C) = a$, then $\ell = b$ in the above discussion, so $c_2(\mathcal{E}(-\ell)) = 0$. Thus $\mathcal{E}$ is a direct sum of line bundles $\mathcal{O}(a) \oplus \mathcal{O}(b)$, and C is a complete intersection. This proves b).

To prove c), suppose that $a = b$ and $s(C) = a - 1$. Then $\mathcal{E}(-a-1)$ has a section corresponding to a curve Y of degree $c_2(\mathcal{E}(-a-1)) = 1$ and $\omega_Y \cong \mathcal{O}_Y(-6)$. There is no such curve, so this case cannot occur. This proves c).

Corollary 1.7. *Let C be an integral subcanonical curve of type (a, b), with $a \leq 3$ or $(a, b) = (4, 4)$. Then C is a complete intersection.*

PROOF. First note that if C is integral and subcanonical of type (a, b), then $a, b > 0$. Indeed, the degree of ω_C is at least -2, so $a, b < 0$ is impossible. Now, by (1,6), if C is not a complete intersection, then $s(C) \leq 2$ and (1.3) applies.

To study subcanonical with larger values of a, b, we will make use of the following result, which generalizes earlier work of [5] for curves on smooth surfaces.

Proposition 1.8. *Let C be an integral curve of degree $d = ab$ and genus $g = \frac{1}{2}ab(a + b - 4) + 1$ on an integral surface X of degree f, and assume that $a, b > f$.*

a) *C is subcanonical (necessarily of type (a, b)) if and only if the divisor $D = C - (a + b - f)H$ on X is effective, where H is the hyperplane section.*

b) *In that case, D is also subcanonical with*

$$\deg D = ab - (a + b - f)f$$
$$\omega_D \cong \mathcal{O}_D(2f - a - b - 4)$$

and the effective divisor D is unique ($h^0(\mathcal{L}(D)) = 1$).

PROOF. When speaking of curves as divisors, we use the theory of generalized divisors [15] which makes sense for any curve on any surface in $\mathbb{P}^3$. We use the exact sequence

$$0 \to \mathcal{O}_X \to \mathcal{L}(C) \to \omega_C \otimes \omega_X^{\vee} \to 0$$

[15, 2.10]. If C is subcanonical, then $\omega_C \cong \mathcal{O}_C(a + b - 4)$, and in any case $\omega_X = \mathcal{O}_X(f - 4)$, so we have

$$0 \to \mathcal{O}_X \to \mathcal{L}(C) \to \mathcal{O}_C(a + b - f) \to 0.$$

Twisting by $-(a + b - f)$ we have

$$0 \to \mathcal{O}_X(f - a - b) \to \mathcal{L}(C - (a + b - f)H) \to \mathcal{O}_C \to 0.$$

Since $a, b > f$, the first sheaf here has neither H^0 nor H^1, so we find

$$h^0(\mathcal{L}(D)) = h^0(\mathcal{O}_C) = 1.$$

Thus D is effective, and is unique.

Conversely, if D is effective, the same argument shows that $h^0(\omega_C(4-a-b)) \neq 0$. Since C is integral, by reason of degree, this shows that $\omega_C(4-a-b) \cong \mathcal{O}_C$, so C is subcanonical, as required.

To prove b), let $\mathcal{E}$ be the rank 2 vector bundle on $\mathbb{P}^3$ associated to C so that

$$0 \to \mathcal{O} \xrightarrow{s} \mathcal{E} \to \mathcal{I}_C(a+b) \to 0.$$

Since $a, b > f$, the surface X is the surface of minimum degree containing C. The corresponding section of $\mathcal{I}_C(f)$ gives rise to a section $t \in H^0(\mathcal{E}(f-a-b))$ and a curve D' satisfying

$$0 \to \mathcal{O} \xrightarrow{t} \mathcal{E}(f-a-b) \to \mathcal{I}_{D'}(2f-a-b) \to 0.$$

Now by construction, the quotient of $\mathcal{E}$ by s and t will give the ideal sheaf of C on X:

$$0 \to \mathcal{O} \oplus \mathcal{O}(a+b-f) \xrightarrow{s,t} \mathcal{E} \to \mathcal{I}_{C,X}(a+b) \to 0.$$

It follows that D' lies on the surface X, and

$$\mathcal{I}_{D',X}(f) \cong \mathcal{I}_{C,X}(a+b).$$

Thus D' is linearly equivalent to $C-(a+b-f)H$, so by uniqueness, $D' = D$. By construction, $D' = D$ is subcanonical and $\omega_D \cong \mathcal{O}_D(2f-a-b-4)$.

Proposition 1.9. *Let C be an integral subcanonical curve of type (a, b) with $4 \leq a \leq b$ on a nonsingular cubic surface X. Then the divisor D of (1.8) is of the form $D = (a+b-3)(L_1 + \cdots + L_k)$, where $1 \leq k \leq 6$ and $L_1, \ldots, L_k$ are skew lines on X. The possible values of k, a, b are as follows*

k	(a,b)
1	$(5,8), (6,6)$
2	$(6,15), (7,10)$
3	$(7,24), (8,15), (9,12)$
4	$(8,35), (9,21), (11,14)$
5	$(9,48), (10,28), (12,18), (13,16)$
6	$(10,63), (11,36), (12,27), (15,18)$.

Furthermore, for each k, a, b above there exist nonsingular such curves C, taking $L_1, \ldots, L_k$ to be $E_1, \ldots, E_k$. In addition for $k = 5$ there are other such curves obtained by taking $L_1, \ldots, L_5 = E_1, E_2, E_3, E_4, F_{56}$. (Here we use the usual notation for the lines on a cubic surface.)

PROOF. First we need a lemma

Lemma 1.10. *Let D be an effective divisor on a normal cubic surface X in $\mathbb{P}^3$ that is rigid, i.e. $h^0(\mathcal{L}(D)) = 1$. Then the support of D is a set of disjoint lines.*

PROOF. Since X is normal, if D is rigid, so is each connected component of D_{red}. So it will be sufficient to show conversely, that if D is a reduced connected divisor of degree $d \geq 2$, then $h^0(\mathcal{L}(D)) \geq 2$. Consider the exact sequence

$$0 \to \mathcal{O}_X \to \mathcal{L}(D) \to \omega_D \otimes \omega_X^{-1} \to 0 \, .$$

Here we have $h^0(\mathcal{O}_X) = 1$, $h^1(\mathcal{O}_X) = 0$, and $\omega_X^{-1} = \mathcal{O}_X(1)$ since X is a cubic surface. Thus it will be sufficient to show $h^0(\omega_D(1)) \geq 1$. By Serre duality on D, we must show $h^1(\mathcal{O}_D(-1)) \geq 1$. Using Riemann-Roch and $h^0(\mathcal{O}_D(-1)) = 0$, since D is reduced, we find

$$h^1(\mathcal{O}_D(-1)) = d - 1 + p_a \, .$$

Now $d \geq 2$ by hypothesis, and $p_a \geq 0$ since D is reduced and connected, so $h^1(\mathcal{O}_D(-1)) \geq 1$ as required.

PROOF OF (1.9). Given C, let D be the associated divisor of (1.8). Then D is rigid, so by the lemma, $D = \sum r_i L_i$ with L_i skew lines. On the cubic surface, if L is one of the 27 lines, the divisor rL is subcanonical with $\omega \cong \mathcal{O}(-r-1)$. In our case the r_i must all be equal to $a + b - 3$, so $D = (a + b - 3)(L_1 + \cdots + L_k)$. There are at most 6 skew lines on X, so $1 \leq k \leq 6$.

Now we compute the degree of D two ways and find

$$ab = (k + 3)(a + b - 3).$$

We look for integral solutions of this equation with $k = 1, \ldots, 6$ and find the possibilities in the above list. For example, if $k = 1$, we have

$$ab = 4(a + b) - 12.$$

Letting $u = a + b$, the integers a, b are roots of the equation

$$x^2 - ux + 4u - 12 = 0.$$

To have integral solutions, the discriminant

$$\Delta = u^2 - 16u + 48$$

must be a square. If $v = u - 8$, then $\Delta = v^2 - 16$. Set $\Delta = y^2$. Then we must solve the equation $v^2 - y^2 = 16$. Factoring gives $(v - y)(v + y) = 16$. The factorization $2 \cdot 8$ and $4 \cdot 4$ of 16 gives rise to $(a, b) = (5, 8)$ and $(a, b) = (6, 6)$. The cases $k = 2, \ldots, 6$ are handled similarly.

For existence, just take D in the form described, let $C = D + (a + b - 3)H$, and observe, by the theory of divisors on the cubic surface, that C is represented by an irreducible nonsingular curve. Then by (1.8a) it will be subcanonical of type (a, b).

Remark 1.11. We have constructed these curves using the internal theory of divisors on the cubic surface. One can also construct them more geometrically as follows.

For $k = 1$, consider the Del Pezzo surface of degree 4 in $\mathbb{P}^4$, which is a complete intersection of two quadric hypersurfaces. Let C_1 be the complete intersection of this surface with a hypersurface of degree 9. Then C_1 is a complete intersection of type $(2, 2, 9)$ in $\mathbb{P}^4$ with $\omega_{C_1} \cong \mathcal{O}_{C_1}(8)$. If C is a smooth projection of C_1 into $\mathbb{P}^3$, then deg $C = 36$ and $\omega_C \cong \mathcal{O}_C(8)$, so C is subcanonical of type (6,6). For a general projection, C will lie on a quartic surface with a double curve. If, however, we project from a general point of the Del Pezzo surface, C will lie on a nonsingular cubic surface. Similarly, a complete intersection (2,2,10) in $\mathbb{P}^4$ gives subcanonical curves of type (5,8) in $\mathbb{P}^3$. This construction was pointed out to us by R. Miro-Roig. Chiantini and Valabrega [6] also noted the subcanonical curves of type (5,8) on a quartic surface, and gave a number of examples of other subcanonical curves of type (a, b) lying on higher degree surfaces.

For $k \geq 2$, consider the Del Pezzo surfaces of degree e in $\mathbb{P}^e$ for $e = 5, \ldots, 9$. Since these surfaces have $K = \mathcal{O}(-1)$, any hypersurface section will be subcanonical, and for suitable choices of e and degree, we obtained the curves listed.

Proposition 1.12. *Let C be an integral subcanonical curve of type (a, b) with $4 \leq a \leq b$ on a rational normal cubic surface X_0 in $\mathbb{P}^3$. Then (a, b) must be one of the pairs listed in (1.9).*

PROOF. Let D_0 be the associated divisor of (1.8). Then D_0 is subcanonical with $\omega_{D_0} \cong \mathcal{O}_{D_0}(2-a-b)$. It is also rigid, and so by (1.10) its support is a set of disjoint lines in X_0.

According to the main theorem of [3], any effective divisor on X_0 is the limit of a flat family of effective divisors on smooth cubic surfaces X_t. So let D_t be a family of divisors on X_t specializing to D_0. Since D_0 is rigid on X_0, it follows by semicontinuity that D_t is rigid on X_t. Therefore D_t is supported on a set of disjoint lines X_t, say $D_t = \sum D_i$ with $D_i = r_i L_i$, $i = 1, \dots, k$.

We claim that all the r_i are equal to $a + b - 3$. Then the same numerical calculations used in the proof of (1.9) will give the result.

Looking at the connected components of D_0, it will be sufficient to prove the following: let L be a line of X_0, let $D_0 = rL$ for some $r \geq 1$, and suppose that D_0 is rigid and subcanonical with $\omega_{D_0} = \mathcal{O}_{D_0}(-\ell)$. Let D_0 be a limit of the family D_t on X_t where $D_t = \sum D_i$, $D_i = r_i L_i$, $i = 1, \dots, k$, and the L_i are disjoint lines. Then we must show all the r_i are equal to $\ell - 1$.

Equating the degrees of D_0 and D_t we have $r = \sum r_i$. For each $D_i = r_i L_i$ on the smooth cubic, we know that $\omega_{D_i} \cong \mathcal{O}_{D_i}(-r_i - 1)$. Equating the arithmetic genus of D_0 and D_t we find $\sum r_i(r_i + 1) = r\ell$.

Next, note that $h^0(\omega_{D_0}(-1)) = 0$ since D_0 is rigid (cf. proof of (1.10)). Since $\omega_{D_0} \cong \mathcal{O}_{D_0}(-\ell)$ this gives $h^0(\mathcal{O}_{D_0}(-\ell+1)) = 0$. By semicontinuity, we must have also $h^0(\mathcal{O}_{D_i}(-\ell+1)) = 0$ for each i. This is dual to $h^1(\omega_{D_i}(\ell - 1)) = h^1(\mathcal{O}_{D_i}(\ell - r_i - 2))$. Considering the map $\mathcal{O}_{D_i} \to \mathcal{O}_{L_i} \to 0$, where L_i is a line, we see that $H^1(\mathcal{O}_{D_i}(-2)) \neq 0$. Hence $\ell - r_i - 2 \geq -1$, so $\ell - r_i - 1 \geq 0$. These inequalities for each i, combined with the equality $\sum r_i(r_i + 1) = r\ell$ imply $r_i = \ell - 1$ for each i, as required.

Proposition 1.13. *Let C be an integral subcanonical curve of type (a, b) with $4 \leq a \leq b$ on the ruled cubic surface X of [15 §6]. The possible forms of D and values of (a, b) are as follows.*

a) *$D = L_2$, a multiplicity two structure on L, and $(a, b) = (4, 5)$.*

b) $D = L_2 + (a+b-3)E$, *and* $(a,b) = (5,10)$ *or* $(6,7)$.

c) $D = (a+b-3)E$, *and* $(a,b) = (5,8)$ *or* $(6,6)$.

Furthermore, the cases a) *and* b) *exist as irreducible nonsingular curves, while case* c) *the curves exist only as integral curves with 10 and 9 nodes, respectively.*

PROOF. Because of the structure of divisors on the ruled cubic surface [15, 6.5, 6.6] any effective divisor which does not move in a linear system must be a sum of copies of E, a suitable multiplicity two structure on L, and rulings of the surface. If rulings were present, then $\omega_D = \mathcal{O}_D(\ell)$ with $\ell \geq -2$, which is impossible if $a, b \geq 4$. Thus we can have only the three types listed. Computation of the degree of D, as in the proof of (1.9) yields the possible values for (a,b) shown.

As for existence, we take D of the form shown, choosing the multiplicity two structure on L_2 to have $\omega \cong \mathcal{O}(2-a-b)$, and let $C = D + (a+b-3)H$. In cases 1) and 2) we then get nonsingular curves [15, 6.6] but in case 3, the multiples of H we add will create nodes along L.

Remark 1.14. One can also show the existence of the curves of types a),b) more geometrically. In the first case, for example, the divisor class of C on X is $(13, 6, \alpha)$ with $h(\alpha) = 13$. So take a plane curve C_1 of degree 13 having a 6-fold point at the point P which is blown up to construct X. Then we get a nonsingular curve C of degree $2 \cdot 13 - 6 = 20$. On the other hand, the adjoint curves are of degree 10, passing 5 times through P, so $\omega_C \cong \mathcal{O}_C(5)$. Thus C is subcanonical of type (4,5).

For type b), take smooth plane curves of degrees 26 and 21 respectively.

Corollary 1.15. *Let* C *be a smooth subcanonical curve of type* (a,b), *with* $a \leq 4$ *or* $(a,b) = (5,5)$. *Then* C *is a complete intersection, except for the case* $(a,b) = (4,5)$ *in which case there exist smooth such curves on a ruled cubic surface which are not complete intersections.*

PROOF. If C is integral subcanonical of type (a,b) with $a \leq 4$ or $(a,b) = (5,5)$, and C is not a complete intersection, then $a, b, \geq 0$ as in the proof of (1.7), and by (1.6), we have $s(C) \leq 3$. We have already discussed the case $s(C) \leq 2$ (1.3), so we may assume $s(C) = 3$. Since C is integral, it lies on an integral cubic surface X.

If X is smooth, there are no such C, by (1.9). If X is normal and rational, there are no such curves, by (1.12).

If X is the general ruled cubic surface we have seen (1.13) the only possible exception, namely $(a, b) = (4, 5)$. If X is the other, more special ruled cubic surface, and if C is smooth, we can lift C to the nonsingular ruled cubic surface S in $\mathbb{P}^4$, then project C by a generic projection to a general ruled cubic surface in $\mathbb{P}^3$, and then apply (1.13).

If X is the cone over a nonsingular cubic curve, any multiplicity structure on a line of degree ≥ 2 moves in a linear system, so there are no possible divisors D as in (1.8).

Finally, if X is the cone over an irreducible singular cubic curve, and if C is smooth, we can lift it to the normalization $\tilde{X}$ of X, which is the cone over a twisted cubic curve. Then A Pic $\tilde{X} = \mathbb{Z}$, and one can check the degree and genus do not allow any subcanonical curves except complete intersections.

Corollary 1.16. *If C is a smooth curve in $\mathbb{P}^3$ which is a specialization of complete intersections of type (a, b) with $a \leq 4$ or $(a, b) = (5, 5)$, then C is a complete intersection, except possibly for the case $(a, b) = (4, 5)$ which remains open.*

Remarks 1.17. We expect (1.15) and (1.16) to remain true for integral curves, but do not have a complete proof.

Let us examine the case (4,5) more closely. One can show that the Hilbert scheme of smooth curves of degree 20 and genus 51 has four irreducible components:

- Y_1 the complete intersection curves of type (4,5). This is a generically smooth component of dimension 85.
- Y_2 curves of type (15;5,4,4,4,4,4) on a nonsingular cubic surface. These form a generically smooth component of dimension 89.
- Y_3 curves of type (16;6,5,5,5,4,3) on a nonsingular cubic surface. These form another generically smooth component of dimension 89.
- Y_4 the subcanonical curves of type (4,5) on a ruled cubic surface, described above. These form a nonreduced component of dimension 96.

For reasons of semicontinuity, the only conceivable specializations could be from Y_i to Y_j with $i < j$. There are none from Y_1 to Y_2 or Y_3 because the curves in Y_2 and Y_3 are not subcanonical (1.9). Conjecture A would say there are no specializations from Y_1 to Y_4. Conjecture B′ below would imply there are no specializations from Y_2 or Y_3 to Y_4. So we may have four irreducible components with no specializations between them.

§2. Families of curves with changing s

In the previous section we discussed the subcanonical curves which might be specializations of complete intersections, but we did not address the question of specialization directly. In this section we focus on possible families of curves in which the invariant s, the least degree of a surface containing the curve, drops.

Conjecture B. *Let $C \subseteq \mathbb{P}^3_T$ be a flat family of curves of degree d over an integral scheme T of dimension ≥ 1, and let $0 \in T$ be a closed point. Assume that for all $t \neq 0$, the fiber C_t lies on a normal surface X_t of degree $s = s(C_t)$, and assume that $s(C_0) < s$. Assume also* char.$k = 0$. *Then $d < s^2$, unless $s = 3$, in which case we allow also $d = s^2$.*

Proposition 2.1. *Conjecture B implies Conjecture A.*

PROOF. Suppose that C_0 is a smooth curve which is a specialization of complete intersection curves C_t of type (a, b). First we note that C_0 is necessarily integral, because a complete intersection is connected, and a limit of connected curves is connected. Secondly, the hypothesis C_0 smooth implies almost all of the C_t smooth, so by shrinking T if necessary, we may assume all C_t smooth.

Let X_t be a surface of degree $s = s(t) = a$ containing C_t, for $t \neq 0$. Then C_t is the complete intersection of X_t with another surface, so X_t must be smooth along C_t. In particular, X_t has isolated singularities, and so is normal. So we are in the situation of Conjecture B. Since $d = ab \geq s^2$, we conclude that $s(C_0) = s = a$. But then it follows from (1.2) and (1.6) that C_0 is a complete intersection. Note that the exception $s=3$, $d=9$ corresponds to $(a, b)=(3, 3)$ in which case Conjecture A is true (1.7).

Remarks 2.2 (a) The hypothesis X_t normal for $t \neq 0$ in Conjecture B is necessary. For example, there exist smooth rational curves C of any degree

$d \geq 5$ in the cubic scroll in $\mathbb{P}^4$. Projecting from a general point P_t in $\mathbb{P}^4$ will give a smooth rational curve C_t in a ruled cubic surface with a double line in $\mathbb{P}^3$, having $s = s(C_t) = 3$. If the point P_t specializes to a general point P_0 in the cubic scroll, the projected curve C_0 will be smooth, rational on a nonsingular quadric surface Q. Thus $s(C_0) = 2$.

(b) According to (2.1), Mohan Kumar's examples (1.1c) will give counterexamples to Conjecture B in characteristic $p > 0$.

(c) We will see below (2.5) an example with $s = 3$ and $d = s^2$.

(d) Conjecture B is true in case $s = 2$. Indeed, a plane curve C_0 has maximum genus for its degree, so can never be a limit of a flat family of curves not contained in a plane.

While we cannot prove any nontrivial case of Conjecture B, there is a certain amount of evidence which supports it. We begin by discussing the case $s = 3$, $s(C_0) = 2$.

Proposition 2.3. *Let C_0 be a curve of bidegree (a, b) on a nonsingular quadric surface Q. If $a, b \geq 3$, then C_0 cannot be a limit of any flat family of curves C with $s(C_t) > 2$ for $t \neq 0$.*

PROOF. We use semicontinuity of $h^0(\mathcal{O}_C(2))$. On the quadric surface we have

$$0 \to \mathcal{I}_{C_0,Q}(2) \to \mathcal{O}_Q(2) \to \mathcal{O}_{C_0}(2) \to 0.$$

Now $\mathcal{I}_{C_0,Q}(2) = \mathcal{O}_Q(2-a, 2-b)$. Since $a, b \geq 3$, this sheaf has neither h^0 nor h^1. So $h^0(\mathcal{O}_{C_0}(2)) = h^0(\mathcal{O}_Q(2)) = 9$.

On the other hand, if C_t is a curve with $s \geq 3$, then $h^0(\mathcal{O}_{C_t}(2)) \geq 10$, so by semicontinuity, there can be no such family.

Remarks 2.4. Curves of type $(1, b)$ on Q are rational curves of degree $d = b + 1$. For $d \geq 5$ these are limits of the general rational curve of degree d, which does not lie on a quadric surface.

For type $(2, b)$ we can make examples as follows. Take the union of a curve C of type $(2, b-1)$ and a line L of type (0,1), meeting C in two points. We can then move L off the quadric, so as to still meet C in two points. This gives flat families whose general curve does not lie on a quadric, for any $b \geq 4$.

It follows from (2.3) that for $a, b \geq 3$, the curves of type (a, b) on quadric surfaces form an (open subset of) an irreducible component of the Hilbert scheme.

Proposition 2.5. *There are families of smooth curves C_t lying on smooth cubic surfaces X_t, whose limit curve is a smooth curve C_0 lying on a non-singular quadric surface Q, for*

$$g = 0, \qquad d = 5, 6, 7, \quad \textit{and}$$
$$g = d - 3, \quad d = 6, 7, 8, 9.$$

PROOF. We use three different methods. The first is the method of the irreducible Hilbert scheme. One knows that the Hilbert scheme $H_{d,g}$ of smooth curves of degree d and genus g, for $g \leq d - 3$, is irreducible [7]. For $g = 0$, $d = 5, 6$, and for $g = d - 3$, $d = 6, 7$ the general such curve lies on a nonsingular cubic surface, while there are such curves (of bidegrees (1,4), (1,5); (2,4); (2,5) respectively) on a nonsingular quadric surface. Since the Hilbert scheme is irreducible, there are flat families as claimed. For larger values of d, this method does not work, because the general curve no longer lies on a cubic surface.

The second method, using the codimension of determinantal varieties, was suggested to us by D. Eisenbud. We illustrate this method with the case $d = 9$, $g = 6$. The Hilbert scheme $H_{9,6}$ is irreducible of dimension $4d = 36$. For a curve C, we consider the sequence

$$0 \to H^0(\mathcal{I}_C(3)) \to H^0(\mathcal{O}_{\mathbb{P}^3}(3)) \xrightarrow{\alpha} H^0(\mathcal{O}_C(3)).$$

The curve C will be contained in a surface of degree 3 if and only if α has a kernel. Now $h^0(\mathcal{O}_{\mathbb{P}^3}(3)) = 20$ and $h^0(\mathcal{O}_C(3)) = 22$. These ranks are constant as C varies, so α gives rise to a map of locally free sheaves of ranks 20 and 22 respectively on the Hilbert scheme. A theorem on determinantal varieties ([8], Ex.10.9, p.244) says that every irreducible component of the locus where α has less than maximal rank must have codimension ≤ 3, i.e. dimension ≥ 33. So we calculate: the dimension of the family of such curves lying on a quadric surface is 32; on a nonsingular cubic surface 33; on a ruled cubic surface 32. We conclude that the families of such curves lying on quadrics or ruled cubic surfaces are in the closure of the family of these curves on nonsingular cubic surfaces. Thus the required family with $(d, g) = (9, 6)$ exists.

This method also shows the existence of the family with $(d, g) = (8, 5)$. For $(d, g) = (7, 0), (8, 0), (9, 0)$, this method shows the existence of a family whose general number has $s(C_t) = 3$, but does not allow us to show the general curve of the family lies on a smooth cubic. Since we know already there exist families coming from the ruled cubic surfaces (2.2a) this gives no new information.

The third method is the explicit construction of families of curves by analysis of the divisor classes on a degenerating family of surfaces, as in the paper [16]. In the family of [16, 3.4] where nonsingular cubic surfaces degenerate to the union of a nonsingular quadric surface Q plus a transversal plane, one can obtain families of smooth curves with $(d, g) = (5, 0), (6, 0), (6, 3)$. For example to get (6,3) take $h = 2$, $a = -2$ and $b_i = -1$, $i = 1, \ldots, 6$. In this family of surfaces the maximum degree of a flat family of smooth curves is 6.

In the family of [16, 3.11] where nonsingular cubic surfaces degenerate to the union of a nonsingular quadric surface Q plus a tangent plane, we can obtain a family of smooth curves with $(d, g) = (7, 4)$ by taking $h = 0$, $e = f_1 = \cdots = f_5 = g = 1$. In this family of surfaces the maximum degree of a flat family of smooth curves is 7.

Going one step further, we construct another family as follows. Let X_1 be the cubic surface obtained by blowing up six points $P_1, \ldots, P_6$ in $\mathbb{P}^2$ with P_1, P_2, P_3 collinear. Then the line F_{123} collapses to make a double point. There are 21 lines on this surface, which we denote by $E_1, \ldots, E_6$, $G_1, \ldots, G_6$, and F_{ij}, $i \in \{1, 2, 3\}$, $j \in \{4, 5, 6\}$, with the usual incidence relations, if we understand that F_{12}, F_{13}, F_{23} are gone and $F_{45} = G_6$, $F_{46} = G_5$, $F_{56} = G_4$. The lines $E_1, E_2, E_3, G_4, G_5, G_6$ pass through the double point.

Let Q be the nonsingular quadric surface containing the three skew lines F_{16}, F_{26}, F_{36}. They meet E_6 and G_6, so these two lines are also in Q. Let H_0 be the plane containing the triangle E_4, G_5, $F_{45} = G_6$. Let $X_0 = Q \cup H_0$, and construct a family by taking $(1 - t)X_0 + tX_1$.

Now one can show, using the same techniques as in [16, §3], that H_0 is tangent to Q. Let O be the point of tangency, and let L_1, L_2 be the two lines in which Q intersects H_0, taking $L_2 = G_6$. Then E_6 lies in the first family on Q and meets L_1 at a point $R \neq O$. The lines $F_{16}, F_{26}, F_{36}, G_4 = F_{56}$ meet L_2 in points P_1, P_2, P_3, P_5 distinct from O, with P_5 being the double point of X_1. The line E_4 lies in H_0, passing through O, and different from L_1 and L_2. Now the divisor class

$$C = E_6 + H - E_4 + F_{16} + F_{26} + F_{36} + 2G_4$$

on the family will give a curve on X_0 of bidegree (2,6) in Q. A general member of this linear system on Q is a smooth curve C_0 of $(d, g) = (8, 5)$. Using [16, 1.6], there is a family of smooth curves C_t restricting to C_0.

The same construction, taking $C' = C - E_6$, gives curves with $(d, g) = (7, 0)$.

Remark 2.6. Combining the previous two results, we see that Conjecture B is true for families of smooth curves with general $s = 3$, except for the case $(d, g) = (9, 6)$ and possibly for the cases $g = 0$, $d \geq 8$ and $g = d - 3$, $d \geq 10$, which remain open.

One approach to proving the conjecture for $s = 3$ would be to extend the methods of [16, §4] to include the behavior of divisors in the neighborhood of the bad point O where Q is tangent to H_0.

Remark 2.7. Looking for families of smooth curves whose general C_t lies on a normal quartic surface, and whose special curve C_0 lies on a cubic surface is much more difficult.

(a) One can generalize (2.3) to say that if C_0 lies on an integral cubic surface, and if $D = C_0 - 3H$ is represented by a connected reduced curve, then C_0 cannot be the specialization of any family with $s(C_t) > 3$ for $t \neq 0$.

(b) The method of irreducible Hilbert schemes (2.4) shows the existence of families with general $s = 4$ lying on smooth quartics, and $s(C_0) = 3$ for $g = 0$, $d = 7, 8$, and for $g = d - 3$, $d = 8, 9, 10$ for example.

The largest d for which we can prove the existence of such a family is 14. An example follows.

Example 2.8. There exists a family C_t of smooth curves with $d = 14$, $g = 23$, whose general member lies on a smooth quartic surface, and whose special member C_0 lies on a smooth cubic surface.

To construct C_0, let D_0 be the disjoint union of a rational quartic curve and a line on a nonsingular cubic surface F. Let $C_0 = D_0 + 3H$. Then C_0 is smooth of $d = 14$ and $g = 23$, and note by construction $h^1(\mathcal{I}_{C_0}(3)) = 1$. The family of all curves constructed by this method, as F varies, has dimension $d + g + 18 = 55$. This is less than $4d = 56$. One knows that every irreducible component of the Hilbert scheme has dimension $\geq 4d$, so we conclude that C_0 is a limit of a family of curves C_t, whose general $s(C_t) \geq 4$.

Next, we study possible smooth curves C with $d = 14$, $g = 23$, and $s(C) \geq 4$. By reason of its degree, $\mathcal{O}_C(4)$ will be nonspecial, so we can compute $h^0(\mathcal{O}_C(4)) = 4d + 1 - g = 34$. This is less than $h^0(\mathcal{O}_{\mathbb{P}^3}(4)) = 35$, so C must be contained in a quartic surface X. Furthermore, supposing C to be a general curve of the family C_t, by semicontinuity we must have $h^1(\mathcal{I}_C(3)) \leq 1$.

Case 1: $h^1(\mathcal{I}_C(3)) = 1$. Then $h^0(\mathcal{O}_C(3)) = 21$ so $h^1(\mathcal{O}_C(3)) = 1$. By duality $h^0(\omega_C(-3)) = 1$. This implies, as in the proof of (1.8), that the linear system $C - 3H$ is effective. If $D \sim C - 3H$, then deg $D = 2$, $g(D) = -1$. We can take for D two skew lines, so such curves C exist and the general one will lie on a nonsingular quartic surface.

Case 2: $h^1(\mathcal{I}_C(3)) = 0$. In any case $C - 2H$ will be effective, giving a curve D' with $\deg(D') = 6$, $g(D') = 3$. Furthermore, $h^1(\mathcal{I}_{D'}(1)) = 0$. It follows from the classification of curves with $(d, g) = (6, 3)$ [1] that either D' is ACM or D' is of bidegree (2,4) on a quadric surface. In either case, such curves C exist and the general ones will lie on nonsingular quartic surfaces.

Thus we have shown that any irreducible component of the Hilbert scheme of curves C, of $(d, g) = (14, 23)$, not lying on a cubic surface, with $h^1(\mathcal{I}_C(3)) \leq 1$, contains curves lying on nonsingular quartic surfaces. Thus there exist families with limit C_0, as claimed. In fact, one can show that the two non-ACM families of Cases 1,2 are in the closure of the ACM family, so there is just one irreducible component of the Hilbert scheme of these curves.

As further evidence for Conjecture B, we give one very special case in which we can prove the result, and an example to show that the bound in this case is sharp.

Proposition 2.9. *Let C_t be a family of smooth curves of degree d, contained in a family X_t of normal surfaces of degree $s = s(C_t)$ for $t \neq 0$. Suppose that $s(C_0) = s - 1$, and that X_0 consists of a normal surface Y of degree $s - 1$, plus a plane H_0, such that Y and H_0 meet in a nonsingular curve Γ.*

(a) *Then Γ will have exactly $s(s-1)$ points of positive index* [16, 4.9], *counted with multiplicity.*

(b) *If furthermore these points all have multiplicity 1, then $d \leq s(s-1)$.*

PROOF. The proof of (a) is the same as [16, 4.11]. To prove (b), we observe, as in the proof of [16, 6.1], that C can meet Γ only at points of positive index. Furthermore, if the point has index 1, then the intersection multiplicity of C with Γ is ≤ 1 at that point [16, 5.6]. It follows that if the index is 1 at every point, C meets Γ with multiplicity $\leq s(s-1)$. But $C.\Gamma = C.H_0 = \deg C$, so $d \leq s(s-1)$.

Example 2.10. For each $s \geq 3$, there is a family of smooth curves C_t of degree $d = s(s-1)$ and genus $g = \frac{1}{2}s(s-1)(2s-5) - s + 3$, of which the general curve C_t lies on a smooth surface X_t of degree s, and the special curve C_0 (also smooth) lies on a smooth surface Y of degree $s - 1$. The curves C_t for $t \neq 0$ are ACM, while C_0 is in the biliaison class of the disjoint union of a plane curve of degree $s - 2$ with a line.

For $s = 3$ this is the family of curves with $d = 6$, $g = 3$ specializing to the curve of bidegree (2,4) on a quadric surface. We obtain our examples for any s by generalizing from this case.

Start with a family of two skew lines $Z_t = L_t \cup M_t$, which specialize to a scheme Z_0 having support equal to two lines L_0, M_0 meeting at a point P_0, and having a nilpotent element at that point. This is a flat family of schemes. The general member is contained in a smooth surface of degree s. The special member is contained in the union of a smooth surface Y containing L_0 and M_0 and a transversal plane H_0 containing P_0. Since $h^0(\mathcal{I}_{Z_t}(n))$ is constant in the family, we can find a family of surfaces X_t of degree s containing the family Z_t, such that

a) X_t is smooth for $t \neq 0$
b) $X_0 = Y \cup H_0$ is the union of a smooth surface Y containing L_0 and M_0, and a plane H_0, and the intersection $\Gamma = Y \cap H_0$ is smooth.

c) The $s(s-1)$ points of positive index on Γ are distinct. Call this set Σ. (This can be accomplished by requiring each X_t to pass through a fixed set of $s(s-1)$ points of Γ.)

Next, as in the proof of [16, 4.11] choose a family of points $x_t \in X_t$ tending to a point $x_0 \in H_0 \backslash \Gamma$. Let H_t be the tangent plane to X_t at x_t, and let D_t be the curve $H_t \cap X_t$. Then H_t tends to H_0, so D_t has as limit a curve D_0 in H_0 with a double point at x_0. Since Γ is irreducible of degree $s-1$ and D_0 is of degree s with a double point at x_0, D_0 must meet Γ properly, namely at the $s(s-1)$ points of Σ. Thus D is a Cohen-Macaulay family of divisors on X.

Note also that even though Z is not a Cohen-Macaulay family, its components L and M are.

Now we define the divisor class $C = sH - D + L - M$ on the family X. We will show that C is effective, and is represented by a family of smooth curves as described.

First we consider C_0. Taking our complete intersections sH on X_0 to pass through Σ, then $sH \cap H_0$ is linearly equivalent to D_0, so that $C_0 \sim sH + L - M$ on Y, where the curves of sH are constrained to pass through Σ. Clearly we can take sH to contain M, so C_0 is effective. Furthermore the divisor class C is still Cohen-Macaulay on X, because L_0 and M_0 pass through the same point P_0 of Σ.

To show that C_0 can be represented by a smooth curve on Y, we blow up the points of Σ to obtain $\tilde{Y}$, and use the criterion of separation of points and tangent vectors to show that the proper transform $\tilde{C}_0$ is very ample on $\tilde{Y}$ except along the curve $\tilde{\Gamma}$. The key point here is the computation $\tilde{C}_0.\tilde{L} = 1$ and that $H^0(\mathcal{O}_{\tilde{Y}}(\tilde{C}_0)) \rightarrow H^0(\mathcal{O}_{\tilde{L}}(\tilde{C}_0))$ is surjective (details left to the reader).

As for C_t for $t \neq 0$, note that $D_t = X_t \cap H_0$, so D_t is linearly equivalent to a hyperplane section H on X_t. Thus $C_t \sim (s-1)H + L_t - M_t$. Clearly C_t is effective, because we can make one of the H's contain M_t.

Now we compute the genus of C_0 and C_t, starting from sH (resp. $(s-1)H$) and subtracting M and adding L. We get the same result $g = \frac{1}{2}s(s-1)(2s-5) - s + 3$ in both cases. Note that here we make essential use of the fact that L_0 and M_0 meet in a point, while L_t and M_t are skew.

Finally, we make a computation which shows that $h^0(\mathcal{O}_{C_t}(n)) = h^0(\mathcal{O}_{C_0}(n))$ for all n. Since the genus is the same, this implies $h^1(\mathcal{O}_{C_t}(n)) = h^1(\mathcal{O}_{C_0}(n))$ for all n, and so by duality $h^0(\omega_{C_t}(n)) = h^0(\omega_{C_0}(n))$ for all n. Now we use the exact sequence [15, 2.10]

$$0 \to \mathcal{O}_X \to \mathcal{L}(C) \to \omega_C \otimes \omega_X^\vee \to 0$$

to conclude also that $h^0(\mathcal{L}_{X_0}(C_0)) = h^0(\mathcal{L}_{X_t}(C_t))$. Then by Grauert's theorem [13, III,12.9] $f_*\mathcal{L}(C)$ is locally free on the parameter scheme T, and commutes with base extension. We conclude, as in [16, 1.6], that there exists a family of effective curves C on X, with C_0, and hence also C_t for $t \neq 0$ nonsingular. This is the required family.

If we write $C_0 \sim (s-1)H + L_0 + (H - M_0)$ on Y and $C_t \sim (s-2)H + L_t + (H - M_t)$ on X_t, then it is clear that C_0 is in the biliaison class of the disjoint union of a plane curve $H - M_0$ of degree $s-2$ and the line L_0; while C_t on X_t is the biliaison class of a plane curve $H - M_t$ and a line L_t attached at one point, which is ACM.

Remark 2.11. One can also obtain this family via the method of triads of [17]. Start with a Koszul module $k[x, y, z, w]/(f_1, f_2, f_3, f_4)$ where the f_i have degree $1, 1, 1, s-2$. Following [17, 5.18.1′] we construct the associated modular triad. According to the calculations there, we obtain a minimal family of curves of degree $s+3$ and genus $\frac{1}{2}(s-1)(s-2)+2$. The general curve is ACM and lies on a surface of degree 3 while the special curve is in the biliaison class of a plane curve of degree $s-2$ plus a disjoint line, and lies on a surface of degree 2. Making a biliaison of height $s-3$ on a family of smooth surfaces of degree $s+1$ will give our family.

§3. Lifting from characteristic p

We say a scheme Y_0, defined over a field k of characteristic $p > 0$, is *liftable to characteristic* 0 if there exists a flat family of schemes Y over an integral scheme T, and a point $0 \in T$ such that Y_0 is the fiber of Y over 0, and the characteristic of the residue field $k(\tau)$ of the generic point $\tau \in T$ is zero. It is known that any nonsingular projective curve lifts (as an abstract variety) to characteristic 0 [10]. Serre [23] gave examples of varieties of dim ≥ 3

which do not lift. There are also surfaces which do not lift, for example, the surfaces of Raynaud [21] or the "generalized" Raynaud surfaces of Lang [18]. In either case, these are surfaces of general type with $c_2 < 0$, which cannot exist in char.0. Oort [20] has studied the problem of lifting curves with their endomorphisms. Here we consider the question of lifting embedded curves.

Conjecture C. *For each $p > 0$, there exist nonsingular curves C_0 in $\mathbb{P}^3_k$, where k is a field of characteristic p, which do not lift (as embedded curves) to any curve in $\mathbb{P}^3_k$ over a field k of characteristic 0. In other words, the Hilbert scheme of curves in $\mathbb{P}^3_{\mathbb{Z}}$ has irreducible components lying over each prime p.*

To approach this conjecture, we formulate another conjecture, closely related to Conjecture B. In that case the surface X_t containing the general curve specialized to a reducible surface X_0. In this case X_0 will be integral but not normal.

Conjecture B′. *Let $C \subseteq \mathbb{P}^3_T$ be a flat family of curves of degree d over an integral scheme T of dimension ≥ 1, and let $0 \in T$ be a closed point. Let C be contained in a flat family of surfaces $X \subseteq \mathbb{P}^3_T$ of degree $s = s(C_t)$. Assume that X_t is normal for $t \neq 0$, that C_0 is nonsingular, and that X_0 is integral but not normal. Then $d \leq \phi(s)$, where $\phi(s)$ is a bound depending only on the degree s of the surfaces.*

We do not yet have enough evidence to make a good guess what the bound $\phi(s)$ should be, but we expect it to be of the order of s^2. For $s = 3$, we expect $\phi(s) = 10$ should do.

Remark 3.1. We believe that Conjecture B' will imply Conjecture C, and offer the following sketch of a proof. Let Y be a nonsingular projective surface over the field k of characteristic p which does not lift to characteristic 0. Embed Y in some $\mathbb{P}^n$ by a sufficiently high multiple of a very ample divisor, so that its generic projection X_0 in $\mathbb{P}^3$ has only ordinary singularities: a double curve with a finite number of pinch points and triple points [22]. Let us assume the double curve is irreducible. Let $s = \deg Y = \deg X_0$.

Now let C' be the intersection of Y with a hypersurface of degree f in $\mathbb{P}^n$, and let $C_0 \subseteq X_0$ be its projection to $\mathbb{P}^3$. For sufficiently general choice of C', the projection C_0 will be nonsingular. Its degree is $d = fs$. Its genus can be computed on Y as

$$g = \tfrac{1}{2}C'(C' + K) + 1.$$

Writing $C' \sim fH$ where H is the hyperplane class, we get

$$g = \tfrac{1}{2}f^2 s + \tfrac{1}{2}fb + 1$$

where $b = (H.K)$ is constant. Substituting $f = d/s$ we can write

$$g = \frac{1}{2s}(d^2 + bd) + 1.$$

Now suppose that C_0 is the limit of a flat family of curves $C \subseteq \mathbb{P}^3_T$ whose general fiber C_t is in characteristic 0. Then C_t will have the same degree and genus. If C_t is not contained in a surface of degree s, then g will be bounded by a certain function $G(d, s)$

$$g \le \frac{1}{2s+2}\, d^2 + \text{lower terms}$$

[11]. For f sufficiently large, the genus of our curve C_t will exceed this bound, so we conclude that $s(C_t) = s$. Let X be a family of surfaces of degree s containing the family C.

If the general surface X_t is not normal, then, since the singular curve of X_0 is irreducible, X_t must have a singular curve of the same degree. Assuming that the normalizations of the surfaces X_t will form a family of smooth surfaces with limit Y, then Y would be liftable to characteristic 0, contrary to hypothesis.

We conclude that the general surface X_t is normal. Now we are in a position to apply Conjecture B$'$, and find that $d < \phi(s)$. For f sufficiently large, d will exceed this bound, so we conclude that the curve C_0 will not be liftable.

Remark 3.2. Concerning evidence for the Conjecture B$'$, we can say something in the case $s = 3$. Already some time ago, Gruson and Peskine [12] noted that there are some curves on the ruled cubic surface which cannot be specializations of curves on a nonsingular cubic surface. Indeed, if C is a curve on the ruled cubic surface whose divisor class is (a, b, α), in the notation of [15, §6], then the dimension of the family of such curves in $\mathbb{P}^3$ is $d+g+a+12$ [12]. On the other hand, the dimension of the family of curves of the same degree and genus on a nonsingular cubic surface is $d+g+18$. Thus, if $a \geq 6$, the general such curve on the ruled cubic cannot be a limit of a family of such curves on smooth cubics. Our conjecture in this case says that for a sufficiently large, no nonsingular curve on the ruled cubic is a limit of curves on the nonsingular cubic.

For evidence, we will show existence of families for $a \leq 5$, and nonexistence for some particular classes of curves with $a \geq 7$.

Proposition 3.3. *Any nonsingular curve C_0 on the ruled cubic surface having divisor class (a, b, α), with $a \leq 5$ is a limit of a family of curves on a nonsingular cubic. The largest degree attained by these curves is 10, for the class $(5, 0, \alpha)$ with genus 6.*

PROOF. These can all be shown by the method of the irreducible Hilbert scheme or by Eisenbud's method, cf. (2.5).

Proposition 3.4. *Let C_0 be a nonsingular curve on the ruled cubic surface with divisor class $(a, 0, \alpha)$, where $a \geq 7$ and $h(\alpha) = a$. Then C_0 cannot be the limit of a flat family of curves whose general member lies on a nonsingular cubic surface. These are curves with $d = 2a$, $g = \frac{1}{2}(a-1)(a-2)$.*

PROOF. Let $D_0 = aH - C_0$ on the ruled cubic surface X_0. Then $D_0 = (a, a, \alpha)$. This is an effective divisor, represented by a disjoint union of a rulings of X_0. It has deg $D_0 = a$, $g(D_0) = -a+1$. Note that $h^0(\mathcal{O}_{D_0}(-1)) = 0$, so $h^1(\mathcal{I}_{D_0,X_0}(-1)) = h^1(\mathcal{L}(C_0 - (a+1)H)) = 0$.

Now suppose that C_0 is the limit of a family of curves C whose general curve C_1 lies on a nonsingular cubic surface X_1. Then $D_1 = aH - C_1$ will be a divisor class of the same degree and genus as D_0, namely deg $D_1 = a$, $g(D_1) = -a+1$. By Riemann-Roch on the surface X_1,

$$h^0(\mathcal{L}(D_1)) \geq \deg D_1 + g(D_1) = 1.$$

Hence D_1 is effective and rigid, and so must be supported on a set of skew lines (1.10). Since X_1 contains at most 6 skew lines, and $a \geq 7$, D must contain at least one line E with multiplicity ≥ 2. Therefore $h^0(\mathcal{O}_D(-1)) \neq 0$. This translates to $h^1(\mathcal{L}(C_1-(a+1)H)) \neq 0$, which contradicts semicontinuity of cohomology in the family. Therefore no such family can exist4

Remark 3.5. Recently Brevik and Mordasini [4] have proved Conjecture B′ for the case of smooth cubic surfaces with limit a general cubic ruled surface. The degree of families of smooth curves in this case is $d \leq 10$. Otherwise Conjecture B′ remains open.

References

[1] Aït-Amrane, S., Sur le schéma de Hilbert $H_{d,(d-3)(d-4)/2}$, in preparation.

[2] Ballico, E., Ciliberto, C., eds., "Algebraic curves and projective geometry," Proceedings Trento 1988, Springer Lecture Notes in Mathematics **1389** (1989).

[3] Brevik, J., Families of curves on families of surfaces degenerating to normal rational cubic surfaces in $\mathbb{P}^3$. Thesis, Berkeley (1996).

[4] Brevik, J., Mordasini, F., in preparation.

[5] Casnati, G., Dolcetti, A., Ellia, Ph., On subcanonical curves lying on smooth surfaces in $\mathbb{P}^3$, *Revue Roumaine Math. Pures. Appl.* **40** (1995), 289–300.

[6] Chiantini, L., Valabrega, P., Subcanonical curves and complete intersections in projective 3-space. *Annali Mat. pura appl.* **138** (1984), 309–330.

[7] Ein, L., Hilbert scheme of smooth space curves, *Ann. Sci. ENS* **19** (1986), 469–478.

[8] Eisenbud, D., *Commutative Algebra, with a view toward algebraic geometry*, Springer, GTM **150** (1995).

[9] Gherardelli, G., Sulle curve sghembe algebriche intersezioni complete di due superficie, *Atti Accad. Italia Rend.* **VII**, ser. 4 (1943), 128–132.

[10] Grothendieck, A., Géométrie formelle et géométrie algébrique. *Séminaire Bourbaki* **182** (1958/59).

[11] Gruson, L., Peskine, C., Genre des courbes de l'espace projectif, Springer LNM **687** (1977), 31–59.

[12] Gruson, L., Peskine, C., Genre des courbes de l'espace projectif, II, *Ann. Sci. ENS* (4) **15** (1982), 401–418.

[13] Hartshorne, R., *Algebraic Geometry*, Springer (1977).

[14] Hartshorne, R., Stable vector bundles of rank 2 on $\mathbb{P}^3$, *Math. Ann.* **238** (1978), 229–280.

[15] Hartshorne, R., Generalized divisors on Gorenstein schemes, *K-Theory* **8** (1994), 287–339.

[16] Hartshorne, R., Families of curves in $\mathbb{P}^3$ and Zeuthen's problem, *Memoirs AMS.*

[17] Hartshorne, R., Martin-Deschamps, M., Perrin, D., Triades et familles de courbes gauches, in preparation.

[18] Lang, W. E., Examples of surfaces of general type with vector fields, in *Arithmetic and Geometry*, dedicated to I.R.Shafarevich, ed. M.Artin and J.Tate, Birkhäuser (1983), vol.II, 167–173.

[19] Mohan Kumar, N., Smooth degeneration of complete intersection curves in positive characteristic, *Invent. Math.* **104** (1991), 313–319.

[20] Oort, F., Lifting algebraic curves, abelian varieties, and their endomorphisms to characteristic 0, in *Algebraic Geometry*. Bowdoin 1985, Proceedings Symp. Pure Math. **46**, American Math Society (1987), vol.2, 165–195.

[21] Raynaud, M., Contre-exemple au "Vanishing Theorem" en caractéristique $p > 0$, in C.P.Ramanujam — A tribute. Springer (1978), 273–278.

[22] Roberts, J., Generic projections of algebraic varieties, *Amer. J. Math.* **93** (1971), 191–214.

[23] Serre, J.-P., Exemples de variétés projectives en caractéristique p non relevable en caractéristique zero. *Proc. Nat. Acad. Sci.* **47** (1961), 108–109.

On Codimension Two k-Buchsbaum Subvarieties of $\mathbf{P}^n$

Ph. Ellia[(1)]-A. Sarti[(2)]

(1) Dipartimento di Matematica
via Machiavelli,35-44100 Ferrara,Italy
email: Phe@dns.unife.it
(2)Mathematisches Institut der Universität,
Bunsen Strasse 3-5,
37073 Göttingen, Germany
email: Sarti@cfgauss.uni-math.gwdg.de

1. Introduction:

In the last few years there has been a great deal of activity on arithmetically Buchsbaum and then on k-Buchsbaum curves in $\mathbf{P}^3$. Recall that a closed subscheme $X \subset \mathbf{P}^n$ is said to be *arithmetically Buchsbaum* if the local ring of the vertex of the affine cone of X is a Buchsbaum ring (see [5]).The property of being arithmetically Buchsbaum is preserved by hyperplane section. A more geometrical definition can be found in [1]: let $X \subset \mathbf{P}^n$ be of codimension two, then X is arithmetically Buchsbaum if $m.H^p_*(\mathcal{I}_{X\cap L}) = 0$ for any linear subspace L of dimension $t, 1 \leq p \leq t-2$(here m denotes the maximal ideal of $S = k[X_0, ..., X_n]$). Anyway, it turns out that a curve $C \subset \mathbf{P}^3$ is arithmetically Buchsbaum if $m.H^1_*(\mathcal{I}_C) = 0$. This notion has been extended in the following way: C is k-Buchsbaum if $m^k.H^1_*(\mathcal{I}_C) = 0$. This motivates the following:

Definition: A closed subscheme, of codimension two, $X \subset \mathbf{P}^n$ is said to be k-Buchsbaum if $m^k.H^p_*(\mathcal{I}_{X\cap L}) = 0$ for every p, $1 \leq p \leq m-2$, where L is a general linear subspace of dimension m.

Note that we consider only general linear subspaces (cp [1]), this avoids problems with non proper intersections (and also doesn't cause any trouble).

Clearly if X is k-Buchsbaum, $X \cap L$ is also k-Buchsbaum; also any X is k-Buchsbaum for some k. Finally if X is arithmetically Buchsbaum then it is 1-Buchsbaum. In this note we prove:

Theorem 1. *Let $X \subset \mathbf{P}^n, n \geq 4$, be a closed, codimension two, subscheme which is subcanonical and k-Buchsbaum with $k \leq 2$, then X is a complete intersection.*

As a corollary we get that a smooth codimension two, $X \subset \mathbf{P}^n$, $n \geq 6$, which is k-Buchsbaum with $k \leq 2$, is a complete intersection. For $k = 1$ we recover a result of Mei-Chu Chang ([1]).

The proof goes as follows: since X is two-codimensional and subcanonical, we may associate a rank two vector bundle, $\mathcal{E}$, to X; now consider a linear section $C = X \cap L$ with $L \simeq \mathbf{P}^3$. The curve $C \subset \mathbf{P}^3$ is k-Buchsbaum and is a section of the rank two vector bundle $E := \mathcal{E}_{|L}$. Since C is k-Buchsbaum, the rank two vector bundle E is "k-Buchsbaum", i.e. it satisfies $m^k.H^1_*(E) = 0$. Now we have (see Thm.11):

Theorem 2. *Let E be a k-Buchsbaum, rank two vector bundle on $\mathbf{P}^3$. If $k \leq 2$, then either $E \simeq \mathcal{O}(a) \bigoplus \mathcal{O}(b)$ $(k = 0)$ or E is a null-correlation bundle (E is stable with $c_1 = 0, c_2 = 1; k = 1$), or E is stable with $c_1 = 0, c_2 = 2$ $(k = 2)$.*

The case $k = 1$ was proved in [2]. Now if E extends to $\mathbf{P}^4$ then $E \simeq \mathcal{O}(a) \bigoplus \mathcal{O}(b)$, because $(c_1, c_2) = (0,1), (0,2)$ do not satisfy the Schwarzenberger condition (S^2_4). It follows that $\mathcal{E} \simeq \mathcal{O}(a) \bigoplus \mathcal{O}(b)$ and thus X is a complete intersection.

With the same approach we get a partial result concerning 3-Buchsbaum subvarieties (see Prop.13).

2. K-BUCHSBAUM RANK 2 VECTOR BUNDLES ON $\mathbf{P}^3$.

In this section we consider k-Buchsbaum rank two vector bundles on $\mathbf{P}^3$, i.e. rank two vector bundles which satisfy: $m^k.H^1_*(E) = 0$.

2.1. 1-Buchsbaum stable bundles. The next proposition is contained in [2] but for the convenience of the reader we include it here.

Proposition 3. *Let E be a stable rank two vector bundle on $\mathbf{P}^3$, then E is 1-Buchsbaum if and only if E is a null-correlation bundle.*

Proof: We assume E normalized. If E is a null-correlation bundle then $H^1_*(E) = k$ (in degree -1) hence E is 1-Buchsbaum.

Now let E be a stable, 1-Buchsbaum rank two vector bundle. The exact sequence of restriction to a plane: $0 \longrightarrow E(-1) \longrightarrow E \longrightarrow E_H \longrightarrow 0$ yields ... $\longrightarrow H^0(E_H) \longrightarrow H^1(E(-1)) \longrightarrow 0$. If E is not a null-correlation bundle, by Barth's restriction theorem $h^0(E_H) = 0$. It follows that $h^1(E(-1)) = 0$. Now, by Riemann-Roch, $\chi(E(-1)) = -c_2$ if $c_1 = 0$ (resp. $-\frac{c_2}{2}$ if $c_1 = -1$). On the other hand $\chi(E(-1)) = h^2(E(-1))$. It follows that $c_2 \leq 0$ which is impossible since E is stable ■

2.2. k-Buchsbaum stable bundles. *Notation:* We will denote by E a k-Buchsbaum, $k \geq 2$, stable, normalized rank two vector bundle with Chern classes c_1, c_2 with $c_2 \geq 2$.

Lemma 4. *(i) With notations as above $h^1(E(t)) = 0$ if $t \leq -k$.*

(ii) $h^2(E(l)) = 0, \forall l \geq k - 4 - c_1$.

Proof: (i) Since E is not a null-correlation bundle ($c_2 > 1$), E_H is stable if H is a general plane. It follows that multiplication by H induces an injective map $\varphi_H(t-1) : H^1(E(t-1)) \hookrightarrow H^1(E(t))$ if $t \leq 0$. Since $\varphi_H(m)\circ\ldots\circ\varphi_H(m-k+1) = 0, \forall m$ because E is k-Buchsbaum, we get the result.

(ii) Follows from (i) by Serre's duality ■

Let $L = H \cap H'$ be a line defined by the planes H, H'. If we consider the Koszul complex of $\mathcal{I}_L$ and the defining sequence for L we get the following diagram (D1):

$$\begin{array}{ccccccccc} & & & & & & 0 & & \\ & & & & & & \downarrow & & \\ 0 & \to & \mathcal{O}(-2) & \to & \mathcal{O}(-1)\bigoplus\mathcal{O}(-1) & \to & \mathcal{I}_L & \to & 0 \\ & & & & & & \downarrow & & \\ & & & & & & \mathcal{O} & & \\ & & & & & & \downarrow & & \\ & & & & & & \mathcal{O}_L & & \\ & & & & & & \downarrow & & \\ & & & & & & 0 & & \end{array}$$

twisting by $E(l+1)$ and taking cohomology we get

$$\psi(l) : H^1(E(l) \bigoplus E(l)) \to H^1(E(l+1)), \psi(l) = (\varphi_H(l), \varphi_{H'}(l)).$$

Lemma 5. *With notations as above and if H and H' are sufficiently general then: $\psi(l) : H^1(E(l) \bigoplus E(l)) \to H^1(E(l+1))$ is surjective for $l \geq k - 3 - c_1$.*

Proof: (1) We may assume that L is a general line for E. By Grauert-Mülich's theorem: $E_{|L} \simeq \mathcal{O}_L(c_1) \bigoplus \mathcal{O}_L$, so $h^1(E_{|L}(l)) = 0$ if $l \geq -1 - c_1$ (*). By 4, $h^2(E(l)) = 0, \forall l \geq k - 4 - c_1$ (**). Taking cohomology in (D.1) twisted by $E(l+1)$ we get:

$$\begin{array}{ccccccccc} \ldots & \to & H^1(E(l-1)) & \to & 2.H^1(E(l)) & \xrightarrow{\alpha_l} & H^1(E(l+1)\bigotimes\mathcal{I}_L) & \to & 0 \\ & & & & & \psi_l \searrow & \downarrow \gamma_l & & \\ & & & & & & H^1(E(l+1)) & & \\ & & & & & & \downarrow & & \\ & & & & & & 0 & & \end{array}$$

By (*), γ_l is surjective for $l+1 \geq -1-c_1$, by (**) α_l is surjective for $l-1 \geq k-4-c_1$; hence ψ_l is surjective for $l \geq k - 3 - c_1$. ■

Lemma 6. *Let H be a general plane then $\varphi_H(l) : H^1(E(l)) \to H^1(E(l+1))$ is identically zero for $l \geq 2k-4-c_1$.*

Proof: We will show, by descending induction on $t, 1 \leq t \leq k-1$, that for every form, S, of degree t, the multiplication map $\varphi_S(l) : H^1(E(l)) \longrightarrow H^1(E(l+t))$ is identically zero if $l \geq 2k-t-3-c_1$.

First consider the case $t = k-1$. The restriction to S yields: $0 \to E(l) \to E(k-1+l) \to E_S(k-1+l) \to 0$. We have:

$$\begin{array}{ccc} & & H^0(E_S(k-1+l)) \\ & & \downarrow r \\ (*) \quad H^1(E(l-1)) & \xrightarrow{\varphi_H(l-1)} & H^1(E(l)) \\ & & \downarrow \varphi_S(l) \\ & & H^1(E(k-1+l)) \end{array}$$

Since E is k-Buchsbaum, $\varphi_S(l) \circ \varphi_H(l-1) = 0$, hence $\mathrm{Im}(\varphi_H(l-1)) \subset Ker(\varphi_S(l))$. By the way this must hold for every plane H. It follows that $\mathrm{Im}(\varphi_H(l-1)) + \mathrm{Im}(\varphi_{H'}(l-1)) \subset Ker(\varphi_S(l))$. Now if H and H' are sufficiently general $\mathrm{Im}(\varphi_H(l-1)) + \mathrm{Im}(\varphi_{H'}(l-1)) = \mathrm{Im}(\psi(l-1)) = H^1(E(l))$ if $l-1 \geq k-3-c_1$ (see 5). We conclude that $\varphi_S(l) \equiv 0$ for $l \geq k-2-c_1$.

Now assume that $\varphi_S(l) : H^1(E(l)) \to H^1(E(t+1+l))$ is zero for every form, S, of degree $t+1$ and if $l \geq 2k-t-4-c_1$. Let S be a degree t form and consider the diagram:

$$\begin{array}{ccc} H^1(E(l-1)) & \xrightarrow{\varphi_H(l-1)} & H^1(E(l)) \\ & & \downarrow \varphi_S(l) \\ & & H^1(E(t+l)) \end{array}$$

By inductive assumption $\varphi_S(l) \circ \varphi_H(l-1) = 0$ if $l-1 \geq 2k-t-4-c_1$. Arguing as before we see that $H^1(E(l)) = \mathrm{Im}(\varphi_H(l-1)) + \mathrm{Im}(\varphi_{H'}(l-1)) \subset Ker(\varphi_S(l))$ if $l \geq \max\{2k-3-t-c_1, k-3-c_1\} = 2k-3-t-c_1$ ■

Proposition 7. *Let E be a normalized rank two vector bundle on $\mathbf{P}^3$ with Chern classes c_1, c_2. Assume E stable and k-Buchsbaum, $k \geq 2$. Let H be a general plane, then :*

a) if $h^0(E_H(2k-3-c_1)) = 0$ then $h^i(E(2k-4-c_1)) = 0, \forall i$, and:

$$c_2 = \begin{cases} \frac{2(k-1)(2k-1)}{3} & \text{if } c_1 = -1 \\ \frac{(2k-3)(2k-1)}{3} & \text{if } c_1 = 0 \end{cases}$$

b) if $h^0(E_H(2k-3-c_1)) \neq 0$ and $h^1(E_H(2k-3-c_1)) = 0$ then:

$$c_2 \leq \begin{cases} \frac{2k(2k-1)}{3} & \text{if } c_1 = -1 \\ \frac{2k(2k-2)}{3} & \text{if } c_1 = 0 \end{cases}$$

c) if $h^i(E_H(2k-3-c_1)) \neq 0, 0 \leq i \leq 1$, then:

$$2(k-c_1-1) \leq c_2 \leq \begin{cases} \frac{2(2k-1)(k-1)(4k+3)}{3(4k-3)} & \text{if } c_1 = -1 \\ \frac{(2k-3)(2k+1)(2k-1)}{3(2k-2)} & \text{if } c_1 = 0 \end{cases}$$

Proof: By 6 we have $H^0(E_H(2k-3-c_1)) \to H^1(E(2k-4-c_1)) \to 0$

A) $h^0(E_H(2k-3-c_1)) = 0$

It follows that $h^1(E(2k-4-c_1)) = 0$. We also have $h^0(E(2k-4-c_1)) = 0$ (because $h^0(E_H(2k-4-c_1)) = 0$ and descending induction), $h^2(E(2k-4-c_1)) = 0$ (4) and $h^3(E(2k-4-c_1)) = 0$ by stability. So $\chi(E(2k-4-c_1)) = 0$ and, since $k \neq 1$, we conclude with Riemann-Roch. This is case a).

B) $h^0(E_H(2k-3-c_1)) \neq 0$

We have

$$h^0(E_H(2k-3-c_1)) = \chi(E_H(2k-3-c_1)) + h^1(E_H(2k-3-c_1))$$

($h^2(E_H(2k-3-c_1)) = 0$ by stability). It follows that:

$\chi(E_H(2k-3-c_1)) + h^1(E_H(2k-3-c_1)) = h^0(E_H(2k-3-c_1)) \geq$
$h^1(E(2k-4-c_1))$ (*)

On the other hand: $h^1(E(2k-4-c_1)) = -\chi(E(2k-4-c_1)) + h^0(E(2k-4-c_1)) \geq$
$-\chi(E(2k-4-c_1))$ (**)

Combining (*) and (**):

$$h^1(E_H(2k-3-c_1)) \geq -\chi(E_H(2k-3-c_1)) - \chi(E(2k-4-c_1)) \ (\#)$$

-If $h^1(E_H(2k-3-c_1)) = 0$, from (#) and Riemann-Roch we get case b).

-Assume $h^1(E_H(2k-3-c_1)) \neq 0$. Set $t := \min\{m/h^0(E_H(m)) \neq 0\}$. Since $k \geq 2$, by Barth's restriction theorem, $t \geq 1$. Moreover $t \leq 2k-3-c_1$ because of assumption B). By [3], Thm. 7.4, $2k-3-c_1 < c_2 - t^2 - c_1 t - 1$, hence: $2(k-c_1-1) \leq c_2$, and: $h^1(E_H(2k-3-c_1)) \leq c_2 - 2k + c_1 - t^2 - c_1 t + 2 \leq c_2 - 2k + 1$. From (#): $c_2 - 2k + 1 \geq -\chi(E_H(2k-3-c_1)) - \chi(E(2k-4-c_1))$ and, using Riemann-Roch, we get case c) ■

From this proposition, proposition 3 and taking into account that c_2 is even if $c_1 = -1$, we get:

Corollary 8. *Let E be a k-Buchsbaum, stable, normalized, rank two vector bundle on $\mathbf{P}^3$.*

(i) If $k = 1$, E is a null-correlation bundle

(ii) If $k = 2$, $c_2 \leq \begin{cases} 4 \text{ if } c_1 = -1 \\ 2 \text{ if } c_1 = 0 \end{cases}$

(iii) If $k = 3$, $c_2 \leq \begin{cases} 10 \text{ if } c_1 = -1 \\ 8 \text{ if } c_1 = 0 \end{cases}$

2.3. k-Buchsbaum unstable bundles.

Proposition 9. *Let E be a k-Buchsbaum, normalized, rank two vector bundle on $\mathbf{P}^3$. Assume E not stable and set $r := \max\{m \geq 0/h^0(E(-m)) \neq 0\}$. We have an exact sequence:*

$$0 \to \mathcal{O} \to E(-r) \to \mathcal{I}_Y(-2r+c_1) \to 0 \ (*)$$

If E doesn't split then the curve Y satisfies $h^0(\mathcal{I}_Y(k-1)) \neq 0$.

Proof: Assume $h^0(\mathcal{I}_Y(k-1)) = 0$. From the exact sequence (*) we get: $h^0(E(l)) = h^0(\mathcal{O}(l+r))$ if $l \leq k+r-1-c_1$ (A).

Restricting (*) to a general surface, S, of degree k yields:

$$0 \to \mathcal{O}_S \to E_S(-r) \to \mathcal{I}_{Y \cap S,S}(-2r+c_1) \to 0$$

The assumption $h^0(\mathcal{I}_Y(k-1)) = 0$ implies $h^0(\mathcal{I}_{Y\cap S,S}(k-1)) = 0$ (otherwise there is a surface, F, of degree $k-1$ intersecting Y in the kd points of $Y \cap S$; but this forces F to contain Y). It follows that: $h^0(E_S(l)) = h^0(\mathcal{O}_S(l+r))$ if $l \leq k+r-1-c_1$ (B).

Now, since E is k-Buchsbaum, for every l we have an exact sequence:

$$0 \to H^0(E(l-k)) \stackrel{.S}{\to} H^0(E(l)) \to H^0(E_S(l)) \to H^1(E(l-k)) \to 0$$

By (A) and (B): $h^0(E(l)) = h^0(E(l-k)) + h^0(E_S(l))$ if $l \leq k+r-1-c_1$. It follows that $h^1(E(t)) = 0$ if $t \leq r-1-c_1$ (C).

By Serre duality: $h^2(E(r-2-c_1)) = h^1(E(-r-2)) = 0$ (use C), and $h^3(E(r-3-c_1)) = h^0(E(-r-1)) = 0$ by definition of r. By Castelnuovo-Mumford's lemma we get: $h^1(E(t)) = 0, t \geq r-1-c_1$. In conclusion $H^1_*(E) = 0$. By Horrocks' theorem E splits ■

Corollary 10. *Let E be a k-Buchsbaum rank two vector bundle on $\mathbf{P}^3$. If E is not stable and if $k \leq 2$ then $E \cong \mathcal{O}(a) \oplus \mathcal{O}(b)$ for suitable integers a, b.*

Proof: The condition $h^0(\mathcal{I}_Y(k-1)) = 0$ of the previous proposition is certainly satisfied if $k = 1$; for $k = 2$, if $h^0(\mathcal{I}_Y(1)) \neq 0$, then $0 = H^1_*(\mathcal{I}_Y) = H^1_*(E)$ and E splits ■

Remark 1. *One could also try an approach similar to the one of the stable case to get vanishings and bounds on c_2.*

2.4. Classification of k-Buchsbaum bundles for k≤ 2. Aim of this section is to prove:

Theorem 11. *Let E be a rank two normalized vector bundle on $\mathbf{P}^3$. If E is k-Buchsbaum, $k \leq 2$, then either: $E \simeq \mathcal{O}(a) \bigoplus \mathcal{O}(b)$ $(k = 0)$, or E is a null-correlation bundle $(k = 1)$, or E is stable with $c_1 = 0, c_2 = 2$ $(k = 2)$.*

Proof: By corollary 10 we may assume E stable. If $c_1 = 0$ then either $k = 1 = c_2$ and E is a null-correlation bundle (Prop.3) or $c_2 = 2 = k$ (Prop.8). Notice that every stable bundle E with $c_1 = 0, c_2 = 2$ is such that $E(1)$ has a section vanishing along three skew lines (see [3], Cor.9.6); it follows that every such bundle is 2-Buchsbaum.

To conclude we have to show that there exists no stable 2-Buchsbaum rank two vector bundle with $c_1 = -1$. By Prop.8, such a bundle would have $c_2 = 2$ or $c_2 = 4$.

(I) $c_1 = -1, c_2 = 2$:

Observe that since $\chi(E(1)) = 0$ by Riemann-Roch, we have $h^1(E(1)) = 0 \Leftrightarrow h^0(E(1)) = 0$.

(α) Assume $h^1(E(1)) = 0$. We have $h^2(E) = 0$ (see Lemma 4) and $h^3(E(-1)) = h^0(E(-2)) = 0$ hence by Castelnuovo-Mumford's lemma, $E(2)$ is generated by global sections. A general section of $E(2)$ yields: $0 \to \mathcal{O} \to E(2) \to \mathcal{I}_X(3) \to 0$ where X is a smooth curve of degree 4 with $\omega_X \simeq \mathcal{O}_X(-1)$. The only possibility is that X is the disjoint union of two conics, but in this case $h^0(\mathcal{I}_X(2)) = h^0(E(1)) \neq 0$, a contradiction.

(β) Assume $h^1(E(1)) \neq 0$. As said before this is equivalent to $h^0(E(1)) \neq 0$. Hence we have $0 \to \mathcal{O} \to E(1) \to \mathcal{I}_Y(1) \to 0$ where Y is a curve of degree 2 with $\omega_Y \simeq \mathcal{O}_Y(-2)$ and we see that Y is a double line with $p_a = -2$. The Rao module of such a curve is a complete intersection module of type $(1,1,2,2)$ (see [4]) so it is concentrated in three degrees $H^1_*(\mathcal{I}_Y) = H^1(\mathcal{I}_Y(-1)) \bigoplus H^1(\mathcal{I}_Y) \bigoplus H^1(\mathcal{I}_Y(1))$ and it is generated by the degree -1 piece in particular $H^1(\mathcal{I}_Y(-1)) \bigotimes S_2 \longrightarrow H^1(\mathcal{I}_Y(1))$ is surjective; in contradiction with the fact that Y is a 2-Buchsbaum curve.

(II) $c_1 = -1, c_2 = 4$.

We have $h^1(E(-2)) = 0$ (Lemma 4), $h^0(E_L) = 1$ if L is a general line (Grauert-Mülich theorem) and by Riemann-Roch $h^1(E(-1)) = 2$. From the diagram (D.1) we get:

$$\begin{array}{ccccc} & & & H^0(E_L) & \\ & & & \downarrow r & \\ 0 & \to & 2.H^1(E(-1)) & \xrightarrow{r_1} & H^1(E \bigotimes \mathcal{I}_L) \\ & & & \searrow \psi & \downarrow r_2 \\ & & & & H^1(E) \end{array}$$

Since $\dim(\mathrm{Im}(r_1) \cap \mathrm{Im}(r)) \leq 1$, we have $\dim(\mathrm{Im}(\psi)) \geq 3$. Now consider the following diagram:

$$\begin{array}{ccc} & & H^0(E_H(1)) \\ & & \downarrow \rho \\ H^1(E(-1)) & \xrightarrow{\varphi_{H'}(-1)} & H^1(E) \\ & & \downarrow \varphi_H \\ & & H^1(E(1)) \end{array}$$

By assumption $\mathrm{Im}(\varphi_{H'}(-1)) \subset Ker\varphi_H = \mathrm{Im}(\rho)$ for any plane H'. It follows that $\mathrm{Im}(\psi) = \mathrm{Im}(\varphi_{H'}(-1)) + \mathrm{Im}(\varphi_{H''}(-1)) \subset \mathrm{Im}(\rho)$ and this implies $h^0(E_H(1)) \geq 3$. Taking a section we get: $0 \to \mathcal{O} \to E_H(1) \to \mathcal{I}_Z(1) \to 0$, where Z is a zero-dimensional subscheme of degree 4 with $h^0(\mathcal{I}_Z(1)) \geq 2$, contradiction ∎

3. Codimension two subvarieties

Now we can prove the results stated in the introduction:

Theorem 12. *Let $X \subset \mathbf{P}^n, n \geq 4$, be a locally complete intersection closed subscheme of codimension two which is subcanonical. If X is k-Buchsbaum with $k \leq 2$,* *then X is a complete intersection.*

In particular any smooth, codimension two subvariety $X \subset \mathbf{P}^n, n \geq 6$, which is k-Buchsbaum with $k \leq 2$ is a complete intersection.

Proof: We may associate a rank two vector bundle $\mathcal{E}$ to X. A general linear section $L \cap X = C$ ($L \simeq \mathbf{P}^3$) is a curve section of the rank two vector bundle $E := \mathcal{E}_{|L}$. Since C is k-Buchsbaum, E is k-Buchsbaum too. By Theorem 11 it follows that if E doesn't split then $(c_1, c_2) = (0,1)$ or $(0,2)$, but the Chern classes of a rank two vector bundle on $\mathbf{P}^4$ have to verify Schwarzenberger's condition (S_4^2): $c_2(c_2 + 1 - 3c_1 - 2c_1^2) \equiv 0(\text{mod}\, 12)$. We conclude that E and hence also $\mathcal{E}$ splits, this implies that X is a complete intersection. The last statement follows from Barth's theorem: every smooth codimension two $X \subset \mathbf{P}^n, n \geq 6$, is subcanonical ■

In the same vein we have also:

Proposition 13. *Let $X \subset \mathbf{P}^n, n \geq 6$, be a smooth subvariety of codimension two. Assume $h^0(\mathcal{I}_X(\varepsilon)) = 0$ ($\varepsilon = \left[\frac{n+1+e}{2}\right]$ where $\omega_X \cong \mathcal{O}_X(e)$). If X is 3-Buchsbaum then X is a complete intersection.*

Proof: The assumption $h^0(\mathcal{I}_X(\varepsilon)) = 0$ implies that the vector bundle, $\mathcal{E}$, associated to X is stable. As before, consider $E = \mathcal{E}_{|L}$, it is a stable, 3-Buchsbaum rank two vector bundle on $\mathbf{P}^3$. By Cor. 8, the possibilities for its Chern classes are: $c_1 = -1, c_2 \in \{2,4,6,8\}$ or $c_1 = 0, 2 \leq c_2 \leq 10$. These Chern classes do not satisfy the Scharzenberger conditions (S_4^2) and (S_6^2) (this latter being expressible as: $\binom{a+5}{6} + \binom{b+5}{6} \in \mathbf{Z}$, where $a + b = c_1, ab = c_2$). It follows that E doesn't extend to $\mathbf{P}^6$ ■

References

[1] Chang, M.C: "Characterization of arithmetically Buchsbaum subschemes of codimension 2 in $\mathbf{P}^n$", *J. Differential Geometry*, **31**, 323-341 (1990)

[2] Ellia, Ph-Fiorentini, M: "Quelques remarques sur les courbes arithmétiquement Buchsbaum de l'espace projectif", *Ann. Univ. Ferrara*, vol. XXXIII, 89-111 (1987)

[3] Hartshorne, R: "Stable vector bundles of rank 2 on $\mathbf{P}^3$", *Math. Ann.*, **238**, 229-280 (1978)

[4] Migliore, J: "On linking double lines", *Trans. Amer. Math. Soc.*, **294**, 177-185 (1986)

[5] Stückrad,J-Vogel,W: *Buchsbaum rings and applications*, Springer, New York (1986)

Curves Contractable in General Surfaces

D. Franco * - A.T. Lascu **

Dipartimento di Matematica, Università di Ferrara

via Machiavelli 35 - 44100 Ferrara, Italy

*e-mail: frv@dns.unife.it

**e-mail: lsl@dns.unife.it

Affectionately dedicated to Mario Fiorentini in appraisal of his work and life.

Abstract. We prove two necessary and sufficient conditions for an integral space curve Y, that is a local complete intersection, to be contractable in *any* surface S of suitable high degree in $\mathbf{P}^3$ containing Y as a Cartier divisor. The first condition is that Y be $\mathbf{Q}$-subcanonical. The second is that Y be contractable in *some* such surface to a $\mathbf{Q}$-Gorenstein singular point.

Introduction.

Let Y be an integral curve, local complete intersection in the projective space $\mathbf{P}^3$ over an algebraically closed field k of arbitrary characteristic. Call Y $\mathbf{Q}$-*subcanonical* if $\omega_Y^m = \mathcal{O}_Y(n)$ for a suitable $m \geq 1$ and n.

Let S be an arbitrary surface in $\mathbf{P}^3$ containing Y as a Cartier divisor. Further, let $f : S \to \overline{S}$ be a *contraction* of Y to a point $\overline{P}$; that means, f is a birational projective morphism such that $f(Y) = \overline{P}$ and f carries $S - Y$ isomorphically onto $\overline{S} - \overline{P}$. Since S is normal at any point of Y, one may assume that $\overline{S}$ is normal at $\overline{P}$. Call f a $\mathbf{Q}$-*Gorenstein* contraction of Y if, additionally to the fact that $\overline{S}$ is normal at $\overline{P}$, there is a non-zero multiple of the canonical sheaf of $\overline{S} - \overline{P}$ which is the restriction of an invertible sheaf on $\overline{S}$. We shall say that Y is *contractable in general surfaces* , if there is s_0 such that for any general surface S of degree at least s_0, Y is contractable in S to a point. Recall that the system of surfaces of degree s containing Y as a Cartier divisor is open; moreover it is non empty for $s >> 0$, because Y is a local complete intersection.

Our main result is the following theorem.

Theorem. *In the situation above the following conditions on Y are equivalent:*

1) The curve Y is contractable in general surfaces.

2) The curve Y is $\mathbf{Q}$-subcanonical.

3) There exists at least one $\mathbf{Q}$-Gorenstein contraction of Y.

Moreover if these equivalent conditions are satisfied, then any contraction of Y is $\mathbf{Q}$-Gorenstein.

Beginning with Castelnuovo's criterion, contractable curves on surfaces have a long history, classical as well as modern, involving global and local aspects. When Y is an integral curve on a surface S, the most general result is Grauert-Artin criterion $(Y, Y) < 0$, for the contractability to a point in an algebraic two dimensional space. Obviously, $(Y, Y) < 0$ is true for any S of degree $s >> 0$, hence any curve is contractable to a point in an *algebraic space* . Nevertheless, in this paper we look for contractions to *algebraic projective surfaces* .

The existence of a projective contraction $f = \phi_{|D|} : S \to \overline{S}$, defined by an effective divisor D, leads naturally to the Picard group of S. Namely, if $Y \subset S$ and S is a general surface of degree $>> 0$, is a fact that $Pic(S)$ is generated by the class of Y and the class of the plane section (see for instance [Lo]). Hence $D \sim aY + bH$ where $(Y, Y) < 0$ and $(D, Y) = 0$. Then $a, b > 0$, $\omega_D \simeq \mathcal{O}_D(\alpha)$ where $b = \alpha - s + 4$ and D is a subcanonical curve on S which is not a complete intersection of S with another surface. In this way our paper is related to the *existence of smooth connected subcanonical curves lying on a smooth surface* studied by Casnati, Dolcetti and Ellia [CDE]. Their paper gives an exaustive solution for particular surfaces (cubics, quartics through a line, surfaces whose Picard group is of rank 2; more generally for surfaces containing a line and for all curves whose class belongs to the Picard subgroup generated by the line and the plane section). Throughout the paper, the method employed relies on the fact ([CDE] Theorem 1.2) that any of the curves looked for is bilinked to a subcanonical curve Y such that $(Y, Y) < 0$. One can easily see that Y is projectively contractable in S to a $\mathbf{Q}$-Gorenstein singularity.

Let us give a short outline of the paper. In §1 we give a contractability criterion, Theorem 1.2, assuming a numerical condition combined with the additional assumption that $\mathcal{O}_Y(D) \sim \mathcal{O}_Y$ with $D \sim aY + bH$, a, $b > 0$. Section §2 is concerned with the singularity $\overline{P} \in \overline{S}$ obtained by the contraction recipe of §1. Specifically (see Theorems 2.1 and 2.2), the problems dealt with are two: normality of $\overline{S}$ and relation between S and $\overline{S}$ blown up at $\overline{P}$. In §3 we look at the algebraic counterparts of the blowing down, the Rees algebra of $\overline{P}$ in an affine neighbourhood and the graded ring of the affine tangent cone of $\overline{S}$ at $\overline{P}$. In §4 we develop an alternative way of contracting a curve, which also points out to a special family of curves, the $\mathbf{Q}$ – *subcanonical* curves (Proposition 4.2). In §5 we prove the aforesaid equivalence between **Q**-subcanonical curves, **Q**-Gorenstein contractable curves and curves which are contractable on general surfaces.

Aknowledgements. We are grateful to Philippe Ellia and Steven L. Kleiman for helpful discussions about the argument. Also many thanks are due to Steven Kleiman and Lucian Badescu for criticism on the preliminary draft, which helped us to write up an improved version.

§1 The existence of contractions

Let Y be an integral curve on a surface S in $\mathbf{P}^3$. In order to contract Y in S by a birational morphism $f : S \to \overline{S}$ where $f(Y) = \overline{P} \in \overline{S}$, the assumption $(Y, Y) < 0$, as well known, is necessary but not sufficient. In general, $\overline{S}$ exists as a two dimensional algebraic space. We are looking for the existence of $\overline{S}$ as a projective surface, hence for the existence of an effective divisor D, very ample on $S - Y$ and such that $D \cdot Y = 0$. If $D \sim aY + bH$, where $a, b > 0$, the the first condition is granted, hence we need only $D \cdot Y = 0$. An extra numerical constraint on $s = degS$ which garantees also $(Y, Y) < 0$ for Y of arithmetic genus > 0 is enough.

Lemma 1.1. *Let Y be an integral curve and S a surface in $\mathbf{P}^3$ of degree s containing Y. Assume that Y is Cartier on S,that $(Y, Y) < 0$ and $4g - 4 < (s-4)d$ where g is the arithmetic genus and d the degree of Y. Let H be a plane section of S, set $D := aY + bH$ with $a > 0$, $b > 0$ and assume $H^0(\mathcal{O}_Y(D)) \neq 0$. Set $E := D - Y$. Then the following homomorphisms are surjective for any $l > 0$:*

$$H^0(\mathcal{O}_S(lD)) \to H^0(\mathcal{O}_Y(lD)),$$

$$H^0(\mathcal{O}_S(lE)) \to H^0(\mathcal{O}_Y(lE)).$$

Proof. We may substitute lD by D, since the assumptions are *a fortiori* preserved. Assume $a \geq r \geq 2$. If we tensor by $\mathcal{O}_S(D)$ over $\mathcal{O}_Y$ the following exact sequence

$$0 \to \frac{\mathcal{I}^{r-1}_{Y,S}}{\mathcal{I}^{r}_{Y,S}} \to \mathcal{O}_{rY} \to \mathcal{O}_{(r-1)Y} \to 0$$

we get

$$0 \to \mathcal{O}_Y((a-r+1)Y + bH) \to \mathcal{O}_{rY}(D) \to \mathcal{O}_{(r-1)Y}(D) \to 0$$

(recall that $\frac{\mathcal{I}^{r-1}_{Y,S}}{\mathcal{I}^{r}_{Y,S}} \simeq \mathcal{I}^{r-1}_{Y,S} \otimes_{\mathcal{O}_S} \mathcal{O}_Y \simeq \mathcal{O}_Y(-(r-1)Y)$). We claim that $H^1(\mathcal{O}_Y((a-r+1)Y+bH)) = 0$. Indeed by Serre's duality we have $H^1(\mathcal{O}_Y((a-r+1)Y+bH)) \simeq H^0(\omega_Y((r-1-a)Y-bH))^*$ and the last vector space is isomorphic to $H^0(\mathcal{O}_Y((r-a)Y+(s-4-b)H))^*$; because we have $\omega_Y \simeq \mathcal{O}_Y(Y+(s-4)H)$ by adjunction. Now, by hypothesis $H^0(\mathcal{O}_Y(D)) \neq 0$, hence $H^0(\mathcal{O}_Y((r-a)Y+(s-4-b)H)) = H^0(\mathcal{O}_Y(rY+(s-4)H-D)) \subset H^0(\mathcal{O}_Y(rY+(s-4)H))$ and the claim follows from $deg(rY+(s-4)H)\mid_Y \leq 2(Y,Y)+(s-4)d = 2g-2+(Y,Y) < 0$, by the numerical constraint. This implies that the map $H^0(\mathcal{O}_{rY}(D)) \to H^0(\mathcal{O}_{(r-1)Y}(D))$ is surjective.

Consider the following commutative diagram

$$\begin{array}{ccccccc}
0 & \to & \mathcal{O}_S(bH) & \to & \mathcal{O}_S(D) & \to & \mathcal{O}_{aY}(D) & \to 0 \\
 & & \downarrow & & \downarrow & & \downarrow & \\
0 & \to & \mathcal{O}_S(bH+Y) & \to & \mathcal{O}_S(D) & \to & \mathcal{O}_{(a-1)Y}(D) & \to 0 \\
 & & \downarrow & & \downarrow & & \downarrow & \\
 & & \vdots & & \vdots & & \vdots & \\
 & & \downarrow & & \downarrow & & \downarrow & \\
0 & \to & \mathcal{O}_S(D-Y) & \to & \mathcal{O}_S(D) & \to & \mathcal{O}_Y(D) & \to 0
\end{array}$$

Then the surjectivity of $H^0(\mathcal{O}_S(D)) \to H^0(\mathcal{O}_{aY}(D))$ is obvious. Combining it with the surjectivity of $H^0(\mathcal{O}_{rY}(D)) \to H^0(\mathcal{O}_{(r-1)Y}(D))$ for $a \geq r \geq 2$, it follows that $H^0(\mathcal{O}_S(D)) \to H^0(\mathcal{O}_Y(D))$ is surjective too.

This accounts for the first statement. The proof of the second one is similar.

Theorem 1.2. *Under the assumptions of the Lemma, assume moreover that* $\mathcal{O}_Y(D) \simeq \mathcal{O}_Y$. *Then* $\mid D \mid$ *has no base points and the morphism* $f : S \to \overline{S}$ *defined by* $\mid D \mid$ *contracts schematically* Y *to a closed point* $\overline{P}$ *i.e.*

i) $Suppf(Y) = \overline{P}$

ii) $\mathcal{I}_Y = \mathcal{I}_{\overline{P}}\mathcal{O}_S$

iii) $f : S - Y \simeq \overline{S} - \overline{P}$

where $\mathcal{I}_Y = \mathcal{I}_{Y,S}$ *and* $\mathcal{I}_{\overline{P}} = \mathcal{I}_{\overline{P},\overline{S}}$

Proof. By Lemma 1.1, taking account of $\mathcal{O}_Y(D) \simeq \mathcal{O}_Y$, the map $H^0(\mathcal{O}_S(D)) \to H^0(\mathcal{O}_Y)$ is surjective. Also $k = H^0(O_Y)$. Then there is $\sigma \in H^0(\mathcal{O}_S(D))$ such that $\sigma \to 1 \in H^0(\mathcal{O}_Y)$, hence σ corresponds to a divisor $D_1 \sim D$ whose support is disjoint of Y. In particular, $\mid D \mid$ has no base points in Y, and f is a morphism, since bH is very ample, hence $\mid D \mid$ is very ample in $S - Y$. These facts account for $i)$ and $iii)$.
It remains only to prove $ii)$: $\mathcal{I}_Y = \mathcal{I}_{\overline{P}}\mathcal{O}_S$. This is due to the Lemma above taking also account that $E_Y = -(Y, Y)$ is non special, hence without base points, since $degE_Y = -(Y, Y) = (s-4)d - 2g + 2 > 2g - 2$.

Remark. *Note that the conditions* $(D, Y) = 0$ *and* $H^0(\mathcal{O}_Y(D)) \neq 0$ *are equivalent to* $\mathcal{O}_Y(D) \simeq \mathcal{O}_Y$. *In §5, Th. 5.2 shows that the condition* Y $\mathbf{Q}$*-subcanonical makes this assumption effective .*

§2 Blowing down and blowing up

In this section Y is supposed irreducible and smooth and S normal. Consider $f : S \to \overline{S} \subset \mathbf{P^n}$ $(n := dim \mid D \mid)$, where f is defined by $\mid D \mid$ as in Theorem 1.2, and let $\pi : \tilde{S} \to \overline{S}$ be the blowing up morphism of $\overline{S}$ in $\overline{P}$. There is an unique birational morphism $g : S \to \tilde{S}$ making commutative the diagram

$$\begin{array}{ccc} S & \xrightarrow{g} & \tilde{S} \\ & f \searrow \quad \swarrow \pi & \\ & \overline{S} & \end{array}$$

by the universal property of π, since $\mathcal{I}_{\overline{P}}\mathcal{O}_S = \mathcal{I}_Y$ by Lemma 1.1 and also Y is Cartier in S. Then $Y = g^{-1}(\tilde{Y})$, where $\tilde{Y} := \pi^{-1}(\overline{P})$. Two natural questions arise:

1) when is $\overline{S}$ normal at $\overline{P}$ i.e. when is $\overline{S}$ normal in a neighbourhood of $\overline{P}$?

2) when is g an isomorphism i.e. when does S coincides with $\overline{S}$ blown-up at $\overline{P}$?

We discuss these questions in terms of $\mid D \mid$. Denote $\overline{D} = f(D)$, $\overline{E} = f(E)$ and let $\Sigma = \{\overline{E} \mid E \in \mid D - Y \mid\}$. Note that $\tilde{S}$ can be identified with the graph $\Gamma \subset \mathbf{P^n} \times \mathbf{P^{n-1}}$ of the rational application $\phi : \overline{S} - - \to \mathbf{P}^{n-1}$ defined by Σ, $\phi := \phi_\Sigma$ and $\pi := pr_1$, taking into account that $\overline{P} = \bigcap_{\overline{E} \in \Sigma} \overline{E}$. Then $\iota := pr_2 \mid_Y$ is the inclusion morphism, defined by the surjective graded homomorphism $\mathcal{R} := k[t_1, ...t_n] = \bigoplus_{r=0}^{\infty} H^0(\mathcal{I}^r_{\overline{P},\mathbf{P^n}}) \to G := Gr_m\mathcal{O} = \bigoplus_{r=0}^{\infty} \frac{m^r}{m^{r+1}}$, with $\mathcal{O} := \mathcal{O}_{\overline{P},\overline{S}}$ and m is its maximal ideal. It follows that $\tilde{Y} = Proj(G)$ and the morphism $\psi : Y \to \tilde{Y}_{red}$, induced by $g \mid_Y$, is defined by $\mid E_Y \mid$ since $H^0(\mathcal{O}_S(E)) \to H^0(\mathcal{O}_Y(E))$ is onto, by Lemma 1.1 (see Preliminary Remark 2 below).

The following Theorem answers the first question above :

Theorem 2.1. *Under the assumptions of Theorem 1.2, let* $t_0, \ldots, t_n \in H^0(\mathcal{O}_S(D))$ *be a base over* k *such that* $H^0(\mathcal{O}_S(E)) = \Sigma_1^n t_i k$, *hence* $\mathcal{T} := im(Sym_k H^0(\mathcal{O}_S(D)) \to H^0_*(\mathcal{O}_S(D))) = k[t_0, \ldots, t_n]$ *and* $I_{\overline{P},\mathbf{P^n}} = (t_1, \ldots, t_n)$.

Denote $U := S - D$, $\overline{U} := \overline{S} - \overline{D}$, *where* $D \cap Y = \emptyset$. *The following conditions are equivalent:*

1) $\overline{S}$ *is normal.*

2) *the cokernel* L *of* $\mathcal{T} \to H^0_*(\mathcal{O}_S(D))$ *is finite dimensional over* k.

3) *the cokernel* M *of* $I_{\overline{P}} \to H^0_*(\mathcal{O}_S(-Y)(D))$ *is finite dimensional over* k.

4) $H^0(\mathcal{O}_U) = H^0(\mathcal{O}_{\overline{U}})$, *equivalently* $f_*\mathcal{O}_S \simeq \mathcal{O}_{\overline{S}}$.

5) $H^0(\mathcal{I}_Y, U) = H^0(\mathcal{I}_{\overline{P}}, \overline{U})$, *equivalently* $f_*\mathcal{I}_Y \simeq \mathcal{I}_{\overline{P}}$.

Proof. By definition of $\overline{S}$ we have a commutative diagram:

$$\begin{array}{ccccc} & & S & & \\ & f \swarrow & & \searrow \phi & := \phi_{|D|} \\ \overline{S} & & \xrightarrow{\overline{\phi}} & & \mathbf{P^n} \end{array}$$

where $\overline{\phi}$ is the inclusion map and $\mathcal{T}$ is the homogeneous coordinates ring of $\overline{S}$. Denote by $k(\overline{S})$ the field of rational functions of $\overline{S}$. One can see that $H^0(\mathcal{O}_S(qD)) = L_q(\overline{S}) = \{\rho \in k(\overline{S}) \mid div_{\overline{S}}(\rho) + q\overline{D} > 0\}$. This is obvious when $\overline{S}$ is normal since $f_*\mathcal{O}_S(qD) = \mathcal{O}_{\overline{S}}(q)$. In the general case one uses the commutative diagram

$$\begin{array}{ccccc} & & S & & \\ & f' \swarrow & & \searrow f & \\ \overline{S}' & & \xrightarrow{\nu} & & \overline{S} \end{array}$$

where ν is the normalization of $\overline{S}$, noticing also that $L_q(\overline{S}') = L_q(\overline{S})$. Then the equivalence 1) $\Leftrightarrow$ 2) is standard ([Z] Proposition 12.10).

2) $\Leftrightarrow$ 3) comes out of the commutative diagram

$$\begin{array}{ccccccccc} & & 0 & & 0 & & 0 & & \\ & & \downarrow & & \downarrow & & \downarrow & & \\ 0 & \to & I_{\overline{P}} & \to & \mathcal{T} & \to & k[t_0] & \to & 0 \\ & & \downarrow & & \downarrow & & \downarrow \| & & \\ 0 & \to & H^0_*(\mathcal{O}_S(-Y)(D)) & \to & H^0_*(\mathcal{O}_S(D)) & \to & \oplus k & \to & 0 \\ & & \downarrow & & \downarrow & & & & \\ & & M & \simeq & L & & & & \\ & & \downarrow & & \downarrow & & & & \\ & & 0 & & 0 & & & & \end{array}$$

taking account of $\mathcal{T} = k[t_0, ..., t_n]$, $I_{\overline{P}} = (t_1, ..., t_n)$ and also of the exact sequence (see Lemma 1.1)

$$0 \to H^0(\mathcal{O}_S(-Y)(rD)) \to H^0(\mathcal{O}_S(rD)) \to H^0(\mathcal{O}_Y) \to 0$$

1) $\Leftrightarrow$ 4) comes out of the fact that $H^0(\mathcal{O}_U)$ is normal, finite over $H^0(\mathcal{O}_{\overline{U}})$ and have the same fraction field.

In order to prove 4) $\Leftrightarrow$ 5) let $\nu : \overline{S}' \to \overline{S}$ be a projective normalization of $\overline{S}$. Then f factors through ν, since S is normal:

$$\begin{array}{ccc} & S & \\ f' \swarrow & & \searrow f \\ \overline{S}' & \xrightarrow{\nu} & \overline{S} \end{array}$$

Also $f' = \phi_{|rD|}$ with $r >> 0$. By restriction, we get a commutative diagram:

$$\begin{array}{ccc} & U & \\ \swarrow & & \searrow \\ \overline{U}' & \to & \overline{U} \end{array}$$

where $\overline{U}' = \overline{S}' - f'(rD)$, $\overline{U}$ is an affine neighbourhood of $\overline{P}$ and similarly $\overline{U}'$ is an affine neighbourhood of $f'(Y) = \overline{P}'$. Let $\mathcal{O} = \mathcal{O}_{\overline{P},\overline{S}}$ as above, $\mathcal{O}' = \mathcal{O}_{\overline{P}',\overline{S}'}$, and m' its maximal ideal. We have $\mathcal{O} \subset \mathcal{O}' \subset k(\overline{S})$ and moreover $\mathcal{O}'$ is the integral closure of $\mathcal{O}$ in $k(\overline{S})$. It follows that $\mathcal{O}'$ is a finite $\mathcal{O}$-module hence, by Nakayama, $\mathcal{O} = \mathcal{O}'$ iff $m\mathcal{O}' = m'$.
Set $\overline{U} = SpecA$ and let I be the maximal ideal corresponding to $\overline{P}$ in A. It follows that $A_I = \mathcal{O}$ and $m = I\mathcal{O}$. Similarly , $m' = I'\mathcal{O}'$ is the maximal ideal of $\mathcal{O}' = A'_{I'}$, where $\overline{U}' = SpecA'$ and I' is the ideal $\overline{P}'$.
We have a commutative diagram

$$\begin{array}{ccc} A & \to & A' \\ & \searrow \quad \swarrow & \\ & H^0(\mathcal{O}_S, U) & \end{array}$$

where $A' \simeq H^0(\mathcal{O}_S, U)$ since $\overline{U}'$ is normal. This implies $I' = H^0(\mathcal{I}_Y, U)$. It is now clear that 5) $\Rightarrow$ 4). The converse is obvious.

In the sequel we consider the second question, that is when is g an isomorphism? Theorem 2.4 below gives a qualitative answer.

Preliminary Remarks.

1) If the homomorphism $\frac{m}{m^2} \to \frac{m'}{m'^2}$ is onto then $\overline{U}$ is normal i.e. $\overline{U} \simeq \overline{U}'$. Indeed in this case $A \to A'$ is unramified since $\Omega^1_{A,k} \otimes_A k = \frac{m}{m^2}$ and $\Omega^1_{A',k} \otimes_{A'} k = \frac{m'}{m'^2}$;

2) The map $\frac{m}{m^2} \to H^0(\mathcal{O}_Y(E))$ is surjective i.e. $< \tilde{Y}_{red} >= \mathbf{P}^{q-1}$ where $q = h^0(\mathcal{O}_Y(E))$; this is true because $H^0(\mathcal{O}_S(E))A = I$, since I is generated by linear forms, also $H^0(\mathcal{O}_S(E)) \to H^0(\mathcal{O}_Y(E))$ is surjective and we have a commutative diagram

$$\begin{array}{ccc} \frac{m}{m^2} & \xrightarrow{\sim} & \frac{I}{I^2} \\ & \searrow \quad \swarrow & \\ & H^0(\mathcal{O}_Y(E)) & \end{array}$$

3) $g : S \to \tilde{S}$ is the normalization of $\tilde{S}$. Indeed S is normal and g is birational and finite.

First let us prove the following result.

Lemma 2.2. *Let A be an integral algebra of finite type over k. $V = SpecA$, $P \in V$ a closed point such that $V - P$ is normal. Assume $dimV > 1$. Let $\mathcal{O} = \mathcal{O}_{P,V}$, m its maximal ideal, $G = Gr_m(\mathcal{O})$ and $B = \bigoplus_0^\infty I^n$ where $I = m \cap A$ is the maximal ideal of A corresponding to P. Consider the blowing-up diagram of V in P*

$$\begin{array}{ccc} \tilde{Y} & \to & \tilde{V} \\ \downarrow & & \downarrow \pi \\ Speck & \to & V \end{array}$$

where $\tilde{Y} = Proj(G)$ is the exceptional divisor and $\tilde{V} = Proj(B)$. Under these assumptions one has the following

1) If $\tilde{Y}$ is integral then $\tilde{V}$ is normal.

2) If moreover $\tilde{Y}$ is linearly normal in $\mathbf{P}(\frac{m}{m^2})$ then one has an exact commutative diagram

$$\begin{array}{ccccccccc}
 & & 0 & & 0 & & & & \\
 & & \downarrow & & \downarrow & & & & \\
0 & \to & I^2 & \to & I & \to & \frac{I}{I^2} & \to & 0 \\
 & & \downarrow & & \downarrow & & \downarrow & & \\
0 & \to & H^0(\mathcal{J}^2)) & \to & H^0(\mathcal{J}) & \to & H^0(\mathcal{O}_{\tilde{Y}}(1)) & \to & 0 \\
 & & \downarrow & & \downarrow & & \downarrow & & \\
 & & L_2 & \to & L_1 & \to & 0 & & \\
 & & \downarrow & & \downarrow & & & & \\
 & & 0 & & 0 & & & &
\end{array}$$

where $\mathcal{J}$ is the ideal sheaf of $\tilde{Y}$.

3) If moreover $\tilde{Y}$ is projectively normal in $\mathbf{P}(\frac{m}{m^2})$ then G is normal, i.e. $G = k[\tilde{Y}]$, thc canonical morphism $B \to H^0_*(\mathcal{J})$ is bijective and $\mathcal{O}$ is normal.

Proof. 1) Is obvious by Serre criterion since $\tilde{Y}$ is Cartier in $\tilde{V}$. Also 2) is obvious since $\frac{m}{m^2} \to H^0(\mathcal{O}_{\tilde{Y}}(1))$ is onto and $\frac{m}{m^2} \simeq \frac{I}{I^2}$.

In order to prove 3) notice first that, for any $n > 0$, one has a commutative exact diagram

$$\begin{array}{ccccccccc}
 & & 0 & & 0 & & K_n & & \\
 & & \downarrow & & \downarrow & & \downarrow & & \\
0 & \to & I^{n+1} & \to & I^n & \to & \frac{I^n}{I^{n+1}} & \to & 0 \\
 & & \downarrow & & \downarrow & & \downarrow & & \\
0 & \to & H^0(\mathcal{J}^{n+1})) & \to & H^0(\mathcal{J}^n) & \to & H^0(\mathcal{O}_{\tilde{Y}}(n)) & \to & 0 \\
 & & \downarrow & & \downarrow & & \downarrow & & \\
K_n & \to & L_{n+1} & \to & L_n & \to & 0 & & \\
 & & \downarrow & & \downarrow & & & & \\
 & & 0 & & 0 & & & &
\end{array}$$

where $K_n \to L_{n+1}$ is injective, since $I \to H^0(\mathcal{O}_{\tilde{Y}}(1))$ and $S^n H^0(\mathcal{O}_{\tilde{Y}}(1)) \to H^0(\mathcal{O}_{\tilde{Y}}(n))$ are both onto, by assumption. Also $I^{n+1} \simeq H^0(\mathcal{J}^{n+1})$ i.e. $L_{n+1} = 0$ for $n >> 0$ by EGA III 2.3.1. Then, by descending induction,

$L_n = 0$ for any n and $G \simeq H^0_*(\mathcal{O}_{\tilde{Y}}(1))$. Finally, G normal implies $\mathcal{O}$ normal, [ZS] VIII, §1, Th. 3.

Proposition 2.3. *Under the conditions above assume moreover that S is normal. Then one has that*

1) The following are equivalent:

a) $\psi : Y \to \tilde{Y}_{red}$ is birational.

b) $\tilde{Y}$ is generically reduced.

c) $\tilde{S}$ is smooth in codimension 1.

2) The following are equivalent:

a) $Y \simeq \tilde{Y}_{red}$.

b) $\tilde{Y} \simeq \tilde{Y}_{red}$.

c) $S \simeq \tilde{S}$.

Proof. 1) Suppose ψ birational. Notice first that $\tilde{Y}$ is irreducible since Y is irreducible by assumption. Let us show that $\tilde{Y}$ is generically reduced. Let $\xi \in Y$ and $\tilde{\xi} \in \tilde{Y}$ be respectively the generic points and $\mathcal{O}_{\tilde{\xi},\tilde{S}} \to \mathcal{O}_{\xi,S}$ the morphism defined by g. Let also $m_{\tilde{\xi}} \subset \mathcal{O}_{\tilde{\xi},\tilde{S}}$ and $m_\xi \subset \mathcal{O}_{\xi,S}$ be the maximal ideals. Then $\mathcal{O}_{\xi,S}$ is a finite $\mathcal{O}_{\tilde{\xi},\tilde{S}}$-module and moreover $m_{\tilde{\xi}}\mathcal{O}_{\xi,S} = m_\xi$ since $m\mathcal{O}_{\tilde{\xi},\tilde{S}} \subset m_{\tilde{\xi}}$ and $m\mathcal{O}_{\xi,S} = m_\xi$. Hence $k(\tilde{Y}_{red}) \to k(Y)$ is defined by $\frac{\mathcal{O}_{\tilde{\xi},\tilde{S}}}{m_{\tilde{\xi}}} \to \frac{\mathcal{O}_{\xi,S}}{m_\xi}$. By Nakayama, it follows that $\mathcal{O}_{\tilde{\xi},\tilde{S}} \simeq \mathcal{O}_{\xi,S}$ iff $k(\tilde{Y}_{red}) \simeq k(Y)$. In particular $a)$ implies $b)$. Also $a)$ implies that $\tilde{S}$ is generically smooth along $\tilde{Y}$, that is $a)$ implies $c)$.

Assume $b)$ i.e. $\tilde{Y}_{red} \simeq \tilde{Y}$ generically. Then $\tilde{S}$ is smooth in $\tilde{\xi}$ since $\tilde{Y}$ is a Cartier divisor, hence $b)$ implies $c)$. Finally, $\tilde{S}$ is smooth in codimension 1 iff $\tilde{S}$ is generically smooth along $\tilde{Y}$, since $\tilde{S} - \tilde{Y} \sim S - Y$ is normal. Then $\mathcal{O}_{\tilde{\xi},\tilde{S}} \simeq \mathcal{O}_{\xi,S}$ i.e. $c)$ implies $a)$.

2) $b) \Leftrightarrow c)$ by Lemma 2.2 and Preliminary Remark 3. Also $c) \Rightarrow a)$ is obvious. Conversely, assume $Y \simeq \tilde{Y}_{red}$ and let $y \in Y$, $\tilde{y} = g(y)$. By

assumption g is bijective, hence $\mathcal{O}_{y,S} =: R$ is finite over $\mathcal{O}_{\tilde{y},\tilde{S}} =: \tilde{R}$. Denote by m_R, $m_{\tilde{R}}$ the maximal ideals of R, $\tilde{R}$ respectively. One has only to show that $R \subset \tilde{R} + m_{\tilde{R}}R$. Notice that $\mathcal{I}_{\tilde{Y},\tilde{S}}\mathcal{O}_S = \mathcal{I}_Y$, since $\mathcal{I}_Y = \mathcal{I}_{\overline{P}}\mathcal{O}_S$ and $\mathcal{I}_{\tilde{Y},\tilde{S}} = \mathcal{I}_{\overline{P}}\mathcal{O}_{\overline{S}}$. Set $I_{\tilde{y}} = \mathcal{I}_{\tilde{Y},\tilde{S},\tilde{y}}$ and $I_y = \mathcal{I}_{Y,y}$. Then $I_{\tilde{y}}R = I_y$ and $\frac{\tilde{R}}{I_y \cap \tilde{R}} \simeq \frac{R}{I_y}$. It follows that $R \subset \tilde{R} + I_y = \tilde{R} + I_{\tilde{y}}R$, hence $R \subset \tilde{R} + m_{\tilde{R}}R$.

Theorem 2.4. *Assume as above Y irreducible, smooth and S normal. If E_Y is very ample, then $S \simeq \tilde{S}$. If additionally E_Y is projectively normal, in particular if $degE_Y > 2g$ and $chark = 0$, then $\tilde{Y}$ is projectively normal, $G = Gr_m(\mathcal{O})$ is normal, in particular $G = k[\tilde{Y}]$, $\overline{S}$ is normal and $B \simeq H^0_*(\mathcal{I})$ where $\mathcal{I} := \mathcal{I}_Y$ is the ideal sheaf of Y in U.*

Proof. The proof is obvious taking account of Lemma 2.2, Prop. 2.3 and of the fact that $degE_Y > 2g$ implies E_Y projectively normal (see, for instance, [L]).

Remark. *Contractions $X \longrightarrow \overline{X}$ to a point $\overline{P}$ which coincide with $\overline{X}$ blown up at $\overline{P}$ have been studied, in a different setting, by L. Badescu and M. Moroianu [BM].*

Let V be an algebraic scheme over k, $x \in V$ a closed point, $\mathcal{O} = \mathcal{O}_{x,V}$, $m = m_{x,V}$ and $G = Gr_m(\mathcal{O})$. We shall use the followig terminology: $Spec(G)$ is the *affine tangent cone* of V at x and $\tilde{Y} = Proj(G)$ is the *projective tangent cone* of V at x.

Theorem 2.4 states that when E_Y is projectively normal, the affine tangent cone of $\overline{S}$ at $\overline{P}$ is normal, hence, *a fortiori* , it coincides with the affine cone over the projective tangent cone. A simple example below illustrates a different situation.

Example. *An isolated singularity with a degenerate projective tangent cone whose affine tangent cone has an embedded component at the vertex.*

Let Y be a nondegenerate, integral, space curve, $A = k[Y] = k[t_0, ..., t_3]$ its homogeneous coordinate ring, $\overline{U} = SpecA$ the affine cone over Y and $\overline{P}$ its vertex. Then $\overline{U} - \overline{P}$ is smooth. Assume Y linearly, but not projectively, normal. Let $\overline{U}'$ be the normalization of $\overline{U}$ i.e. $\overline{U}' = SpecA'$, where A' is the integral closure of A in its fraction field. Then $A' = k[t_0, ..., t_3, u_1, ..., u_r]$ where $r \geq 1$ and $u_i \notin A$ is homogeneous, $degu_i = a_i > 1$, since Y is linearly normal. We can suppose that $\{t_0, ..., t_3, u_1, ..., u_r\}$ is a minimal homogeneous system of generatorsof A' over k. This implies $(\Sigma_0^{r+3} kt_i) \cap I'^2 = 0$ where $t_{i+3} = u_i$ for $1 \leq i \leq r$, i.e. $dim_k \frac{I'}{I'^2} = r + 4$. Geometrically, I' defines the unique point $\overline{P}' \in \overline{U}'$ corresponding to $\overline{P}$. Then $G' = Gr_{I'}(A')$ gives the affine tangent cone $\Gamma = Spec(G')$ of $\overline{U}'$ at $\overline{P}'$. It follows that the embedding $\Gamma \subset A^{r+4}$ defined by $\{t_0, ..., t_{r+3}\}$ is nondegenerate. On the other hand, $\{t_0, ..., t_{r+3}\}$ defines also an embedding $Y' = Proj(G')$ in $\mathbf{P}^{r+3}$ of the projective tangent cone Y' of $\overline{U}'$ at $\overline{P}'$. Next we shall show that Y' is degenerate in $\mathbf{P}^{r+3}$, it belongs to the subspace defined by $t_4 = ... = t_{r+3} = 0$. This implies that Γ has a 0-dimensional embedded component at its vertex $\overline{P}'$, since Γ is nondegenerate in A^{r+4}. The fact that Y' is degenerate comes out of the assumption that Y is linearly normal, as follows. One has that A' is also a graded ring and $A_n = A'_n$ for $n >> 0$; then there is $s > 0$ such that $u_i^s = Q_i \in A$, $1 \leq i \leq r$. It follows that Q_i is homogeneous, $degQ_i = sa_i > s$, since $a_i > 1$. For any $Q \in A'$, $Q \neq 0$, let $\overline{Q} \in G'$ be its initial form. Then $\overline{u}_i^s = 0$, since $deg\overline{u}_i = 1$ and $degu_i^s = sa_i > s$. It follows that $\overline{u}_i$ belongs to any homogeneous prime ideal of $Proj(G') = Y'$ i.e. $\overline{u}_i = \overline{t}_{i+3} = 0$ for $1 \leq i \leq r$ since Y' is integral.

§3 The local structure of the blowing down

Assume as in §2, Y irreducible and smooth and S normal. With the notations of §2, let $\overline{U} = SpecA$ hence $H^0(\mathcal{O}_{\overline{U}}) = A$, $I \subset A$ be the ideal of

$\overline{P}$ and $\mathcal{I} \subset \mathcal{O}_U$ be the ideal sheaf of Y. Assume moreover $\mathcal{O}_Y(D) \simeq \mathcal{O}_Y$ as in Theorem 1.2. Denote by $B := B(A, I) = A \oplus I \oplus I^2 \oplus \ldots$ the Rees algebra of A with respect to I and by B_+ its irrelevant ideal. Set $G := \frac{B}{IB} = Gr_I(A) = Gr_m(\mathcal{O})$. We have, for any $n >> 0$, the commutative rows exact diagram

$$\begin{array}{ccccccccc}
0 & \to & I^{n+1} & \to & I^n & \to & \frac{m^n}{m^{n+1}} & \to & 0 \\
 & & \downarrow & & \downarrow & & \downarrow & & \\
0 & \to & H^0(\mathcal{I}^{n+1})) & \to & H^0(\mathcal{I}^n) & \to & H^0(\mathcal{O}_Y(nE)) & \to & 0 \\
 & & \uparrow & & \uparrow & & \uparrow & & \\
0 & \to & H^0(\mathcal{O}_S((n+1)E))) & \to & H^0(\mathcal{O}_S(nE))) & \to & H^0(\mathcal{O}_Y(nE)) & \to & 0
\end{array}$$

taking account of Lemma 1.1, in the last row. Also we have an inclusion $B \hookrightarrow H^0_*(\mathcal{I})$ because $I^n \hookrightarrow H^0(\mathcal{I}^n)$.

Remarks. *The diagram above implies the following*

1) $\frac{m}{m^2} \to H^0(\mathcal{O}_Y(E))$ *is surjective, as already seen in the preliminary Remark, §2;*

2) E_Y *very ample* $\Rightarrow Coker(\frac{m^n}{m^{n+1}} \to H^0(\mathcal{O}_Y(nE))) = 0$ *for* $n >> 0$;

3) E_Y *projectively normal* $\Rightarrow Coker(\frac{m^n}{m^{n+1}} \to H^0(\mathcal{O}_Y(nE))) = 0\ \forall n$;

4) $Ker(\frac{m^n}{m^{n+1}} \to H^0(\mathcal{O}_Y(nE))) = 0 \Leftrightarrow I^n \cap H^0(\mathcal{I}^{n+1}) = I^{n+1}$ *and moreover these equivalent conditions are satisfied for* $n >> 0$ *iff* $\tilde{Y} = \tilde{Y}_{red}$;

5) $H^0(\mathcal{I}^n) = I^n \Rightarrow Sym^n_A H^0(\mathcal{I}) \to H^0(\mathcal{I}^n)$ *is onto and when* $\overline{U}$ *is normal these conditions are equivalent. Also* $H^0(\mathcal{I}^n) = I^n\ \forall n$ *means that* $B = H^0_*(\mathcal{I})$;

6) *when* E_Y *is very ample one has an exact sequence*

$$0 \to k[\tilde{Y}_{red}] \to H^0_*(\mathcal{O}_Y(E)) \to M^1(\tilde{Y}_{red}) \to 0$$

where $M^1(\tilde{Y}_{red})$ *is the Rao-module of* $\tilde{Y}_{red}$.

Proposition 3.1. *Under the assumptions above i.e.* Y *irreducible, smooth and* S *normal, assume moreover* $chark = 0$, $\overline{S}$ *normal,* E_Y *very ample* *and* $\frac{m}{m^2} \simeq H^0(\mathcal{O}_Y(E))$. *Assume also that* $(s-4)d > 4g-2$. *Then* $\tilde{Y}$ *is Buchsbaum.*

Proof. In the exact commutative diagram

$$\begin{array}{ccccccccc} 0 & \to & I^2 & \to & I & \to & \frac{m}{m^2} & \to & 0 \\ & & \downarrow & & \downarrow & & \downarrow & & \\ 0 & \to & H^0(\mathcal{I}^2)) & \to & H^0(\mathcal{I}) & \to & H^0(\mathcal{O}_Y(E)) & \to & 0 \end{array}$$

we have $\frac{m}{m^2} \simeq H^0(\mathcal{O}_Y(E))$ and additionally $I = H^0(\mathcal{I})$ since $\overline{S}$ is normal. Then $I^2 = H^0(\mathcal{I}^2)$, hence the map $\frac{m^2}{m^3} \to H^0(\mathcal{O}_Y(2E))$ is surjective because $H^0(\mathcal{I}^2) \to H^0(\mathcal{O}_Y(2E))$ is onto. Notice that $\tilde{Y} \simeq \tilde{Y}_{red}$, by Prop. 2.3, since E_Y is very ample i.e. $Y \simeq \tilde{Y}_{red}$. It follows that $\phi_{|2E_Y|}(Y) = \mathcal{V}_2(\tilde{Y})$ where $\mathcal{V}_2$ is the Veronese embedding of $\tilde{Y}$. Notice also that $2E_Y$ is projectively normal, since $deg(2E_Y) > 2g$ [L]. Then $(k[\tilde{Y}]_{2p} = H^0(\mathcal{O}_Y(2pE))$ hence $M^1(\tilde{Y})_{2p} = 0 \; \forall p$ (see the last Remark). Finally, $\frac{m}{m^2} M^1(\tilde{Y}) = 0$ i.e. $\tilde{Y}$ is Buchsbaum.

Proposition 3.2. *Under the assumptions above i.e. Y irreducible and smmoth and S normal, the following are equivalent:*

i) G is normal.

ii) E_Y is projectively normal.

Furthermore, if these equivalent conditions are satisfied, then $B \simeq H^0_*(\mathcal{I})$ *and* $\mathcal{O}$ *is normal.*

Proof. *ii)* imlies *i)* by 3) of Lemma 2.2 for $V = \overline{U}$. The converse is obvious, taking also account of Remark 1) above.

Lemma 3.3. *Let A be an algebra of finite type over k, $I \subset A$ a maximal ideal, $B = B(A, I)$, $\mathcal{O} = A_I$ and m its maximal ideal. Assume A normal nd $dimA = 2$. The following conditions are equivalent:*

i) The Rees algebra B is Cohen Macaulay.

ii) $\mathcal{B} := B_I = \mathcal{O} \oplus m \oplus m^2 \oplus \ldots$ *is Cohen Macaulay.*

Proof. $i) \Leftarrow ii)$ is obvious. In order to prove that $ii) \Leftarrow i)$ let $P \subset B$ be a prime ideal and $P \cap A = p$. It is enough to see that $B_p = B(A_p, I_p)$ is CM. When $I \subset P$ then $p = I$, hence $A_p = 0$, $I_p = m$, $B_p = B_I$ which is CM by assumption. If $I \not\subset P$ then $I_p = A_p$ hence $B_p = B(A_p, A_p) = A_p[t]$ where t is an indeterminate and also A_p is CM since it is normal of dimension ≤ 2. Hence $B_p = A_p[t]$ is CM.

Proposition 3.4. *Assume Y integral and E_Y projectively normal. Then the following conditions are equivalent:*

i) The Rees algebra B is Cohen Macaulay

ii) $\mathcal{B} := B_I = \mathcal{O} \oplus m \oplus m^2 \oplus \ldots$ *is Cohen Macaulay*

iii) Y is smooth and rational i.e. f is a rational contraction.

Proof. In view of Lemma 3.3, one has only to show that $ii) \Leftrightarrow iii)$. Theorem 2.4 implies $\frac{m}{m^2} \simeq H^0(\mathcal{O}_Y(E))$. By [HIO] Prop. (48.1) B is Cohen Macaulay iff $e(m) = embeddim(\mathcal{O}) - 1$. Further $e(m) = (Y, E) = -(Y, Y) = (s-4)d - 2g + 2 = degE_Y$. Also $embeddim(\mathcal{O}) = dim_k(\frac{m}{m^2}) = h^0(\mathcal{O}_Y(E))$. It follows that $embeddim(\mathcal{O}) - 1 = h^0(\mathcal{O}_Y(E)) - 1 = degE_Y - g + h^1(\mathcal{O}_Y(E))$ where $h^1(\mathcal{O}_Y(E)) = h^0(\omega_Y(-E)) < g$ if $g \neq 0$.

Remark. *By Prop. 3.2, it follows that B C.M. $\Rightarrow$ the affine tangent cone of $\overline{S}$ at $\overline{P}$ is normal since, for Y smooth and rational, E_Y is projectively normal.*

Proposition 3.4. *Assume $char k = 0$, Y integral and smooth, and $degE_Y > 2g$. Then $R^1 f_* \mathcal{I}^n = 0 \ \forall n > 0$*

Proof. By Lemma 1.1 we have an exact sequence

$$0 \to H^0(\mathcal{I}^{n+1}) \to H^0(\mathcal{I}^n) \to H^0(\mathcal{O}_S(nE)) \to 0$$

Then

$$0 \to R^1 f_* \mathcal{I}^{n+1}) \to R^1 f_* \mathcal{I}^n \to R^1 f_* \mathcal{O}_S(nE) \to 0$$

or else

$$0 \to H^1(\mathcal{I}^{n+1}) \to H^1(\mathcal{I}^n) \to H^1(\mathcal{O}_S(nE)) \to 0$$

where also $H^1(\mathcal{O}_Y(nE)) = 0$ for $n \geq 1$, since $degE_Y > 2g$.
Then $H^1(\mathcal{I}^{n+1}) = H^1(\mathcal{I}^n)$ and the conclusion comes out from the fact that $R^1 f_* \mathcal{I}^n = 0$ for $n >> 0$, since f is the blowing up morphism of $\overline{S}$ in $\overline{P}$ (EGA III (2.4.1)).

§4 Q-subcanonical space curves and linkage: an alternative recipe for contraction

Lemma 4.1. *Let Y be a curve which is a Cartier divisor on a normal surface $S \subset \mathbf{P}^3$. Assume that $\alpha Y + \beta H \sim \Delta$ where $\alpha, \beta > 0$ and $Y \cdot \Delta = 0$. Then αY is subcanonical and there is an effective Cartier divisor $\Delta_0 \sim \Delta$, in particular $(SuppY) \cap (Supp\Delta_0) = \emptyset$ since $Y \cdot \Delta_0 = 0$.*

Proof. The assumption implies that $\alpha Y^2 + \beta H_Y \sim 0$ hence $\alpha^2 Y^2 + \beta H_{\alpha Y} \sim 0$. By adjunction $\alpha^2 Y^2 + (s-4) H_{\alpha Y} \sim K_{\alpha Y}$. Then $(s - 4 - \beta) H_{\alpha Y} \sim K_{\alpha Y}$, which shows that αY is subcanonical. Let $m_1 = s$, $F_1 = S$ and $m_2 \geq m_1$ be such that there exists a geometric linkage $F_1 \cap F_2 = \alpha Y + C$ with $m_2 = degF_2$ and C integral. Set $Z = F_1 \cap F_2$. By linkage we have the exact sequence

$$0 \to \mathcal{I}_{Z,S} \to \mathcal{I}_{C,S} \to \omega_{\alpha Y}(-m_1 - m_2 + 4) \to 0$$

Then $\omega_{\alpha Y}(-m_1 - m_2 + 4) = \mathcal{O}_{\alpha Y}(-\beta - m_2)$ since $m_1 = s$ and $\omega_{\alpha Y} = \mathcal{O}_{\alpha Y}(s - 4 - \beta)$. It follows that

$$0 \to \mathcal{I}_{Z,S}(m_3) \to \mathcal{I}_{C,S}(m_3) \to \mathcal{O}_{\alpha Y} \to 0$$

where $m_3 = m_2 + \beta > m_2$, hence $C = F_1 \cap F_2 \cap F_3$, i.e. C is a quasi complete intersection of type (m_1, m_2, m_3). Also $m_3 > m_2$ implies that it exists F_3 which gives a geometric linkage $F_1 \cap F_3 = C + \Delta_0$. Notice also that Δ_0 and Y have no common points, since $C = F_1 \cap F_2 \cap F_3$ and $Y \subset F_2$. In terms of Cartier divisors of S

1) $m_2 H \sim \alpha Y + C$

2) $m_3 H \sim C + \Delta_0$ where $m_3 = m_2 + \beta$.

It follows that $\alpha Y + \beta H \sim \Delta_0$.

Remark. *Notice that $\mid \Delta \mid$ is cut out on S by the surfaces of degree m_3 through C.*

Definition. *Let Y be a local complete intersection space curve. Then Y is called* **Q***-subcanonical if $\omega_Y^{\otimes m} = \mathcal{O}_Y(n)$ where $m \geq 1$ and n are integers.*

Proposition 4.2 *Under the assumptions of the Lemma 4.1 assume moreover Y connected. Then*

i) Y is **Q***-subcanonical*

ii) There is a contraction of Y $f : S \to \overline{S}$ to a normal point $\overline{P} \in \overline{S}$ defined by a convenient multiple of $\mid \Delta \mid$.

Proof. i) one has $\alpha Y + \beta H \sim \Delta$ and $Y \cdot \Delta = 0$ which implies $\mathcal{O}_Y(\alpha Y + \beta H) \simeq \mathcal{O}_Y$ i.e. $\alpha Y^2 + \beta H_Y \sim 0$ where $Y^2 = K_Y - (s-4)H_Y$ hence $\alpha K_Y \sim (\alpha(s-4) - \beta)H_Y$.

ii) $\mid \Delta \mid$ is very ample in $S - Y$ and moreover it has no base points, because of the existence of $\Delta_0 \in \mid \Delta \mid$ as in Lemma 4.1. Hence $\phi_{\mid \Delta \mid} : S \to \overline{S}$ is a birational projective morphism which contracts Y to a point, since Y is connected.

Corollary 4.3. *Assume* $char k = 0$. *Let* Y *be a connected curve which is local complete intersection and a Cartier divisor on a normal surface* S. *Assume moreover* Y *subcanonical,* $\omega_Y = \mathcal{O}_Y(\alpha)$ *and* $\beta = s - 4 - \alpha > 0$, *where* $s = degS$. *Set* $D = Y + \beta H$. *Then* $\phi_{|D|} = f : S \to \overline{S}$ *is a birational morphism which contracts* Y *to a point* $\overline{P}$. *Moreover* $E = D - Y$ *is projectively normal. If additionally* Y *is integral and smooth and* $(s-4)d > 4g-2$, *where* $d = degY$, *then* E_Y *is projectively normal,* $\overline{S}$ *is normal and* S *is isomorphic to* $\overline{S}$ *blown up at* $\overline{P}$.

Proof. The statement about f is already in the Proposition above. Also $E = \beta H$ is very ample since $\beta > 0$. The final statement comes out of Theorem 2.4.

Proposition 4.4. *Let* Y *be a curve which is a Cartier divisor on a surface* $S \subset \mathbf{P^3}$. *Assume* S *normal and* $D = aY + bH$ *with* $a, b > 0$ *defining a contraction* $f = \phi_{|D|} : S \to \overline{S}$ *of* Y *to a point* $\overline{P}$. *Assume moreover that* $E = \mid D - Y \mid$ *is very ample. Then* S *is isomorphic to* $\overline{S}$ *blown up at* $\overline{P}$.

Proof. Let $\Sigma := \{\overline{E} \mid \overline{E} = f(E),\ E \in \mid D - Y \mid\}$, $\psi = \phi_{|E|} : S \to \mathbf{P^n}$ and $\overline{\psi} = \phi_\Sigma : \overline{S} - - \to \mathbf{P^n}$, hence $\psi = \overline{\psi} \cdot f$. Let $\Gamma \subset \overline{S} \times \mathbf{P^n}$ be the graph of $\overline{\psi}$ and π_i its projections. Then one has a canonical $\overline{S}$-isomorphism

$$\begin{array}{ccc} \tilde{S} & \simeq & \Gamma \\ \pi \searrow & & \swarrow \pi_1 \\ & \overline{S} & \end{array}$$

where π is the blowing up morphism of $\overline{S}$ at $\overline{P}$. Hence one can identify $\tilde{S}$ with Γ. This gives a commutative diagram

$$\begin{array}{ccc} S & \xrightarrow{g} & \tilde{S} \\ f \downarrow & & \downarrow \pi_2 \\ \overline{S} & \overset{\overline{\psi}}{- \to} & \mathbf{P^n} \end{array}$$

where $\psi = \pi_2 \cdot g$ is an isomorphism.

§5 Q-Gorenstein contractions, Q-subcanonical space curves and curves contractable on general surfaces

Definition 1. *Let X be a normal surface. Then we shall say that X have* $\mathbf{Q}$*-Gorenstein singularities if the canonical (Weil) divisor K_X is such that rK_X is Cartier for some $r > 0$.*

Let $S \subset \mathbf{P}^3$ be a normal surface and Y in S be an *integral curve* , which is a Cartier divisor. Assume that Y is contractable in S to a point $P_0 \in S_0$ where S_0 is a *normal surface with* $\mathbf{Q}$*-Gorenstein singularities* i.e. there exists a birational morphism $\psi : S \to S_0$ such that S_0 is normal, $\psi(Y) = P_0$, $S - Y \simeq S_0 - P_0$ and there is an integer $r > 0$ such that rK_{S_0} is a Cartier divisor, i.e. S_0 has $\mathbf{Q}$-Gorenstein singularities.

Definition 2. *Under the assumptions above we shall call ψ a* $\mathbf{Q}$*-Gorenstein contraction of Y and we shall also say that Y is* $\mathbf{Q}$*-Gorenstein contractable in S.*

Definition 3 *A local complete intersection curve $Y \subset \mathbf{P}^3$ is contractable on general surfaces in $\mathbf{P}^3$ iff for any surface S of degree $s >> 0$, containing Y as a Cartier divisor, Y is contractable on S.*

Example. Let $f : S \to \overline{S}$ with $f = \psi_{|D|}$, $D = aY + bH$ be as in sections 1-3, where we assume moreover that $\overline{S}$ is normal. Then f is a semi-canonical contraction, since $K_{\overline{S}} = (s-4)\overline{H}$ and $b\overline{H} \sim \overline{E} \sim \overline{D}$, i.e. $bK_{\overline{S}} \sim (s-4)\overline{D}$. Hence $bK_{\overline{S}}$ is Cartier at $\overline{P}$ because $\overline{S}$ is normal and we may suppose that $\overline{P} \notin \overline{D}$.

The next Proposition shows that any semi-canonical contraction is of this type.

Proposition 5.1. *Let* Y *be an integral curve which is a Cartier divisor on a normal surface* S *and let* $\psi : S \to S_0$ *be a* $\mathbf{Q}$*-Gorenstein contraction of* Y. *Then there is an effective divisor* $\alpha Y + \beta H =: D$ *with* $\alpha > 0$, $\beta > 0$, *such that* $\phi_{|D|} : S \to \overline{S}$ *contracts* Y *to a* $\mathbf{Q}$*-Gorenstein singularity. Moreover, there is an unique isomorphism* u *making commutative the diagram*

$$\begin{array}{ccccc} & & S & & \\ & f \swarrow & & \searrow \phi & \\ \overline{S} & & \xrightarrow{u} & & S_0 \end{array}$$

Proof. One has that $K_{S_0} = (s-4)H_0$ where $H_0 = \psi(H)$ since $K_S = (s-4)H$. By assumption $rK_{S_0} = r(s-4)H_0$ is Cartier, hence $r(s-4)H_0 \sim \Delta_0$ where Δ_0 is a Cartier divisor of $S_0 - P_0$. Then $\Delta_0 = \psi(\Delta)$ where Δ is Cartier on S. It follows that $r(s-4)H + \mu Y \sim \Delta$ where $\mu > 0$, since $P_0 \in SuppH_0$. Also $\Delta \cdot Y = 0$ since $Supp\Delta \subset S - Y$. Then $\phi := \phi_{|\Delta|} : S \to \overline{S}$ contracts Y to a point $\overline{P}$. Moreover, substituting Δ by $l\Delta$, $l >> 0$ one has that $\overline{S}$ is normal, hence one obtains $D := l\Delta \sim \alpha Y + \beta H$ as required. Finally, u is a morphism by Zariski's Main Theorem.

Theorem 5.2. *Let* Y *be an integral local complete intersection space curve. Then the following are equivalent:*

1) Y *is* $\mathbf{Q}$*-subcanonical.*
2) Y *is* $\mathbf{Q}$*-Gorenstein contractable to a point, in a suitable normal surface* $S_1 \subset \mathbf{P}^3$ *containing* Y *as a Cartier divisor.*
3) Y *is contractable on general surfaces.*

Proof. 1) $\Rightarrow$ 2). Assume $mK_Y \sim nH_Y$ with $m > 0$. For $s >> 0$, there is a surface S, of degree s, containing Y as a Cartier divisor. Then

$Y + (s-4)H = \Gamma$ is a Cartier divisor such that $\Gamma \cdot Y \sim K_Y$. It follows that $\mathcal{O}_Y(\alpha Y + \beta H) \simeq \mathcal{O}_Y$ where $\alpha = m > 0$, $\beta = m(s-4) - n > 0$ provided $m(s-4) > n$. Then, by Proposition 4.2, the linear system $\mid l\Delta \mid$ contracts Y to a $\mathbf{Q}$-Gorenstein singularity in $\overline{S}$, since $D = l\Delta = l(\alpha Y + \beta H)$ implies $\overline{D} = l\beta\overline{H}$, hence $(l\beta)K_{\overline{S}} = (s-4)\overline{D}$.

2) $\Rightarrow$ 1). By Proposition 5.1 above $(\alpha Y + \beta H) \cdot Y \sim 0$, hence $mK_Y \sim nH_Y$ with $m = \alpha$ and $n = \alpha(s-4) - \beta$, as in the proof of Prop. 4.2.

1) $\Rightarrow$ 3). As already seen in the proof of 1) $\Rightarrow$ 2), Y is $\mathbf{Q}$-Gorenstein contractable on any surface S of degree s such that $m(s-4) > n$, containing Y as a Cartier divisor.

3) $\Rightarrow$ 1). For $s >> 0$ and S general, $Pic(S)$ is generated by Y and H [Lo], hence the contraction $f : S \to \overline{S}$ is given by $D \sim \alpha Y + \beta H$ with $\alpha, \beta > 0$.

REFERENCES

[A] Artin, M.: *Algebraization of formal moduli II,* Annals of Math., **91** (1970) 88-136

[BM] Badescu, L. - Moroianu, M.: *Algebraic Contractions and Complete Intersections,* Rend. Acc. Lincei, **52** (1972) 884-892

[HIO] Herrmann, M. - Ikeda, S. - Orbanz, U.: *Equimultiplicity and Blowing up,* Springer (1988)

[CDE] Casnati, G. - Dolcetti, A. - Ellia, Ph.: *On subcanonical curves lying on smooth surfaces in* $\mathbf{P}^3$, Rev. Roumaine Math. Pures Appl., **40** (1995) 289-300

[H] Hartshorne, R.: *Algebraic Geometry,* Springer GTM 52 (1977)

[L] Lazarsfeld, R.: *A sampling of vector bundles techniques in the study of linear series*, Lectures on Riemann surfaces, M. Cornalba et al. (eds.), World Scientific Press, (1989) 500-599

[Lo] Lopez, A. F.: *Noether Lefschetz theory and the Picard group of projective surfaces,* Mem. of AMS, **438**

[Z] Zariski, O. : *An Introduction to the Theory of Algebraic Surfaces,* Springer LNM 83 (1969)

[ZS] Zariski, O. - Samuel, P. : *Commutative Algebra,* Springer GTM 28-29 (1975)

On Nagata's Theorem for the Class Group, II

Stefania Gabelli

Dipartimento di Matematica, Università di Roma "La Sapienza",
Piazzale A. Moro 2, 00185 Roma, Italy.
e-mail: gabelli@mat.uniroma1.it

Introduction

The general form of Nagata's Theorem for the class group states that, if D is a Krull domain and $T = \cap\{D_P ; P \in \Lambda\}$, with Λ a set of height-one prime ideals of D, then the natural map of divisor class groups $\mathcal{C}(D) \longrightarrow \mathcal{C}(T)$, $[I] \longrightarrow [(IT)_t]$, is a well defined surjective homomorphism and its kernel is generated by the classes of the height-one primes which are not in Λ (see [16, Theorem 7.1]). Each divisorial ideal of a Krull domain is t-invertible and in [11] and [12] the notion of class group was generalized to any domain as the quotient group $\mathcal{C}(D) := \mathcal{T}(D)/\mathcal{P}(D)$, where $\mathcal{T}(D)$ denotes the group of t-invertible fractional t-ideals of D and $\mathcal{P}(D)$ the group of principal fractional ideals. During the last ten years several attempts have been made to generalize Nagata's theorem relaxing the Krull assumption on D and various different sufficient conditions have been given for the natural homomorphism $\mathcal{C}(D) \longrightarrow \mathcal{C}(T)$ to be surjective in the case when T is an intersection of localizations, in particular a ring of fractions of D (see for example [1, 2, 3, 6, 7, 9, 10, 13, 19, 27]). However it seems rather difficult, if not impossible, to find out a satisfactory characterization of domains satisfying Nagata's Theorem. Indeed, even if D has very good ideal theoretical properties and/or T is a ring of fractions of D with respect to a multiplicative part generated by one (prime) element, the homomorphism $\mathcal{C}(D) \longrightarrow \mathcal{C}(T)$ need not be surjective [1, 7, 19].

Partially supported by research funds of Ministero dell'Università e della Ricerca Scientifica e Tecnologica and NATO Grant 970140.

In this paper we consider generalized rings of fractions $D_{\mathcal{F}}$ of D (intersections of localizations are generalized ring of fractions). In this setting, we give general sufficient conditions for the surjectivity of the natural homomorphism of class groups $\mathcal{C}(D) \longrightarrow \mathcal{C}(D_{\mathcal{F}})$, which include as special cases most of those already known; thus providing a unifying interpretation of several results appeared in the literature.

In particular we deal with the cases when D has t-finite character (that is each element of D is contained at most in finitely many t-maximal ideals), T has t-dimension one (that is each t-maximal ideal of T has height one) or D is a v-coherent domain.

We recall that a domain D is v-coherent if (D:I) is a v-finite divisorial ideal for each finitely generated ideal I [13, Section 3]. The class of v-coherent domains is very large, including quasi-coherent domains (domains such that (D:I) is finitely generated for each finitely generated ideal I, for example coherent domains), Mori domains (domains satisfying the ascending chain conditions on divisorial ideals, for example Noetherian and Krull domains) and Prüfer v-multiplication domains, in short PvMDs (domains in which each v-finite divisorial ideal is t-invertible, for example Prüfer domains). An example of a v-coherent domain which is not in the union of these classes of domains is given in [28, Example 2.8].

A central role in this paper is played by the v-*finite multiplicative systems* of ideals of D, which we introduce and study in Section 1 and characterize in Theorem 3.3 in the case when D is v-coherent. Of particular interest are those (v-finite) saturated multiplicative systems $\mathcal{F}$ of D such that each t-ideal in $\mathcal{F}$ is t-invertible, characterized in Theorem 1.10. A particular example of a multiplicative system of this type is given by the saturation of $\mathcal{F} := \{sD; s \in S\}$, where S is a multiplicative part of D generated by primes.

It turns out that, if $\mathcal{F}$ is v-finite, then $D_{\mathcal{F}}$ is a t-subintersection of D, more precisely, $D_{\mathcal{F}} = \cap\{D_P ; P \in \Lambda\}$, where Λ is a suitable set of pairwise incomparable t-prime ideals of D (Proposition 1.9). In addition, denoting by t(D) the semigroups of all fractional t-ideals of D, the natural homomorphism $t(D) \longrightarrow t(D_{\mathcal{F}})$, $I \longrightarrow (ID_{\mathcal{F}})_t$ is surjective (Proposition 1.8). If moreover D has t-dimension one, also the induced homomorphism $\phi_{\mathcal{F}}: \mathcal{T}(D) \longrightarrow \mathcal{T}(D_{\mathcal{F}})$ is surjective (Theorem 2.3). This last statement allows us to give a partial answer to a problem raised in [3] and generalize some of the results proved there.

Theorem 1.13 shows that, if D is any domain and $\mathcal{F}$ is a saturated multiplicative system of ideals of D such that each t-ideal in $\mathcal{F}$ is t-invertible, then $\mathrm{Ker}(\phi_{\mathcal{F}})$ is generated by the prime t-ideals of D which are in $\mathcal{F}$ and the natural homomorphism $\phi_{\mathcal{F}}$: $\mathcal{T}(D) \longrightarrow \mathcal{T}(D_{\mathcal{F}})$ is surjective whenever $P_{\mathcal{F}}$ is a t-ideal for each prime t-ideal P of D. This last condition is verified when D has t-dimension one (Proposition 2.1) or when D is v-coherent (Proposition 3.2). Under a slightly different condition on $\mathcal{F}$, the natural homomorphism $\phi_{\mathcal{F}} : \mathcal{T}(D) \longrightarrow \mathcal{T}(D_{\mathcal{F}})$ is also surjective when D has t-finite character (Theorem 1.18).

We observe that a Krull domain has t-finite character, t-dimension one and is v-coherent and that each t-ideal of a Krull domain is t-invertible. Thus the classical Nagata's Theorem can be reobtained as a consequence of the results mentioned above. In a similar way, we can also recover and improve several results known for Prüfer domains [15, Section 5], PvMDs [3], Mori domains [9, 19] and v-coherents domains [27].

In the last Section of the paper we prove some splitting properties of the group $\mathcal{T}(D)$. We show for example that, if D has t-finite character, then there exist two overrings D_1 and D_2 of D such that D_1 is Krull, $D = D_1 \cap D_2$ and $\mathcal{T}(D) = \mathcal{T}(D_1) \oplus \mathcal{T}(D_2)$ (Theorem 4.4). For Mori domains this was already proved in [9]. Under a weaker hypothesis of t-finiteness, a similar statement holds when D has t-dimension one or is v-coherent (Theorem 4.3); in this case D_2 is also of t-dimension one or v-coherent.

We recall some notation and definitions. For general properties of divisorial ideals, t-ideals and class group we refer the reader to [1, 6, 17, 21, 24]. If I is a fractional ideal of a domain D with quotient field K, then (D:I) denotes the fractional ideal $\{x \in K ; xI \subseteq D\}$. The *divisorial closure* of I is given by $I_v := (D:(D:I))$ and the t-*closure* of I by $I_t := \cup\{J_v ; J$ is a finitely generated ideal of D and $J \subseteq I\}$. The fractional ideal I is *divisorial* (respectively a t-*ideal*) if $I = I_v$ (respectively $I = I_t$). The v-closure and the t-closure coincide on finitely generated fractional ideals. We have $I \subseteq I_t \subseteq I_v$, so that a fractional divisorial ideal is a fractional t-ideal. The set of all fractional t-ideals of D is a semigroup under t-*multiplication*, defined by $I*J = (IJ)_t$, and will be benoted by t(D). Note that, if $I = H_t$ and $J = L_t$, then $(IJ)_t = (HL)_t$.

The set of (integral) t-ideals has maximal elements under inclusion, called t-*maximal ideals*, and these ideals are prime. A t-ideal which is prime is also called a t-*prime ideal* (or simply a t-*prime*). We denote by t-Spec(D) the set of t-prime ideals of D and by t-Max(D) the set of t-maximal ideals of D. We will often use the properties that a minimal prime of a t-ideal, in particular a height-one prime, is a t-prime and that $I_t = \cap\{I_tD_M ; M \in \text{t-Max}(D)\}$ for any fractional ideal I of D; in particular $D = \cap\{D_M ; M \in \text{t-Max}(D)\} = \cap\{D_P ; P \in \text{t-Spec}(D)\}$. If each t-maximal ideal of D has height one, we say that D has t-*dimension* one.

A v-*finite* fractional ideal is an ideal of type J_v, with J a finitely generated fractional ideal. We denote by $\mathcal{D}_f(D)$ the set of v-finite fractional ideals of D. The set $\mathcal{D}_f(D)$ is a subsemigroup of t(D). A fractional t-ideal I is t-*invertible* if it is invertible in the semigroup t(D). In this case the t-*inverse* of I is (D:I). A t-invertible fractional ideal is divisorial and v-finite, so that the group $\mathcal{T}(D)$ of fractional t-invertible t-ideals of D is the largest subgroup of $\mathcal{D}_f(D)$. A t-ideal I is t-invertible if and only if I is v-finite and ID_M is principal for all t-maximal ideals M of D. A t-invertible t-prime is t-maximal.

The (t-)*class group* of D is defined as the quotient group $\mathcal{C}(D) := \mathcal{T}(D)/\mathcal{P}(D)$, where $\mathcal{P}(D)$ is the group of fractional principal ideals of D. If D is a Krull domain, then $\mathcal{C}(D)$ coincides with the divisor class group of D and, if D is a Prüfer domain, $\mathcal{C}(D)$ coincides with the class group of invertible ideals (or Picard group) of D.

1. v-finite multiplicative systems of ideals

Let D be any commutative ring. If $\mathcal{F}$ is a multiplicative system of ideals of D, that is a family of nonzero integral ideals of D closed under multiplication, the overring $D_{\mathcal{F}} := \cup\{(D:J) ; J \in \mathcal{F}\}$ of D is called the *generalized ring of fractions* of D with respect to $\mathcal{F}$. We refer to [8, Section 1] and [23, Section 4] for basic properties of multiplicative systems of ideals and generalized quotient rings (see also [15, Section 5.1]). If I is a fractional ideal of D, then $I_{\mathcal{F}} := \cup\{(I:J) ; J \in \mathcal{F}\}$ is a fractional ideal of $D_{\mathcal{F}}$ and $ID_{\mathcal{F}} \subseteq I_{\mathcal{F}}$.

We denote by $\mathrm{Sat}(\mathcal{F})$ the *saturation* of $\mathcal{F}$, that is the set of ideals of D containing some ideal in $\mathcal{F}$. Then $\mathrm{Sat}(\mathcal{F})$ is a multiplicative system of ideals of D and clearly $I_{\mathcal{F}} = I_{\mathrm{Sat}(\mathcal{F})}$ for each fractional ideal I of D. We say that $\mathcal{F}$ is *saturated* if $\mathcal{F} = \mathrm{Sat}(\mathcal{F})$.

If S is a multiplicative part of D and $\mathcal{F} := \{sD ; s \in S\}$, then $D_{\mathcal{F}} = D_S$ is a ring of fractions of D. If P is a prime ideal of D, we set $\mathcal{F}(P) := \{ I ; I$ integral ideal of D such that $I \not\subseteq P\}$. Then $\mathcal{F}(P)$ is a saturated multiplicative system of ideals and $D_{\mathcal{F}(P)} = D_P$. Moreover, if Λ is a nonempty family of nonzero prime ideals of D, setting $\mathcal{F}(\Lambda) := \cap\{ \mathcal{F}(P) ; P \in \Lambda\}$, we have that $\mathcal{F}(\Lambda)$ is a saturated multiplicative system of ideals and $D_{\mathcal{F}(\Lambda)} = \cap\{D_P ; P \in \Lambda\}$ [15, Proposition 5.1.4]. In this case, it is also easy to verify that $I_{\mathcal{F}(\Lambda)} = \cap\{ID_P ; P \in \Lambda\}$, for each fractional ideal I of D. In particular, if Λ is a set of pairwise incomparable primes and $P \in \Lambda$, then $P_{\mathcal{F}(\Lambda)} = PD_P \cap D_{\mathcal{F}(\Lambda)}$.

Proposition 1.1. Let D be any commutative ring and $\mathcal{F}$ a multiplicative system of ideals of D. Then:

(a) If I is an ideal of D, then $I_{\mathcal{F}} = D_{\mathcal{F}}$ if and only if $I \in \mathrm{Sat}(\mathcal{F})$.

(b) The map $P \longrightarrow P'$ is a one-to-one inclusion preserving correspondence between the set of prime ideals P of D such that $P \notin \mathrm{Sat}(\mathcal{F})$ and the set of prime ideals P' of $D_{\mathcal{F}}$ such that $JD_{\mathcal{F}} \not\subseteq P'$ for any $J \in \mathcal{F}$, whose inverse is defined by $P' \longrightarrow P' \cap D$. In addition $(D_{\mathcal{F}})_{P_{\mathcal{F}}} = D_P$ for each $P \notin \mathrm{Sat}(\mathcal{F})$.

Proof. (a) Let $I \in \mathrm{Sat}(\mathcal{F})$. If $x \in D_{\mathcal{F}}$, then $xH \subseteq D$ for some $H \in \mathcal{F}$. Hence $xHI \subseteq I$ and, since $IH \in \mathrm{Sat}(\mathcal{F})$, then $x \in I_{\mathrm{Sat}(\mathcal{F})} = I_{\mathcal{F}}$. Whence $I_{\mathcal{F}} = D_{\mathcal{F}}$.

Conversely, if $I_{\mathcal{F}} = D_{\mathcal{F}}$, then $1 \in I_{\mathcal{F}}$ and so $1H = H \subseteq I$ for some $H \in \mathcal{F}$. It follows that $I \in \mathrm{Sat}(\mathcal{F})$.

(b) is [8, Theorem 1.1].

We are interested in the behaviour of t-operation under the extension $D \subseteq D_{\mathcal{F}}$. The following properties are easy to prove (cf. [29, Lemma 1, Section 3] and [25, Lemma 3.4]).

Proposition 1.2. Let D be any commutative ring and $\mathcal{F}$ a multiplicative system of ideals of D. If I, J are fractional ideals of D, then:

(a) $(I:J)_{\mathcal{F}} \subseteq (I_{\mathcal{F}}:J_{\mathcal{F}})$ and, if J is finitely generated or v-finite, then $(I:J)_{\mathcal{F}} = (I_{\mathcal{F}}:J_{\mathcal{F}}) = (I_{\mathcal{F}}:JD_{\mathcal{F}})$.

(b) If J is finitely generated, then $J_vD_{\mathcal{F}} \subseteq (J_v)_{\mathcal{F}} \subseteq (J_{\mathcal{F}})_v = (JD_{\mathcal{F}})_v$; thus $(J_vD_{\mathcal{F}})_v = (J_{\mathcal{F}})_v = (JD_{\mathcal{F}})_v$. If in addition (D:J) is also v-finite, then $(JD_{\mathcal{F}})_v = (J_v)_{\mathcal{F}}$.

(c) $I_{\mathcal{F}}J_{\mathcal{F}} \subseteq (IJ)_{\mathcal{F}} \subseteq I_{\mathcal{F}} \cap J_{\mathcal{F}} = (I \cap J)_{\mathcal{F}}$.

Proof. (a) Let $x \in (I:J)_{\mathcal{F}}$ and $H \in \mathcal{F}$ be such that $xH \subseteq (I:J)$. If $y \in J_{\mathcal{F}}$ and $H' \in \mathcal{F}$ is such that $yH' \subseteq J$, then $xyHH' \subseteq (I:J)J \subseteq I$. Since $HH' \in \mathcal{F}$, then $xy \in I_{\mathcal{F}}$. Whence $x \in (I_{\mathcal{F}}:J_{\mathcal{F}})$. Assume now that $J = x_1D + \ldots + x_nD$ is finitely generated. We have $(I:J)_{\mathcal{F}} \subseteq (I_{\mathcal{F}}:J_{\mathcal{F}}) \subseteq (I_{\mathcal{F}}:JD_{\mathcal{F}})$ and have to show that $(I_{\mathcal{F}}:JD_{\mathcal{F}}) \subseteq (I:J)_{\mathcal{F}}$. Let $y \in K$ be such that $yJD_{\mathcal{F}} \subseteq I_{\mathcal{F}}$. Hence $yJ \subseteq I_{\mathcal{F}}$. If $H_i \in \mathcal{F}$ is such that $yx_iH_i \subseteq I$, $i = 1, \ldots, n$ and $H = H_1 \ldots H_n$, then $H \in \mathcal{F}$ and $yJH \subseteq I$. Thus $yH \subseteq (I:J)$ and so $y \in (I:J)_{\mathcal{F}}$. Finally, set $J = F_v$, with F finitely generated. Then $(I_{\mathcal{F}}:J_{\mathcal{F}}) \subseteq (I_{\mathcal{F}}:JD_{\mathcal{F}}) \subseteq (I_{\mathcal{F}}:FD_{\mathcal{F}}) = (I_{\mathcal{F}}:F_{\mathcal{F}}) = (I:F)_{\mathcal{F}} = (I:J)_{\mathcal{F}} \subseteq (I_{\mathcal{F}}:J_{\mathcal{F}})$ and equalities hold.

(b) By applying (a), $J_vD_{\mathcal{F}} \subseteq (J_v)_{\mathcal{F}} = (D:(D:J))_{\mathcal{F}} \subseteq (D_{\mathcal{F}}: (D:J)_{\mathcal{F}}) = (D_{\mathcal{F}}: (D_{\mathcal{F}}:J_{\mathcal{F}})) = (J_{\mathcal{F}})_v = (D_{\mathcal{F}}: (D_{\mathcal{F}}: JD_{\mathcal{F}}) = (JD_{\mathcal{F}})_v$. Now assume that (D:J) is v-finite. Then $(J_v)_{\mathcal{F}} = (D:(D:I))_{\mathcal{F}} = (D_{\mathcal{F}}: (D:J)_{\mathcal{F}}) = (D_{\mathcal{F}}: (D_{\mathcal{F}}:J_{\mathcal{F}})) = (D_{\mathcal{F}}: (D_{\mathcal{F}}: JD_{\mathcal{F}}) = (JD_{\mathcal{F}})_v$.

(c) is immediate.

By Proposition 1.2(b), if J is finitely generated or v-finite, then $J_{\mathcal{F}} \subseteq (JD_{\mathcal{F}})_v = (JD_{\mathcal{F}})_t$. Thus, in this case, if $J \in \mathrm{Sat}(\mathcal{F})$, it results $(JD_{\mathcal{F}})_t = D_{\mathcal{F}}$. However, in general it may happen that $(JD_{\mathcal{F}})_t \neq J_{\mathcal{F}} = D_{\mathcal{F}}$ for some J in $\mathrm{Sat}(\mathcal{F})$ (see Proposition 1.5 below).

Proposition 1.3. Let D be any domain and $\mathcal{F}$ a multiplicative system of ideals of D. Then the following conditions hold and are equivalent:

(i) $J_v \subseteq (JD_{\mathcal{F}})_v$, for each finitely generated (fractional) ideal J of D;

(ii) $(I_tD_{\mathcal{F}})_t = (ID_{\mathcal{F}})_t$, for each (fractional) ideal I of D;

(iii) If I' is a t-ideal of $D_{\mathcal{F}}$ and $I' \cap D \neq (0)$, then $I' \cap D$ is a t-ideal of D.

Proof. (i) holds by Proposition 1.2. The equivalences (i) $\Leftrightarrow$ (ii) $\Leftrightarrow$ (iii) are proved in [10, Proposition 1.1].

By Proposition 1.3, the natural map $t(D) \longrightarrow t(D_{\mathcal{F}})$, $I \longrightarrow (ID_{\mathcal{F}})_t$ is a well defined homomorphism of semigroups, which induces homomorphisms $\mathcal{D}_f(D) \longrightarrow \mathcal{D}_f(D_{\mathcal{F}})$, $\mathcal{T}(D) \longrightarrow \mathcal{T}(D_{\mathcal{F}})$ and $\mathcal{C}(D) \longrightarrow \mathcal{C}(D_{\mathcal{F}})$ [10, Proposition 1.1] (see also [6, Section 4]). Observe that, if $I \in \mathcal{T}(D)$, then $(ID_{\mathcal{F}})_t = (ID_{\mathcal{F}})_v = I_{\mathcal{F}}$ by Proposition 1.2 (b).

Proposition 1.4. Let D be any domain and $\mathcal{F}$ a multiplicative system of ideals of D. Then the natural homomorphism of semigroups $\mathcal{D}_f(D) \longrightarrow \mathcal{D}_f(D_{\mathcal{F}})$, $I \longrightarrow (ID_{\mathcal{F}})_t$, is surjective.

Proof. Since D and $D_{\mathcal{F}}$ have the same quotient field, if $I' = a_1D_{\mathcal{F}} + \dots + a_nD_{\mathcal{F}}$ is a finitely generated fractional ideal of $D_{\mathcal{F}}$, then $I = a_1D + \dots + a_nD$ is a finitely generated fractional ideal of D and $I' = ID_{\mathcal{F}}$. In addition $I'_v = (ID_{\mathcal{F}})_v = (I_vD_{\mathcal{F}})_v$ by Proposition 1.2 (b).

We give in Proposition 1.7 below sufficient conditions for the surjectivity of the natural homomorphism $t(D) \longrightarrow t(D_{\mathcal{F}})$.

Proposition 1.5. Let D be any domain and $\mathcal{F}$ a multiplicative system of ideals of D. Then the following are equivalent:

(i) $(ID_{\mathcal{F}})_t = (I_{\mathcal{F}})_t$, for each (fractional) ideal I of D;

(ii) $(JD_{\mathcal{F}})_t = D_{\mathcal{F}}$, for each ideal $J \in \mathrm{Sat}(\mathcal{F})$;

(iii) If $P' \in t\text{-Spec}(D_{\mathcal{F}})$, then $P' = P_{\mathcal{F}}$ for some $P \in t\text{-Spec}(D) \setminus \mathrm{Sat}(\mathcal{F})$;

(iv) If I' is a t-ideal of $D_{\mathcal{F}}$ and $I := I' \cap D$, then $I \in t(D)$ and $I' = (ID_{\mathcal{F}})_t = I_{\mathcal{F}}$.

Moreover, under these conditions, $D_{\mathcal{F}} = \cap\{D_P \,;\, P \in t\text{-Spec}(D) \setminus \mathrm{Sat}(\mathcal{F})\}$.

Proof. (i) $\Rightarrow$ (iv). If $I := I' \cap D$, then $I \in t(D)$ by Proposition 1.3 and we have $(ID_{\mathcal{F}})_t \subseteq I' \subseteq I_{\mathcal{F}} \subseteq (I_{\mathcal{F}})_t$. Hence $(ID_{\mathcal{F}})_t = I' = I_{\mathcal{F}}$.

(iv) $\Rightarrow$ (iii) is clear.

(iii) $\Rightarrow$ (ii). If $J \in \mathrm{Sat}(\mathcal{F})$, then $J \not\subseteq P := P' \cap D$ for each $P' \in t\text{-Spec}(D_{\mathcal{F}})$; otherwise $D_{\mathcal{F}} = J_{\mathcal{F}} \subseteq P_{\mathcal{F}} = P'$. Hence $(JD_{\mathcal{F}})_t = D_{\mathcal{F}}$.

(ii) $\Rightarrow$ (i). Always $(ID_{\mathcal{F}})_t \subseteq (I_{\mathcal{F}})_t$. Conversely, let $x \in I_{\mathcal{F}}$ and $J \in \mathcal{F}$ such that $xJ \subseteq I$. Then $(xJD_{\mathcal{F}})_t = xD_{\mathcal{F}} \subseteq (ID_{\mathcal{F}})_t$. Whence $(I_{\mathcal{F}})_t \subseteq (ID_{\mathcal{F}})_t$.

Now we prove that, under these conditions, $D_{\mathcal{F}} = \cap\{D_P \,;\, P \in t\text{-Spec}(D) \setminus \mathrm{Sat}(\mathcal{F})\}$. Always we have $\mathrm{Sat}(\mathcal{F}) \subseteq \cap\{\mathcal{F}(P) \,;\, P \in \mathrm{Spec}(D) \setminus \mathrm{Sat}(\mathcal{F})\} \subseteq \cap\{\mathcal{F}(P) \,;\, P \in t\text{-Spec}(D) \setminus \mathrm{Sat}(\mathcal{F})\}$. Thus $D_{\mathcal{F}} \subseteq \cap\{D_P \,;\, P \in t\text{-Spec}(D) \setminus \mathrm{Sat}(\mathcal{F})\}$. The opposite inclusion follows from condition (iii), because $D_{\mathcal{F}} = \cap\{(D_{\mathcal{F}})_Q \,;\, Q \in t\text{-Spec}(D_{\mathcal{F}})\} = \cap\{(D)_{Q \cap D} \,;\, Q \in t\text{-Spec}(D_{\mathcal{F}})\}$.

Since $D = \cap\{D_P ; P \in \text{t-Spec}(D)\}$, if $\Lambda \subseteq \text{t-Spec}(D)$ we say that $D_{\mathcal{F}(\Lambda)} = \cap\{D_P ; P \in \Lambda\}$ is a t-*subintersection* of D. We have seen that, under the equivalent conditions of Proposition 1.5, setting $\Sigma := \{\text{t-Spec}(D)\backslash\text{Sat}(\mathcal{F})\}$, $D_{\mathcal{F}} = D_{\mathcal{F}(\Sigma)}$ is a t-subintersection of D.

If $P_{\mathcal{F}}$ is a t-ideal for each prime t-ideal P of D (not belonging to $\text{Sat}(\mathcal{F})$), then condition (iii) of Proposition 1.5 is equivalent to the equality $\text{t-Spec}(D_{\mathcal{F}}) = \{P_{\mathcal{F}}; P \in \text{t-Spec}(D)\backslash\text{Sat}(\mathcal{F})\}$. However, $P_{\mathcal{F}}$ need not be a t-ideal of $D_{\mathcal{F}}$. Indeed, if P is a non-divisorial t-maximal ideal of D, then PD_P may not be a t-ideal of D_P (see [30, Proposition 4.3] and [19, Example 2.2]).

Proposition 1.6. Let D be any domain, $\mathcal{F}$ a multiplicative system of ideals of D and $P \in \text{t-Spec}(D)\backslash\text{Sat}(\mathcal{F})$. Then:

(a) $P_{\mathcal{F}}$ is a t-ideal of $D_{\mathcal{F}}$ under one of the following conditions:

(1) P has height one.

(2) P and (D:P) are divisorial v-finite, in particular P is t-invertible.

(b) Set $\Lambda := \{P \in \text{t-Spec}(D) ; P \notin \text{Sat}(\mathcal{F})$ and P is maximal with respect to this property$\}$. If $P_{\mathcal{F}}$ is a t-ideal for each $P \in \text{t-Spec}(D)$, then the following conditions are equivalent:

(i) $\text{t-Spec}(D_{\mathcal{F}}) = \{P_{\mathcal{F}}; P \in \text{t-Spec}(D)\backslash\text{Sat}(\mathcal{F})\}$;

(ii) Λ is not empty and $\text{t-Max}(D_{\mathcal{F}}) = \{M_{\mathcal{F}}; M \in \Lambda\}$.

Proof. (a) If P has height one, then $P_{\mathcal{F}}$ has height one by Proposition 1.1 (b). Hence it is a t-ideal. If condition (2) is verified, $P_{\mathcal{F}} = (PD_{\mathcal{F}})_t$ a t-ideal by Proposition 1.2(b).

(b) (i) $\Rightarrow$ (ii) It is enough to observe that, if (i) holds, then $\Lambda = \{M := M' \cap D ; M' \in \text{t-Max}(D_{\mathcal{F}})\}$.

(ii) $\Rightarrow$ (i). If $J \in \text{Sat}(\mathcal{F})$, then $J \not\subset M := M' \cap D$ for all $M' \in \text{t-Max}(D_{\mathcal{F}})$; otherwise $D_{\mathcal{F}} = J_{\mathcal{F}} \subseteq M_{\mathcal{F}} = M'$. Hence $(JD_{\mathcal{F}})_t = D_{\mathcal{F}}$ and so $\text{t-Spec}(D_{\mathcal{F}}) \subseteq \{P_{\mathcal{F}}; P \in \text{t-Spec}(D)\backslash\text{Sat}(\mathcal{F})\}$ by Proposition 1.5. Since $P_{\mathcal{F}}$ is a t-ideal for any $P \in \text{t-Spec}(D)$, the equality holds.

Proposition 1.7. Under the equivalent conditions of Proposition 1.5, $D_{\mathcal{F}}$ is a t-subintersection of D and the natural homomorphism of semigroups $t(D) \longrightarrow t(D_{\mathcal{F}})$, $I \longrightarrow (ID_{\mathcal{F}})_t$, is surjective.

Proof. Let $I' \in t(D_{\mathcal{F}})$ and set $I' = x J'$, with $x \in K$ and $J' \subseteq D_{\mathcal{F}}$. J' is at-ideal, then the ideal $I := x(J' \cap D) \in t(D)$ by Proposition 1.3 (c) and, by condition (iv) of Proposition 1.5, it results $(ID_{\mathcal{F}})_t = xJ' = I'$.

The converse of Proposition 1.7 does not hold. For example, let V be a valuation domain, assume that the maximal ideal M of V is idempotent and consider the multiplicative system of ideals $\mathcal{F} := \{V, M\}$. Then $V = V_{\mathcal{F}}$, because $(V:M) = (M:M) = V$, and the natural map $t(V) \longrightarrow t(V_{\mathcal{F}})$ is the identity. However $M = MV_{\mathcal{F}} = (MV_{\mathcal{F}})_t \neq (M_{\mathcal{F}})_t = V$ and condition (i) of Proposition 1.5 is not satisfied.

A multiplicative system of ideals $\mathcal{F}$ of D is *finitely generated* if each ideal $I \in \mathcal{F}$ contains a finitely generated ideal J which is still in $\mathcal{F}$. Clearly if $\mathcal{F}$ is finitely generated then Sat($\mathcal{F}$) is finitely generated. If D is noetherian, each multiplicative system of ideals $\mathcal{F}$ of D is finitely generated and, if D is any domain, a multiplicative system of principal ideals $\mathcal{F} := \{sD \,;\, s \in S\}$, S a multiplicative part of D, is also finitely generated. The multiplicative system of ideals $\mathcal{F} := \{V, M\}$ considered above is not finitely generated, because M is not finitely generated.

Denote by t-Sat($\mathcal{F}$) the set of all t-ideals in Sat($\mathcal{F}$). Since $(D:I) = (D:I_t)$ for all fractional ideals I of D, we get that $D_{\mathcal{F}} = \cup\{(D:J) \,;\, J \in \text{t-Sat}(\mathcal{F})\}$. Note that t-Sat($\mathcal{F}$) is closed under t-multiplication (because, if $I, J \in \text{Sat}(\mathcal{F})$, then $(IJ)_t \in \text{t-Sat}(\mathcal{F})$) and is t-saturated (that is each t-ideal containing an ideal in t-Sat($\mathcal{F}$) is in t-Sat($\mathcal{F}$)). Hence in order to study the domain $D_{\mathcal{F}}$, and in particular the behaviour of the homomorphism $t(D) \longrightarrow t(D_{\mathcal{F}})$, we might just assume that $\mathcal{F}$ be a t-multiplicative t-saturated family of t-ideals. This fact motivates the following definition.

We say that $\mathcal{F}$ is v-*finite* if each ideal $I \in \text{t-Sat}(\mathcal{F})$ contains a finitely generated ideal J such that $J_v \in \text{Sat}(\mathcal{F})$. It is immediate to check that $\mathcal{F}$ is v-finite if and only if Sat($\mathcal{F}$) is v-finite and that, if each t-ideal in $\mathcal{F}$ is v-finite, then $\mathcal{F}$ is v-finite. Since in a Mori domain each t-ideal is v-finite, then each multiplicative system of a Mori domain is v-finite.

Each finitely generated multiplicative system of ideals is v-finite. The converse does not hold. For example take the Krull domain $D := F[\underline{X}]$, where F is a field and $\underline{X}$ is an infinite set of indeterminates. Let M be the maximal ideal of D generated by $\underline{X}$ and consider the multiplicative system of ideals $\mathcal{F} := \{M^n \,;\, n \geq 0\}$. Then $\text{t-Sat}(\mathcal{F}) = \{D\}$, because $(M^n)_t = D$, for all n. Hence $\mathcal{F}$ is trivially v-finite. However $\mathcal{F}$ is not finitely generated, because no proper ideal I in $\mathcal{F}$, being M- primary, is finitely generated. In fact, if I is a finitely generated ideal of D, then $I = JD$, for some ideal J of $F[X_1, \ldots, X_n]$, $\{X_1, \ldots, X_n\} \subseteq \underline{X}$. Thus I is contained in some finitely generated prime ideal of D.

If D is a Prüfer domain, then each ideal is a t-ideal, thus in this case $\mathcal{F}$ is v-finite if and only if is finitely generated.

Proposition 1.8. Let D be any domain and $\mathcal{F}$ a v-finite multiplicative system of ideals of D. Then $(I_t)_{\mathcal{F}} \subseteq (ID_{\mathcal{F}})_t = (I_{\mathcal{F}})_t$ for all fractional ideal I of D. Hence all the conditions of Proposition 1.5 are satisfied. In particular $D_{\mathcal{F}}$ is a t-subintersection of D and the natural homomorphism $t(D) \longrightarrow t(D_{\mathcal{F}})$ is surjective.

Proof. Let $x \in (I_t)_{\mathcal{F}}$. Then there exists a finitely generated ideal J of D such that $J_v \in Sat(\mathcal{F})$ and $xJ_v \subseteq I_t$. Since $(J_vD_{\mathcal{F}})_v = (J_v)_{\mathcal{F}} = D_{\mathcal{F}}$ (Proposition 1.2 (b)), then $xD_{\mathcal{F}} = x(J_vD_{\mathcal{F}})_v \subseteq (I_tD_{\mathcal{F}})_t = (ID_{\mathcal{F}})_t$. Whence $(I_t)_{\mathcal{F}} \subseteq (ID_{\mathcal{F}})_t \subseteq (I_{\mathcal{F}})_t$. We now show that $(ID_{\mathcal{F}})_t = (I_{\mathcal{F}})_t$. Let $JD_{\mathcal{F}} \subseteq I_{\mathcal{F}}$, with $J = x_1D + \ldots + x_nD$ a finitely generated ideal of D. If $H_i \in \mathcal{F}$ is such that $x_iH_i \subseteq I$, then $JH \subseteq I$, where $H := H_1 \ldots H_n \in \mathcal{F}$, and so $JH_t \subseteq (JH)_t \subseteq I_t$. Since $H_t \in Sat(\mathcal{F})$, then there exists a finitely generated ideal H' of D such that $H'_v \subseteq H_t$ and $H'_v \in Sat(\mathcal{F})$. By Proposition 1.2 (b), we have $D_{\mathcal{F}} = (H'_v)_{\mathcal{F}} \subseteq (H'_vD_{\mathcal{F}})_v = (H'D_{\mathcal{F}})_v = (H'D_{\mathcal{F}})_t \subseteq (H_tD_{\mathcal{F}})_t$. Hence $(H_tD_{\mathcal{F}})_t = D_{\mathcal{F}}$ and $(JD_{\mathcal{F}})_v = (JD_{\mathcal{F}})_t = (JH_tD_{\mathcal{F}})_t \subseteq (I_tD_{\mathcal{F}})_t$. It follows that $(I_{\mathcal{F}})_t \subseteq (I_tD_{\mathcal{F}})_t = (ID_{\mathcal{F}})_t$.

Proposition 1.9. Let D be any domain, $\mathcal{F}$ a v-finite multiplicative system of ideals of D and $\mathcal{S} := t(D) \setminus Sat(\mathcal{F})$. Then:

(a) Each subset $\mathcal{T}$ of $\mathcal{S}$ has maximal elements and each maximal element of $\mathcal{T}$ is prime.

(b) If Λ is the set of the maximal elements of $\mathcal{S}$, then $t\text{-}Sat(\mathcal{F}) = \mathcal{F}(\Lambda) \cap t(D)$. In particular $\mathcal{F}(\Lambda)$ is v-finite and $D_{\mathcal{F}} = D_{\mathcal{F}(\Lambda)}$.

(c) If $P_{\mathcal{F}}$ is a t-ideal for each $P \in t\text{-}Spec(D)$, then $t\text{-}Max(D_{\mathcal{F}}) = \{M_{\mathcal{F}};\ M \in \Lambda\}$.

Proof. (a). Let $I_0 \subseteq I_1 \subseteq \ldots \subseteq I_n \subseteq \ldots$ be an ascending chain of ideals of $\mathcal{T}$ and $I := \cup\{I_j\ ;\ j \geq 0\}$. If H is a finitely generated ideal contained in I, then $H \subseteq I_n$ for some $n \geq 0$ and, since I_n is a t-ideal, then $H_v \subseteq I_n \subseteq I$. Hence I is a t-ideal and $I \notin Sat(\mathcal{F})$. Otherwise $I_n \in Sat(\mathcal{F})$ for some $n \geq 0$, because $\mathcal{F}$ is v-finite; a contradiction. By Zorn's Lemma, the set $\mathcal{T}$ has maximal elements and, since $Sat(\mathcal{F})$ is multiplicatively closed, a standard argument shows that each maximal element of $\mathcal{T}$ is prime.

(b) Always we have $Sat(\mathcal{F}) \subseteq \cap\{\mathcal{F}(P)\ ;\ P \in Spec(D) \setminus Sat(\mathcal{F})\}$. Hence $t\text{-}Sat(\mathcal{F}) \subseteq \mathcal{F}(\Lambda) \cap t(D)$. The opposite inclusion follows from (a). In fact, if I is a t-ideal and $I \notin Sat(\mathcal{F})$, then $I \in M$ for some $M \in \Lambda$. Hence $I \notin \mathcal{F}(\Lambda) \cap t(D)$.

The last two statements are immediate consequences of the equality just proved.

(c) If $\mathcal{F}$ is v-finite and $P_{\mathcal{F}}$ is a t-ideal for each $P \in t\text{-}Spec(D)$, then $t\text{-}Spec(D) = \{P_{\mathcal{F}};\ P \in Spec(D) \setminus Sat(\mathcal{F})\}$ (Propositions 1.8 and 1.5) and Λ is the set of maximal elements in $Spec(D) \setminus Sat(\mathcal{F})$.

When $\mathcal{F}$ is finitely generated, an analogue of Proposition 1.9 (a) and (b) has been proved in [15, Lemma 5.1.5].

We do not know whether, under the equivalent conditions of Proposition 1.5, t-Sat($\mathcal{F}$) always coincides with $\mathcal{F}(\Sigma) \cap t(D)$, where $\Sigma := \{t\text{-Spec}(D) \setminus \text{Sat}(\mathcal{F})\}$. However, the same argument used in the proof of [23, Proposition 4.2] shows that, with no assumption on $\mathcal{F}$, if each t-ideal of $\mathcal{F}(\Sigma)$ has finitely many minimal (t-)primes and contains a power of its radical then t-Sat($\mathcal{F}$) = $\mathcal{F}(\Sigma) \cap t(D)$.

If each t-ideal in Sat($\mathcal{F}$) is t-invertible, then clearly $\mathcal{F}$ is v-finite. In the next Theorem we characterize these multiplicative systems of ideals of relevant interest.

Theorem 1.10. Let D be any domain and $\mathcal{F}$ a multiplicative system of ideals of D. Then the following statements are equivalent:

(i) t-Sat($\mathcal{F}$) $\subseteq \mathcal{T}(D)$;

(ii) Each t-prime ideal P in Sat($\mathcal{F}$) is t-invertible;

(iii) If I $\in$ t-Sat($\mathcal{F}$), each minimal prime of I is t-invertible;

(iv) If I $\in$ t-Sat($\mathcal{F}$), then $I = (P_1^{n_1} \dots P_k^{n_k})_v$ for some t-invertible t-prime ideals $P_1, \dots, P_k$ and $n_i \geq 1$;

(v) t-Sat($\mathcal{F}$) = $\{(P_1^{n_1} \dots P_k^{n_k})_v\ ;\ P_1, \dots, P_k \in \mathcal{T}(D) \cap \text{Sat}(\mathcal{F}),\ n_i \geq 1\}$.

Proof. (i) $\Rightarrow$ (ii) and (v) $\Rightarrow$ (iv) $\Rightarrow$ (i) are clear.

(ii) $\Rightarrow$ (iii) because a prime minimal over a t-ideal is a t-ideal.

(iii) $\Rightarrow$ (i). By [20, Proposition 1.5], if each minimal prime of a t-ideal I is t-invertible, then I is t-invertible.

(i) $\Rightarrow$ (iv) is [20, Proposition 1.6].

(iv) $\Rightarrow$ (v). If $I = (P_1^{n_1} \dots P_k^{n_k})_v \in$ t-Sat($\mathcal{F}$), then $P_1, \dots, P_k \in \mathcal{T}(D) \cap \text{Sat}(\mathcal{F})$. Conversely, if $I = (P_1^{n_1} \dots P_k^{n_k})_v$ with $P_1, \dots, P_k \in \mathcal{T}(D) \cap \text{Sat}(\mathcal{F})$, then I is t-invertible and so $I_{\mathcal{F}} = (ID_{\mathcal{F}})_v = ((P_1^{n_1} \dots P_k^{n_k})_v D_{\mathcal{F}})_v$. But $(P_i D_{\mathcal{F}})_v = (P_i)_{\mathcal{F}} = D_{\mathcal{F}}$ for each i. Hence $I_{\mathcal{F}} = D_{\mathcal{F}}$ and so I $\in$ t-Sat($\mathcal{F}$).

Since a t-invertible t-prime is t-maximal [G, Corollary 1.8], under the equivalent conditions of Theorem 1.10, all the t-primes in t-Sat($\mathcal{F}$) are t-maximal.

Let S be a multiplicative part of D generated by primes. Then each t-prime ideal P intersecting S contains a prime element p $\in$ S. Since a principal prime ideal is t-maximal, then P = pD is principal, whence t-invertible. Therefore the multiplicative system of ideals $\mathcal{F} := \{sD;\ s \in S\}$ satisfies all the equivalent conditions of Theorem 1.10.

When t-Sat($\mathcal{F}$) $\subseteq \mathcal{T}(D)$, we are able to give a description of the kernel of the natural homomorphism $\phi_{\mathcal{F}} : \mathcal{T}(D) \longrightarrow \mathcal{T}(D_{\mathcal{F}})$, $I \longrightarrow (ID_{\mathcal{F}})_t = I_{\mathcal{F}}$.

Proposition 1.11. Let D be any domain and $\mathcal{F}$ a multiplicative system of ideals of D such that each $I \in$ t-Sat($\mathcal{F}$) contains a t-invertible t-ideal which is still in Sat($\mathcal{F}$). Then $I_{\mathcal{F}} = D_{\mathcal{F}}$ if and only if $I = (J(D:H))_v$, with $J, H \in$ t-Sat($\mathcal{F}$).

Proof. If J is a finitely generated fractional ideal of D and $(J_v)_{\mathcal{F}} = D_{\mathcal{F}}$, then also $(D:J)_{\mathcal{F}} = D_{\mathcal{F}}$. In fact, by Proposition 1.2 (a), we have $D_{\mathcal{F}} = (D_{\mathcal{F}}:(J_v)_{\mathcal{F}}) \subseteq (D_{\mathcal{F}}:J_{\mathcal{F}}) = (D:J)_{\mathcal{F}} = (D:J_v)_{\mathcal{F}} \subseteq (D_{\mathcal{F}}:(J_v)_{\mathcal{F}}) = D_{\mathcal{F}}$. Therefore, if $J, H \in$ t-Sat($\mathcal{F}$) and $I := (J(D:H))_v$ we have $I_{\mathcal{F}} = (J_{\mathcal{F}}(D:H)_{\mathcal{F}})_v = D_{\mathcal{F}}$. Conversely, assume that $I \in \mathcal{T}(D)$ is such that $I_{\mathcal{F}} = D_{\mathcal{F}}$ and consider the ideal $I' := (D:I) \cap D$. Hence I' is a divisorial ideal of D and, since $(D:I)_{\mathcal{F}} = D_{\mathcal{F}}$, we have $I'_{\mathcal{F}} = (D:I)_{\mathcal{F}} \cap D_{\mathcal{F}} = D_{\mathcal{F}}$ (Proposition 1.2 (c)). Whence $I' \in$ Sat($\mathcal{F}$). Then there exists a t-invertible t-ideal H of D such that $H \subseteq I'$ and $H \in$ Sat($\mathcal{F}$). Set $J := (IH)_v$. Thus $J \subseteq D$, $J \in \mathcal{T}(D)$ and $J_{\mathcal{F}} = (H_{\mathcal{F}} I_{\mathcal{F}})_v = D_{\mathcal{F}}$. It follows that $I = (IH(D:H))_v = (J(D:H))_v$, with $J, H \in$ t-Sat($\mathcal{F}$).

We observe that Proposition 1.11 applies to the case when D is a PvMD (that is each v-finite ideal of D is t-invertible) and $\mathcal{F}$ is a v-finite multiplicative system of ideals of D (see Corollary 3.7 below).

Corollary 1.12. Let D be any domain and $\mathcal{F}$ a multiplicative system of ideals of D such that t-Sat($\mathcal{F}$) $\subseteq \mathcal{T}(D)$. Let $\phi_{\mathcal{F}}: \mathcal{T}(D) \longrightarrow \mathcal{T}(D_{\mathcal{F}})$ be the natural homomorphism. Then

$$\mathrm{Ker}(\phi_{\mathcal{F}}) = \{(P_1^{e_1} \dots P_k^{e_k})_v \ ; P_i \in \text{t-Sat}(\mathcal{F}),\ e_i \in \mathbb{Z} \text{ and } k \geq 1\},$$

where, if $n > 0$, we denote by P^{-n} the ideal $(D:P^n)$.

Proof. If $I \in \mathcal{T}(D)$ and $I \subseteq D$, then $I \in \mathrm{Ker}(\phi_{\mathcal{F}})$ if and only if $I \in$ t-Sat($\mathcal{F}$) $= \{(P_1^{n_1} \dots P_k^{n_k})_v \ ; P_i \in \text{t-Sat}(\mathcal{F}),\ n_i \in \mathbb{N} \text{ and } k \geq 1\}$ (Theorem 1.10). We conclude by Proposition 1.11.

Theorem 1.13. Let D be any domain and $\mathcal{F}$ a multiplicative system of ideals of D such that t-Sat($\mathcal{F}$) $\subseteq \mathcal{T}(D)$. If t-Spec($D_{\mathcal{F}}$) $= \{P_{\mathcal{F}};\ P \in \text{t-Spec}(D)\backslash\text{Sat}(\mathcal{F})\}$, then the natural homomorphism $\phi_{\mathcal{F}}: \mathcal{T}(D) \longrightarrow \mathcal{T}(D_{\mathcal{F}})$ is surjective and $\mathrm{Ker}(\phi_{\mathcal{F}})$ is generated by the prime ideals in t-Sat($\mathcal{F}$).

Proof. Let $I' \in \mathcal{T}(D_{\mathcal{F}})$. By Proposition 1.4, we have $I' = (J_t D_{\mathcal{F}})_t$ with J a finitely generated fractional ideal of D. We show that J_t is t-invertible. By Proposition 1.2 (a), $(D_{\mathcal{F}}:I') = (D_{\mathcal{F}}:JD_{\mathcal{F}}) = (D:J)_{\mathcal{F}}$. Thus $D_{\mathcal{F}} = (I'(D_{\mathcal{F}}:I'))_t = (JD_{\mathcal{F}}(D:J)_{\mathcal{F}})_t \subseteq (J_{\mathcal{F}}(D:J)_{\mathcal{F}})_t \subseteq ((J(D:J))_{\mathcal{F}})_t \subseteq D_{\mathcal{F}}$. Whence $((J(D:J))_{\mathcal{F}})_t = D_{\mathcal{F}}$. It follows that $(J(D:J))_{\mathcal{F}}$ is not contained in any t-prime ideal of $D_{\mathcal{F}}$, that is $J(D:J)$ is not contained in any ideal $P \in$ t-Spec(D)\Sat($\mathcal{F}$). Therefore, if $(J(D:J))_t \neq D_{\mathcal{F}}$ and P is a t-prime minimal over $(J(D:J))_t$, then $P \in$ t-Sat($\mathcal{F}$). Thus P is t-invertible. By [20, Proposition 1.5], $(J(D:J))_t$ is t-invertible and so also J_t is t-invertible. Finally, Ker($\phi_{\mathcal{F}}$) is generated by the t-prime ideals in t-Sat($\mathcal{F}$) by Corollary 1.12.

In the statement of Theorem 1.13, the hypothesis that t-Spec($D_{\mathcal{F}}$) = {$P_{\mathcal{F}}$; P ∈ t-Spec(D)\Sat($\mathcal{F}$)} is equivalent to the hypothesis that $P_{\mathcal{F}}$ be a t-ideal for each P ∈ t-Spec(D)\Sat($\mathcal{F}$). This follows from Propositions 1.8 and 1.5, because $\mathcal{F}$ is v-finite. Using Proposition 1.6 (a), we then obtain the following Corollary, generalizing [19, Theorem 1.11, (4) and (5)].

Corollary 1.14. Let D be any domain and $\mathcal{F}$ a multiplicative system of ideals of D such that t-Sat($\mathcal{F}$) ⊆ $\mathcal{T}$(D). Then the natural homomorphism $\phi_{\mathcal{F}}$: $\mathcal{T}$(D) —> $\mathcal{T}$($D_{\mathcal{F}}$) is surjective under either one of the following conditions:

(1) Each P ∈ t-Spec(D)\Sat($\mathcal{F}$) has height one.

(2) D is a Mori domain.

Proof. Recalling that each t-ideal of a Mori domain is divisorial v-finite, in either case t-Spec($D_{\mathcal{F}}$) = {$P_{\mathcal{F}}$; P ∈ t-Spec(D)\Sat($\mathcal{F}$)} by Proposition 1.6 (a). We conclude by applying Theorem 1.13.

From Corollary 1.14, we get that, when t-Sat($\mathcal{F}$) ⊆ $\mathcal{T}$(D), the natural homomorphism $\phi_{\mathcal{F}}$: $\mathcal{T}$(D) —> $\mathcal{T}$($D_{\mathcal{F}}$) is surjective in case D has t-dimension one (see also Corollary 2.4). A Mori domain is v-coherent; we will extend Corollary 1.14 to v-coherent domains in Corollary 3.6.

Corollary 1.15. Let S be a multiplicative part of D generated by primes. If P_S is a t-ideal for each t-prime ideal P not intersecting S, then the natural homomorphism ϕ_S : $\mathcal{T}$(D) —> $\mathcal{T}$(D_S) is surjective.

We recall that, if S is generated by primes, then ϕ_S need not be surjective (cf. [19, Section 2]).

To finish, we exhibit a particular class of v-finite (actually finitely generated) multiplicative systems of ideals.

If Λ is a family of nonzero prime ideals of D, the intersection $D_{\mathcal{F}(\Lambda)} = \cap\{D_P ; P \in \Lambda\}$ has *finite character* if each (noninvertible) element x ∈ D is contained at most in finitely many ideals P ∈ Λ. We say that D has t-*finite character* if the intersection D = ∩{D_M ; M ∈ t-max(D)} has finite character.

Lemma 1.16. Let D be any domain, Λ ⊆ Spec(D) and $\mathcal{F}$:= $\mathcal{F}$(Λ). If the intersection $D_{\mathcal{F}} = \cap\{D_P ; P \in \Lambda\}$ has finite character, then $\mathcal{F}$ is finitely generated.

Proof. Let I be an ideal of D which is in $\mathcal{F}$ and let $x \in I$. Since $D_{\mathcal{F}}$ has finite character, there are at most finitely many ideals $P_1, \ldots, P_s$ in Λ containing x. Let $y_i \in I \setminus P_i$, $i = 1, \ldots, s$. Then the ideal $J := (x, y_1, \ldots, y_s)$ is contained in I and it is not contained in any prime ideal P of Λ. Hence $J \in \mathcal{F}$ and so $\mathcal{F}$ is finitely generated.

Proposition 1.17. Let D be a domain with t-finite character, $\Lambda \subseteq$ t-Max(D) and $\mathcal{F} := \mathcal{F}(\Lambda)$. Then:

(a) $\mathcal{F} := \mathcal{F}(\Lambda)$ is finitely generated.

(b) $\text{t-Max}(D_{\mathcal{F}}) = \{M_{\mathcal{F}}; M \in \Lambda\}$.

Proof. (a) follows from Lemma 1.16.

(b) Since D has t-finite character, then MD_M is a t-ideal for all $M \in$ t-Max(D) [5, Lemma 3.3]. Hence $M_{\mathcal{F}} = MD_M \cap D_{\mathcal{F}}$ is a t-ideal (Proposition 1.3). On the other hand, since $\mathcal{F}$ is v-finite, by Propositions 1.8 and 1.5, we have that each t-maximal ideal M' of $D_{\mathcal{F}}$ is of type $P_{\mathcal{F}}$, for some $P \in$ t-Spec(D))\Sat($\mathcal{F}$). Hence $\text{Max}(D_{\mathcal{F}}) = \{M_{\mathcal{F}}; M \in \Lambda\}$.

We remark that, if D is a domain such that $\mathcal{F}(\Lambda)$ is finitely generated for each $\Lambda \subseteq$ t-Max(D), then D need not have t-finite character. For example, the domain $D := \mathbb{Z}+X\mathbb{Q}[X]$ does not have t-finite character, even though $\mathcal{F}(\Lambda)$ is finitely generated for each $\Lambda \subseteq$ Spec(D) [15, Example 8.4.7].

Theorem 1.18. Let D be a domain with t-finite character, $\Lambda \subseteq$ t-Max(D) and $\mathcal{F} := \mathcal{F}(\Lambda)$. If each (prime) t-ideal in $\mathcal{F}$ is t-invertible, then the natural homomorphism $\phi_{\mathcal{F}}$: $\mathcal{T}(D) \longrightarrow \mathcal{T}(D_{\mathcal{F}})$ is surjective.

Proof. By Proposition 1.17, $\text{t-Max}(D_{\mathcal{F}}) = \{M_{\mathcal{F}}; M \in \Lambda\}$. As in the proof of Theorem 1.13, if $I' \in \mathcal{T}(D_{\mathcal{F}})$ and $I' = (J_tD_{\mathcal{F}})_t$ with J a finitely generated fractional ideal of D, then $((J(D:J))_{\mathcal{F}})_t = D_{\mathcal{F}}$. It follows that $(J(D:J))_{\mathcal{F}}$ is not contained in any t-maximal ideal of $D_{\mathcal{F}}$, that is $J(D:J)$ is not contained in any ideal $M \in \Lambda$. Whence $(J(D:J))_t \in$ t-Sat($\mathcal{F}$) $= \mathcal{F} \cap$ t(D). We conclude that $(J(D:J))_t$ is t-invertible and so J_t is t-invertible. If each t-prime in $\mathcal{F}$ is t-invertible, then each t-ideal in $\mathcal{F}$ is t-invertible by Theorem 1.10.

2. The map $\mathcal{T}(D) \longrightarrow \mathcal{T}(D_{\mathcal{F}})$ when $D_{\mathcal{F}}$ has t-dimension one

In [3] the authors call a domain of t-dimension 1 with t-finite character a *weakly Krull domain* and, among other results, they prove that, if D is weakly Krull and $D_{\mathcal{F}}$ is a t-subintersection of D, then $D_{\mathcal{F}}$ is weakly Krull and the natural homomorphism $\phi_{\mathcal{F}}$: $\mathcal{T}(D) \longrightarrow \mathcal{T}(D_{\mathcal{F}})$ is surjective [3, Proposition 4.7 and Theorem 4.8]. In this contest, they ask what happens relaxing the hypothesis of t-finiteness on D and show that, if D has t-dimension one, the natural homomorphism $\phi_S : \mathcal{T}(D) \longrightarrow \mathcal{T}(D_S)$ is surjective [3, Theorem 4.5].

We start this Section by proving that, if D has t-dimension one, the homomorphism $\phi_{\mathcal{F}}$ is surjective when $\mathcal{F}$ is v-finite, thus generalizing at the same time the two results in [3] mentioned above.

Proposition 2.1. Let D be a domain of t-dimension one, $\mathcal{F}$ a multiplicative system of ideals of D and $\Lambda = \text{t-Max}(D) \setminus \text{Sat}(\mathcal{F})$. If $\mathcal{F}$ is v-finite, then $\text{t-Spec}(D_{\mathcal{F}}) = \text{t-Max}(D_{\mathcal{F}}) = \{P_{\mathcal{F}};\, P \in \Lambda\}$. Hence $D_{\mathcal{F}} = D_{\mathcal{F}(\Lambda)}$ is a t-subintersection of D of t-dimension one.

Proof. If $P \in \Lambda$, then $P_{\mathcal{F}}$ is a t-prime of $D_{\mathcal{F}}$ by Proposition 1.6 (a). By Proposition 1.9 (c), it follows that $\text{t-Spec}(D_{\mathcal{F}}) = \text{t-Max}(D_{\mathcal{F}}) = \{P_{\mathcal{F}};\, P \in \Lambda\}$. Thus $D_{\mathcal{F}} = \cap\{D_P;\, P \in \Lambda\}$ is a t-subintersection of D of t-dimension one.

Corollary 2.2. Let D be a weakly Krull domain. Then D' is a t-subintersection of D if and only if there exists a v-finite multiplicative system of ideals $\mathcal{F}$ of D such that $D' = D_{\mathcal{F}}$. In this case D' is weakly Krull.

Proof. If $\mathcal{F}$ is v-finite, then $D_{\mathcal{F}}$ is a t-subintersection of D by Proposition 1.8. Conversely, assume that $D' = \cap\{D_P;\, P \in \Lambda\}$ for some subset Λ of $\text{t-Max}(D_{\mathcal{F}})$. By Proposition 1.17, $\mathcal{F}(\Lambda)$ is v-finite and $D' = D_{\mathcal{F}(\Lambda)}$. To finish, $D' = D_{\mathcal{F}(\Lambda)}$ is weakly Krull because it has t-finite character and t-dimension one by Proposition 2.1.

The proof of the following Theorem is a modification of the proof given for rings of fractions in [3, Theorem 4.5], which was in turn inspired by the proof given in [19, Theorem 1.18] in the one-dimensional case.

Theorem 2.3. Let D be a domain of t-dimension one and $\mathcal{F}$ a v-finite multiplicative system of ideals of D. Then $D_{\mathcal{F}}$ is a t-subintersection of D of t-dimension one and the natural homomorphism $\phi_{\mathcal{F}}: \mathcal{T}(D) \longrightarrow \mathcal{T}(D_{\mathcal{F}})$ is surjective.

Proof. Set $\Lambda := \text{t-Max}(D) \setminus \text{Sat}(\mathcal{F})$. By Proposition 2.1, we have that $D_{\mathcal{F}} = D_{\mathcal{F}(\Lambda)}$ is a t-subintersection of D of t-dimension one and $\text{t-Max}(D_{\mathcal{F}}) = \{P_{\mathcal{F}};\, P \in \Lambda\}$. Moreover $\mathcal{F}(\Lambda)$ is v-finite by Proposition 1.9 (b).

If I' is a t-invertible t-ideal of $D_{\mathcal{F}}$, by Proposition 1.4, there are a finitely generated ideal $J \subseteq D$ and an element $x \in K$ such that $\phi_{\mathcal{F}}(xJ_t) = (xJD_{\mathcal{F}})_t = I'$. We can assume that $J \notin \text{Sat}(\mathcal{F})$, otherwise $I' = (xD_{\mathcal{F}})_t = \phi_{\mathcal{F}}(xD)$. As in the proof of Theorem 1.13, it results $((J(D{:}J))_{\mathcal{F}})_t = D_{\mathcal{F}}$. Consider the ideal $L := J(D{:}J)$. If $L \notin \mathcal{F}(\Lambda)$, then $L \subseteq P$ for some $P \in \Lambda$; whence $L_{\mathcal{F}} \subseteq P_{\mathcal{F}}$ and so $(L_{\mathcal{F}})_t \neq D_{\mathcal{F}}$. A contradiction. Thus $L \in \mathcal{F}(\Lambda)$ and, since $\mathcal{F}(\Lambda)$ is v-finite, there exists a finitely generated ideal $H' \subseteq L_t$ such that $H := H'_t \in \mathcal{F}(\Lambda)$. Let $\mathcal{H} = \{H^n\}_{n\geq 0}$. Since $\mathcal{H}$ is trivially v-finite, by Proposition 2.1, $D_{\mathcal{H}}$ is a t-subintersection of D of t-dimension one and t-Max($D_{\mathcal{H}}$) = $\{P_{\mathcal{H}}$; P ∈ t-Spec(D) and $H \not\subseteq P\}$. In addition, $D_{\mathcal{F}}$ is a t-subintersection of $D_{\mathcal{H}}$ because if $P \in \Lambda$, then $H \not\subseteq P$ (otherwise $P \in \mathcal{F}(\Lambda)$). Hence $((JD_{\mathcal{H}})_t D_{\mathcal{F}})_t = (JD_{\mathcal{F}})_t \neq D_{\mathcal{F}}$. In particular $(JD_{\mathcal{H}})_t \neq D_{\mathcal{H}}$. Consider the ideal $J_1 = JD_{\mathcal{H}} \cap D$. Any minimal prime of J_1 is the contraction of a minimal prime of $JD_{\mathcal{H}}$ and any t-ideal containing J_1 is minimal because D has t-dimension one. Hence the only t-prime ideals of D containing J_1 are contractions of t-prime ideals of $D_{\mathcal{H}}$ (containing $(JD_{\mathcal{H}})_t$) and so do not contain H . It follows that $(J_1, H)_t = (J_1', H')_t = D$, for some finitely generated ideal $J_1' \subseteq J_1$. Now let $I := (J, J_1')$. Then I is finitely generated and $\phi_{\mathcal{F}}(xI_t) = I'$. In fact, $(xJ_1'D_{\mathcal{F}})_t \subseteq (xJ_1D_{\mathcal{F}})_t = (x(J_1D_{\mathcal{H}})D_{\mathcal{F}})_t = (x(JD_{\mathcal{H}})D_{\mathcal{F}})_t = (xJD_{\mathcal{F}})_t = I'$. We are left to show that I_t is t-invertible.

Since H' is finitely generated and $H' \subseteq L_t$, then it is possible to find finitely generated ideals $A \subseteq J$ and $B \subseteq (D:J)$ such that $H \subseteq (AB)_t$. Since J_1' is finitely generated and $J_1' \subseteq JD_{\mathcal{H}} \subseteq J_{\mathcal{H}}$, there is a $k \geq 0$ such that $H^kJ_1' \subseteq J$, so that $H^kJ_1'B \subseteq D$. Since $H^kJB \subseteq D$, then $H^kIB \subseteq D$ and so $H^kB \subseteq (D:I)$. Finally, $H^{k+1} \subseteq H^k(AB)_t \subseteq (AH^kB)_t \subseteq (I(D:I))_t$. Therefore $D = (J_1, H)_t = (J_1, H^{k+1})_t \subseteq (I, H^{k+1})_t \subseteq (I(D:I))_t \subseteq D$ and so $(I(D:I))_t = D$.

Since a Krull domain has t-dimension one and each t-ideal of a Krull domain is t-invertible, the classical Nagata's Theorem can be reobtained by the following Corollary.

Corollary 2.4. Let D be a domain of t-dimension one and $\mathcal{F}$ a multiplicative system of ideals of D such that t-Sat($\mathcal{F}$) $\subseteq \mathcal{T}(D)$. Then $D_{\mathcal{F}}$ is a t-subintersection of D of t-dimension one and the natural homomorphism $\phi_{\mathcal{F}}: \mathcal{T}(D) \longrightarrow \mathcal{T}(D_{\mathcal{F}})$ is surjective.

Proof. By Theorem 2.3, it enough to observe that $\mathcal{F}$ is v-finite.

A different proof of Corollary 2.4 can be given by using Corollary 1.14.

If D is weakly Krull , by Corollary 2.2 and Theorem 2.3 we recover [3, Theorem 4.8]. But indeed a slightly stronger result holds.

Lemma 2.5. Let D be any domain, Λ a set of height-one t-maximal ideals of D and $\mathcal{F} := \mathcal{F}(\Lambda)$. Then the following conditions are equivalent:

(i) The intersection $D_{\mathcal{F}} = \cap\{D_P ; P \in \Lambda\}$ has finite character;

(ii) $D_{\mathcal{F}}$ is a weakly Krull domain and t-Max($D_{\mathcal{F}}$) = $\{P_{\mathcal{F}}; P \in \Lambda\}$.

Under (any one of) these conditions, $\mathcal{F}$ is v-finite.

Proof. (i) ⇒ (ii). If $D_{\mathcal{F}} = \cap\{D_P ; P \in \Lambda\}$ has finite character, then $\mathcal{F}$ is v-finite by Lemma 1.16. But, if $P \in \Lambda$, then P is an height-one prime. Whence $P_{\mathcal{F}}$ is a t-ideal by

Proposition 1.6 (a). It follows that t-Spec($D_{\mathcal{F}}$) = t-Max($D_{\mathcal{F}}$) = $\{P_{\mathcal{F}}; P \in \Lambda\}$ (Proposition 1.9 (c)). Hence $D_{\mathcal{F}}$) has t-dimension one and so $D_{\mathcal{F}}$ is weakly Krull.

(ii) $\Rightarrow$ (i) is clear.

Corollary 2.6. Let D be a domain of t-dimension one and D' a t-subintersection of D with finite character. Then D' is a weakly Krull domain and the natural homomorphism $\mathcal{T}(D) \longrightarrow \mathcal{T}(D')$ is surjective.

Proof. Let $D' = D_{\mathcal{F}(\Lambda)} = \cap\{D_P ; P \in \Lambda\}$, Λ a set of t-prime ideals of D . Since this intersection has finite character, then D' is weakly Krull by Lemma 2.5. Since in addition $\mathcal{F}(\Lambda)$ is v-finite, we conclude by Proposition 2.3.

We now wish to give another generalization of [3, Proposition 4.7 and Corollary 4.8] relaxing the hypothesis of one-t-dimensionality on D .

Theorem 2.7. Let D be a domain with t-finite character, Λ a set of height-one t-maximal ideals of D and $\mathcal{F} := \mathcal{F}(\Lambda)$. Then $D_{\mathcal{F}}$ is a weakly Krull domain and t-Max($D_{\mathcal{F}}$) = $\{P_{\mathcal{F}}; P \in \Lambda\}$. Moreover the natural homomorphism $\phi_{\mathcal{F}}: \mathcal{T}(D) \longrightarrow \mathcal{T}(D_{\mathcal{F}})$ is surjective.

Proof. $D_{\mathcal{F}}$ is a weakly Krull domain, t-Max($D_{\mathcal{F}}$) = $\{P_{\mathcal{F}}; P \in \Lambda\}$ and $\mathcal{F}$ is v-finite by Lemma 2.5. By Proposition 1.8, all the conditions of Proposition 1.5 are satisfied. In particular, if I' is a t-invertible t-ideal of $D_{\mathcal{F}}$, then $I := I' \cap D$ is a t-ideal of D and $I' = (ID_{\mathcal{F}})_t = I_{\mathcal{F}}$. To show that $I \in \mathcal{T}(D)$, it is enough to show that ID_M is principal for all t-maximal ideals M of D [4, Lemma 2.2]. Since rad(I) = rad(I')$\cap$D and any minimal prime of I is a t-ideal, the only t-maximal ideals of D containing I are the t-maximal ideals P in Λ such that $I' \subseteq P_{\mathcal{F}}$. Since I' is t-invertible and $P_{\mathcal{F}} \in$ t-Max($D_{\mathcal{F}}$) , then $ID_P = I(D_{\mathcal{F}})_{P_{\mathcal{F}}} = I'(D_{\mathcal{F}})_{P_{\mathcal{F}}}$ is principal [17, Proposition 1.1].

Since any t-invertible t-prime is t-maximal [17, Corollary 18], Theorem 2.7 holds in particular when Λ is a set of height-one t-invertible t-prime ideals of D. In this case it can be actually improved weakening the hypothesis of t-finiteness.

For the following Lemma see also [26, Theorem 3.6, (1) $\Leftrightarrow$ (5)].

Lemma 2.8. Let D be any domain. Then D is a Krull domain if and only if each t-prime of D is t-invertible.

Proof. If each t-prime of D is t-invertible, then each t-prime is t-maximal [17, Corollary 1.8] and so of height one. Then D_P is a DVR [17, Remark 1.2 (2)]. Hence D is completely integrally closed and all its t-maximal ideals are divisorial. Thus D is Krull by [17, Theorem 2.6]. The convere is well known.

Theorem 2.9. Let D be any domain, Λ a set of height-one t-invertible t-prime ideals of D and $\mathcal{F} := \mathcal{F}(\Lambda)$. Then, the following are equivalent:

(i) The intersection $D_{\mathcal{F}} = \cap\{D_P ; P \in \Lambda\}$ has finite character;

(ii) $D_{\mathcal{F}}$ is a Krull domain and $\text{t-Max}(D_{\mathcal{F}}) = \{P_{\mathcal{F}}; P \in \Lambda\}$;

(iii) $D_{\mathcal{F}}$ is a weakly Krull domain and $\text{t-Max}(D_{\mathcal{F}}) = \{P_{\mathcal{F}}; P \in \Lambda\}$.

In addition, under (any one of) these conditions, the natural homomorphism $\phi_{\mathcal{F}}: \mathcal{T}(D) \longrightarrow \mathcal{T}(D_{\mathcal{F}})$ is surjective.

Proof. (i) $\Leftrightarrow$ (iii) by Lemma 2.5.

(iii) $\Rightarrow$ (ii). Since any $P \in \Lambda$ is t-invertible, then $P_{\mathcal{F}} = (PD_{\mathcal{F}})_t$ is t-invertible. Whence $D_{\mathcal{F}}$ is Krull by Lemma 2.8.

(ii) $\Rightarrow$ (iii) is trivial.

Under these conditions, the natural homomorphism $\phi_{\mathcal{F}}: \mathcal{T}(D) \longrightarrow \mathcal{T}(D_{\mathcal{F}})$ is surjective because $\mathcal{T}(D_{\mathcal{F}})$ is generated by the set $\{P_{\mathcal{F}}; P \in \Lambda\}$ and $P_{\mathcal{F}} = \phi_{\mathcal{F}}(P)$, with $P \in \Lambda \subseteq \mathcal{T}(D)$.

Corollary 2.10. Let D be any domain, Λ a set of height-one t-invertible t-prime ideals of D and $\mathcal{F} := \mathcal{F}(\Lambda)$. If $D_{\mathcal{F}}$ has t-finite character, then $D_{\mathcal{F}}$ is a Krull domain and the natural homomorphism $\phi_{\mathcal{F}}: \mathcal{T}(D) \longrightarrow \mathcal{T}(D_{\mathcal{F}})$ is surjective.

In general, if $D_{\mathcal{F}}$ is a Krull domain, the homomorphism $\phi_{\mathcal{F}}: \mathcal{T}(D) \longrightarrow \mathcal{T}(D_{\mathcal{F}})$ need not be surjective. In fact in [19, Example 2.2] it is given an example of a domain D and a multiplicative part S of D generated by primes such that D_S is Krull, though the natural homomorphism $\phi_S : \mathcal{T}(D) \longrightarrow \mathcal{T}(D_S)$ is not surjective.

3. The map $\mathcal{T}(D) \longrightarrow \mathcal{T}(D_{\mathcal{F}})$ when D is v-coherent

We recall that a domain D is v-*coherent* if (D:I) is a v-finite divisorial ideal for each finitely generated ideal I [13, Section 3]. The class of v-coherent domains includes Mori domains (domains satisfying the ascending chain conditions on divisorial ideals) and PvMDs (domains in which each v-finite divisorial ideal is t-invertible).

When $t(D) = \mathcal{D}_f(D)$, each multiplicative system of ideals of D is trivially v-finite. In this case, the surjectivity of the map $t(D) \longrightarrow t(D_{\mathcal{F}})$ (Corollary 1.9) implies that $t(D_{\mathcal{F}}) = \mathcal{D}_f(D_{\mathcal{F}})$. Since this map is inclusion preserving, we get that , if D is a Mori domain, then $D_{\mathcal{F}}$ is a Mori domain (cf. [29, Corollaire 1]).

When D is a PvMD, then $\mathcal{D}_f(D) = \mathcal{T}(D)$. In this case, the surjection $\mathcal{D}_f(D) \longrightarrow \mathcal{D}_f(D_{\mathcal{F}})$ (Proposition 1.4) implies that $\mathcal{D}_f(D_{\mathcal{F}}) = \mathcal{T}(D_{\mathcal{F}})$. Hence $D_{\mathcal{F}}$ is a PvMD (cf. [22, Proposition 1.8]).

The first result in this Section shows that, more in general, if D is v-coherent, then $D_{\mathcal{F}}$ is also v-coherent. This has been proved for rings of fractions in [28, Proposition 2.4].

Proposition 3.1. Let D be a v-coherent domain and $\mathcal{F}$ a multiplicative system of ideals of D. Then $D_{\mathcal{F}}$ is v-coherent.

Proof. Let $I' = (JD_{\mathcal{F}})_v \in \mathcal{D}_f(D_{\mathcal{F}})$, with J a finitely generated fractional ideal of D (Proposition 1.4). We have $(D_{\mathcal{F}}:I') = (D_{\mathcal{F}}:JD_{\mathcal{F}}) = (D:J)_{\mathcal{F}}$ (Proposition 1.2 (a)). Set $(D:J) = H_v$, with H a finitely generated fractional ideal of D. By Proposition 1.2 (b), we have that $(D_{\mathcal{F}}:I') = (D:J)_{\mathcal{F}} = (H_v)_{\mathcal{F}} = (HD_{\mathcal{F}})_v$ is v-finite. It follows that $D_{\mathcal{F}}$ is v-coherent.

Proposition 3.2. Let D be a v-coherent domain, $\mathcal{F}$ a multiplicative system of ideals of D and I a fractional ideal of D. Then:

(a) $(I_{\mathcal{F}})_t \subseteq (I_t)_{\mathcal{F}}$. In particular, if $I \in t(D)$, then $I_{\mathcal{F}} \in t(D_{\mathcal{F}})$.

(b) If $\mathcal{F}$ is v-finite, then $(I_t)_{\mathcal{F}} = (ID_{\mathcal{F}})_t = (I_{\mathcal{F}})_t$.

Proof. (a) Let $JD_{\mathcal{F}} \subseteq I_{\mathcal{F}}$, with J a finitely generated fractional ideal of D. We have to show that $(JD_{\mathcal{F}})_t \subseteq (I_t)_{\mathcal{F}}$. But $(JD_{\mathcal{F}})_t = (J_t)_{\mathcal{F}}$ by Proposition 1.2 (b), because (D:J) is v-finite. As in the proof of Proposition 1.8, there exists $H \in \mathcal{F}$ such that $(JH)_t \subseteq I_t$. Since $J_tH_t \subseteq (JH)_t$ and $(H_t)_{\mathcal{F}} = D_{\mathcal{F}}$, by Proposition 1.2 (c), we have $(J_t)_{\mathcal{F}}(H_t)_{\mathcal{F}} = (J_t)_{\mathcal{F}} \subseteq (J_tH_t)_{\mathcal{F}} \subseteq ((JH)_t)_{\mathcal{F}} \subseteq (I_t)_{\mathcal{F}}$. Hence $(J_{\mathcal{F}})_t = (JD_{\mathcal{F}})_t \subseteq (I_t)_{\mathcal{F}}$.

(b) By Proposition 1.8, we have $(I_t)_{\mathcal{F}} \subseteq (ID_{\mathcal{F}})_t = (I_{\mathcal{F}})_t$. Thus we conclude by (a).

The following Theorem characterizes v-finite multiplicative systems of ideals of v-coherent domains. It shows in particular that, when D is v-coherent, a multiplicative system of ideals satisfying the equivalent conditions of Proposition 1.5 is v-finite.

Theorem 3.3. Let D be a v-coherent domain, $\mathcal{F}$ a multiplicative system of ideals of D and $\Lambda := \{P \in \text{t-Spec}(D)\ ;\ P \notin \text{Sat}(\mathcal{F})$ and P is maximal with respect to this property$\}$. Then the following are equivalent:

(i) $\mathcal{F}$ is v-finite;

(ii) $(ID_{\mathcal{F}})_t = (I_t)_{\mathcal{F}}$ for each (fractional) ideal I of D;

(iii) $(JD_{\mathcal{F}})_t = D_{\mathcal{F}}$, for each $J \in \text{Sat}(\mathcal{F})$;

(iv) $\text{t-Sat}(\mathcal{F}) = \{J \in t(D)\ ;\ (JD_{\mathcal{F}})_t = D_{\mathcal{F}}\}$;

(v) $\text{t-Spec}(D_{\mathcal{F}}) = \{P_{\mathcal{F}};\ P \in \text{t-Spec}(D) \setminus \text{Sat}(\mathcal{F})\}$;

(vi) Λ is not empty and $\text{t-Max}(D_{\mathcal{F}}) = \{P_{\mathcal{F}};\ P \in \Lambda\}$;

(vii) If I' is a t-ideal of $D_{\mathcal{F}}$ and $I := I' \cap D$, then $I \in t(D)$ and $I' = (ID_{\mathcal{F}})_t = I_{\mathcal{F}}$.

Moreover, under (any one of) these conditions, t-Sat($\mathcal{F}$) = $\mathcal{F}(\Lambda)\cap$t(D). Hence $D_{\mathcal{F}}= \cap\{D_P \,;\, P \in \Lambda\}$.

Proof. (i) $\Rightarrow$ (ii) $\Leftrightarrow$ (iii) $\Leftrightarrow$ (vii) and (v) $\Rightarrow$ (iii) by Propositions 1.5, 1.8 and 3.2 (a).

(v) $\Leftrightarrow$ (vi) by Propositions 1.6 (b) and 3.2. (iv) $\Rightarrow$ (iii) is clear.

(ii) $\Rightarrow$ (iv). If $(JD_{\mathcal{F}})_t = D_{\mathcal{F}}$, then $J \in$ t-Sat($\mathcal{F}$) because $(JD_{\mathcal{F}})_t \subseteq (J_{\mathcal{F}})_t = I_{\mathcal{F}}$ (Proposition 3.2 (a)). The opposite inclusion follows from (ii).

(ii) $\Rightarrow$ (i). Let $J_t \in$ Sat($\mathcal{F}$). Then $(JD_{\mathcal{F}})_t = (J_t)_{\mathcal{F}} = D_{\mathcal{F}}$ and $1 \in (J'D_{\mathcal{F}})_v$ for some finitely generated ideal $J' \subseteq J$. Since (D:J') is finitely generated, then $(J'D_{\mathcal{F}})_v = (J'_v)_{\mathcal{F}} = D_{\mathcal{F}}$ (Proposition 1.2 (b)) and so $J'_v \in$ Sat($\mathcal{F}$).

(iii) $\Rightarrow$ (v). By Proposition 1.5, t-Spec($D_{\mathcal{F}}$) $\subseteq \{P_{\mathcal{F}};\, P \in$ t-Spec(D)\Sat($\mathcal{F}$)$\}$. Conversely, if $P \in$ t-Spec(D)\Sat($\mathcal{F}$), then $P_{\mathcal{F}}$ is a t-prime by Propositions 1.1(b) and 3.2(a).

The last assertion holds by Proposition 1.9 (b).

When D is Prüfer, any ideal of D is a t-ideal. Thus Theorem 3.3 recovers the characterization of finitely generated multiplicative systems of ideals of Prüfer domains given in [18,Theorem 1.3].

Each multiplicative system of ideals of a Mori domain D is trivially v-finite. Therefore, if D is a Mori domain, for each multiplicative system of ideals $\mathcal{F}$ of D, we have that $D_{\mathcal{F}} = \cap\{D_P \,;\, P \in \Lambda\}$, where Λ is the set of maximal elements of t-Spec(D)\Sat($\mathcal{F}$) [10, Theorem 5.1].

Corollary 3.4. Let D be a v-coherent domain, Λ a set of pairwise incomparable t-primes and $\mathcal{F} := \mathcal{F}(\Lambda)$. Then $\mathcal{F}$ is v-finite if and only if t-Max($D_{\mathcal{F}}$) = $\{P_{\mathcal{F}} \,;\, P \in \Lambda\}$. In this case, if $P \in \Lambda$ is a-t-maximal v-finite prime ideal, then P is t-invertible if and only if $P_{\mathcal{F}}$ is t-invertible.

Proof. The first part follows directly by Theorem 3.3, observing that in this case Λ is the set of maximal elements of t-Spec(D)\Sat($\mathcal{F}$). If P is t-invertible, then clearly $P_{\mathcal{F}} = (PD_{\mathcal{F}})_t$ is t-invertible. Conversely, since $P \subseteq (P(D:P))_t \subseteq D$, if P is not t-invertible, we have (P:P) = (D:P). Hence, by Proposition 1.2 (a), $(P_{\mathcal{F}}:P_{\mathcal{F}}) = (P:P)_{\mathcal{F}} = (D:P)_{\mathcal{F}} = (D_{\mathcal{F}}:P_{\mathcal{F}})$. It follows that $P_{\mathcal{F}}$ is not t-invertible.

We remark that, if D is v-coherent and $\mathcal{F}$ is a v-finite multiplicative system of ideals of D, then the natural homomorphism $\phi_{\mathcal{F}} : \mathcal{T}(D) \longrightarrow \mathcal{T}(D_{\mathcal{F}})$ need not be surjective. In fact, when D is Mori, the homomorphism $\phi_S : \mathcal{T}(D) \longrightarrow \mathcal{T}(D_S)$ may not be surjective, even when $S = \{s^n\}_{n\geq 0}$, $s \in D$, is generated by one element [1]. Next we give a sufficient condition for the surjectivity of the natural homomorphism $\phi_{\mathcal{F}}$.

Theorem 3.5. Let D be a v-coherent domain and $\mathcal{F}$ a multiplicative system of ideals of D such that each v-finite ideal $I \in Sat(\mathcal{F})$ is t-invertible, then

(a) The natural homomorphism of groups $\phi_{\mathcal{F}} : \mathcal{T}(D) \longrightarrow \mathcal{T}(D_{\mathcal{F}})$ is surjective.

(b) If $\mathcal{F}$ is v-finite then $Ker(\phi_{\mathcal{F}}) = \{(J(D:H))_v \ ; J, H \in \mathcal{T}(D) \cap Sat(\mathcal{F})\}$.

Proof. (a) Let $I' \in \mathcal{T}(D_{\mathcal{F}})$ and $I' = (ID_{\mathcal{F}})_v$, with I a v-finite fractional ideal of D. We show that $I \in \mathcal{T}(D)$. As in the proof of Theorem 1.13, we have $((I(D:I))_{\mathcal{F}})_v = D_{\mathcal{F}}$. Since D is v-coherent, H := (D:I) is v-finite and (D:IH) is also v-finite. Then $((IH)_{\mathcal{F}})_v = ((IH)_v)_{\mathcal{F}} = D_{\mathcal{F}}$ (Proposition 1.2(b)). It follows that $(IH)_v \in Sat(\mathcal{F})$ (because $IH \subseteq D$). Since $(IH)_v$ is t-invertible, also I is t-invertible.

(b) follows from Proposition 1.11.

Corollary 3.6. Let D be a v-coherent domain and $\mathcal{F}$ a multiplicative system of ideals of D such that $t\text{-}Sat(\mathcal{F}) \subseteq \mathcal{T}(D)$. Then the natural homomorphism $\phi_{\mathcal{F}} : \mathcal{T}(D) \longrightarrow \mathcal{T}(D_{\mathcal{F}})$ is surjective and $Ker(\phi_{\mathcal{F}})$ is generated by the prime ideals in $t\text{-}Sat(\mathcal{F})$.

Proof. By Theorem 3.5 and Corollary 1.12.

The surjectivity of $\phi_{\mathcal{F}}$ when D is v-coherent and $t\text{-}Sat(\mathcal{F}) \subseteq \mathcal{T}(D)$ follows also from Theorem 1.13, taking in account that $t\text{-}Spec(D_{\mathcal{F}}) = \{P_{\mathcal{F}} ; P \in t\text{-}Spec(D) \setminus Sat(\mathcal{F})\}$ by Theorem 3.3. As a particular case of Corollary 3.6, we get that, if D is a v-coherent domain and S is a multiplicative part of D generated by primes, then the natural homomorphism $\phi_S : \mathcal{T}(D) \longrightarrow \mathcal{T}(D_S)$ is surjective [27, Théorème 1]. Since a Krull domain is v-coherent and each t-ideal of a Krull domain is t-invertible, from Corollary 3.6 we reobtain the classical Nagata's Theorem.

Corollary 3.7. Let D be a PvMD and $\mathcal{F}$ a multiplicative system of ideals of D . Then the natural homomorphism $\phi_{\mathcal{F}} : \mathcal{T}(D) \longrightarrow \mathcal{T}(D_{\mathcal{F}})$ is surjective. If, in addition, $\mathcal{F}$ is v-finite, then $Ker(\phi_{\mathcal{F}}) = \{(J(D:H))_v ; J, H \in \mathcal{T}(D) \cap Sat(\mathcal{F})\}$.

Proof. By Theorem 3.5.

Corollary 3.7 has been proved for Prüfer domains in [15, Proposition 5.8.1].

If D is a PvMD, the surjectivity of $\phi_{\mathcal{F}}$ follows also from the surjectivity of the homomorphism $\mathcal{T}(D) = \mathcal{D}_f(D) \longrightarrow \mathcal{D}_f(D_{\mathcal{F}})$ (Proposition 1.4).

In [3, Theorem 4.4] the authors prove that $\phi_{\mathcal{F}}$ is surjective if D is a PvMD and $D_{\mathcal{F}}$ is a t-subintersection of D . This is enough to prove the surjectivity of $\phi_{\mathcal{F}}$ for any multiplicative system of ideals $\mathcal{F}$, because a generalized ring of fractions of a PvMD is always a t-subintersection [25, Theorem 3.11].

When D is a PvMD, the next Proposition gives in particular an explicit representation of $D_{\mathcal{F}}$ as a t-subintersection of D .

Proposition 3.8. Let D be a PvMD and $\mathcal{F}$ a multiplicative system of ideals of D. If $\Lambda := \{M \cap D, M \in \text{t-Max}(D_{\mathcal{F}})\}$, then $\mathcal{F}(\Lambda)$ is v-finite and $D_{\mathcal{F}} = D_{\mathcal{F}(\Lambda)}$.

Proof. Set $D_{\mathcal{F}} := D'$ and let M be a t-maximal ideal of D'. The ideal $P := M \cap D$ is a t-ideal (Proposition 1.3) and so D_P is a valuation domain. We have $D_P \subseteq D'_M$. But, since D' is a PvMD [22, Proposition 1.8], then also D'_M is a valuation domain. Hence we have $D_P = D'_M$ and $M = PD_P \cap D' = P_{\mathcal{F}(\Lambda)}$. It follows that $D' = \cap\{D_M ; M \in \text{t-Max}(D')\} = \cap\{D_P ; P \in \Lambda\} = D_{\mathcal{F}(\Lambda)}$. Since Λ is a set of pairwise incomparable t-primes and $\text{t-Max}(D') = \{P_{\mathcal{F}(\Lambda)} ; P \in \Lambda\}$, then $\mathcal{F}(\Lambda)$ is v-finite by Corollary 3.4.

Corollary 3.9. Let D be a PvMD and $\mathcal{F}$ a multiplicative system of ideals of D. Then the natural homomorphism $\phi_{\mathcal{F}} : \mathcal{T}(D) \longrightarrow \mathcal{T}(D_{\mathcal{F}})$ is surjective and $\text{Ker}(\phi_{\mathcal{F}}) = \{(J(D:H))_v ; J, H \in \mathcal{T}(D)$ and $J, H \not\subseteq M$ for each $M \in \text{t-Max}(D_{\mathcal{F}})\}$.

Proof. By Corollary 3.7 and Proposition 3.8.

4. Splitting properties of $\mathcal{T}(D)$

In this Section, if D is any domain, we denote by Λ_1 a set of height-one t-invertible t-prime ideals of D and by Λ_2 the complement of Λ_1 in t-Max(D), i.e. $\Lambda_2 := \text{t-max}(D)\backslash\Lambda_1$. We set $\mathcal{F}_i := \mathcal{F}(\Lambda_i)$ and $D_i := D_{\mathcal{F}_i}$ for $i = 1, 2$, so that $D = D_1 \cap D_2$. If Λ_i is empty, we set $D_i := K$. Observe that $\mathcal{F}_i$ is saturated and so $\text{t-Sat}(\mathcal{F}_i) = \mathcal{F}_i \cap t(D)$.

Since any prime in Λ_1 has height one, a t-prime ideal P of D is in $\mathcal{F}_2$ if and only if $P \in \Lambda_1$ and, by Theorem 1.10,

$$\mathcal{F}_2 \cap t(D) = \{(P_1^{n_1} \dots P_k^{n_k})_v \; ; \; P_1, \dots, P_k \in \Lambda_1 \text{ and } n_i \geq 0\} \subseteq \mathcal{T}(D),$$

in particular $\mathcal{F}_2$ is v-finite. If, in addition, $\text{t-Spec}(D_2) = \{Q_{\mathcal{F}_2} ; Q \in \text{t-Spec}(D)\backslash\Lambda_1\}$, then the natural homomorphism $\phi_2 : \mathcal{T}(D) \longrightarrow \mathcal{T}(D_2)$ is surjective by Theorem 1.13.

Moreover, if the intersection $D_1 = \cap\{D_P ; P \in \Lambda_1\}$ has finite character, then D_1 is Krull and the natural homomorphism $\phi_1 : \mathcal{T}(D) \longrightarrow \mathcal{T}(D_1)$ is also surjective (Theorem 2.9).

Proposition 4.1. Let D be any domain and assume that the intersection $D_1 = \cap\{D_P; P \in \Lambda_1\}$ has finite character. Then D_1 is Krull and there is an exact sequence:

$$(1) \longrightarrow \mathcal{T}(D_1) \xrightarrow{\alpha} \mathcal{T}(D) \xrightarrow{\phi_2} \mathcal{T}(D_2),$$

where α is defined on the set of height-one prime ideals of D_1 by $P' \longrightarrow P' \cap D$ and ϕ_2 is the natural homomorphism.

Moreover, if ϕ_2 is surjective, the sequence splits and so, in this case, $\mathcal{T}(D)$ is isomorphic to $\mathcal{T}(D_1) \oplus \mathcal{T}(D_2)$.

Proof. If the intersection $D_1 = \cap\{D_P ; P \in \Lambda_1\}$ has finite character, then D_1 is Krull and $\text{t-Max}(D_1) = \{P_{\mathcal{F}_1}, P \in \Lambda_1\}$ (Theorem 2.9). Hence α is well defined and

$\mathrm{Im}(\alpha) = \{(P_1^{e_1} \cdots P_k^{e_k})_v;\ P_1, \ldots, P_k \in \Lambda_1$ and $e_i \in \mathbb{Z}\}$. On the other hand, $\mathrm{Ker}(\phi_2) = \mathrm{Im}(\alpha)$ by Corollary 1.12. Since $\phi_1 \circ \alpha$ is the identity on $\mathcal{T}(D_1)$, α is injective; hence, if ϕ_2 is surjective, the sequence splits.

Proposition 4.2. If the intersection $D_1 = \cap\{D_P\,;\ P \in \Lambda_1\}$ has finite character and $\text{t-Spec}(D_2) = \{Q_{\mathcal{F}_2}\,;\ Q \in \text{t-Spec}(D)\backslash\Lambda_1\}$, then D_1 is a Krull domain and there is a splitting short exact sequence:

$$(1) \longrightarrow \mathcal{T}(D_1) \xrightarrow{\ \alpha\ } \mathcal{T}(D) \xrightarrow{\ \phi_2\ } \mathcal{T}(D_2) \longrightarrow (1)\,.$$

Proof. If $\text{t-Spec}(D_2) = \{Q_{\mathcal{F}_2}\,;\ Q \in \text{t-Spec}(D)\backslash\Lambda_1\}$, then ϕ_2 is surjective by Theorem 1.13. We conclude by proposition 4.1.

Theorem 4.3. Assume that D has t-dimension one or, respectively, that D is a v-coherent domain. If the intersection $D_1 = \cap\{D_P\,;\ P \in \Lambda_1\}$ has finite character, then D_1 is a Krull domain and there is a splitting short exact sequence:

$$(1) \longrightarrow \mathcal{T}(D_1) \xrightarrow{\ \alpha\ } \mathcal{T}(D) \xrightarrow{\ \phi_2\ } \mathcal{T}(D_2) \longrightarrow (1)\,.$$

In addition, D_2 has t-dimension one or, respectively, D_2 is a v-coherent domain.

Proof. If D has t-dimension one, then D_2 has also dimension one and $\text{t-Spec}(D_2) = \{Q_{\mathcal{F}_2}\,;\ Q \in \text{t-Spec}(D)\backslash\Lambda_1\}$ by Proposition 2.1. If D is a v-coherent domain, then D_2 is a v-coherent domain by Proposition 3.1 and $\text{t-Spec}(D_2) = \{Q_{\mathcal{F}_2}\,;\ Q \in \text{t-Spec}(D)\backslash\Lambda_1\}$ by Proposition 3.3. Hence we can apply Proposition 4.2.

Theorem 4.4. If D is a domain with t-finite character, then D_1 is a Krull domain and there is a splitting short exact sequence:

$$(1) \longrightarrow \mathcal{T}(D_1) \xrightarrow{\ \alpha\ } \mathcal{T}(D) \xrightarrow{\ \phi_2\ } \mathcal{T}(D_2) \longrightarrow (1)\,.$$

Proof. D_1 is a Krull domain by Theorem 2.9 and ϕ_2 is surjective by Theorem 1.18. Hence the sequence splits by Proposition 4.1.

Theorem 4.4 applies to the case when D is a weakly Krull domain. In this case D_2 is a weakly Krull domain by Lemma 2.5 (or [3, Proposition 4.7]).

A Mori domain has t-finite character [9, Theorem 2.1]. Hence, by Theorem 4.4, when D is Mori, $\mathcal{T}(D)$ is isomorphic to $\mathcal{T}(D_1) \oplus \mathcal{T}(D_2)$. In this case, D_2 is Mori by the discussion at the beginning of Section 3 or by [29, Corollaire 1] (cf. [9, Theorem 2.6]).

Now assume that, under the hypotheses of Proposition 4.1, there is a splitting short exact sequence:

$$(1) \longrightarrow \mathcal{T}(D_1) \xrightarrow{\ \alpha\ } \mathcal{T}(D) \xrightarrow{\ \phi_2\ } \mathcal{T}(D_2) \longrightarrow (1)\,.$$

A discussion similar to the one carried out in [9, Remark 2.7 (2)] shows that we have also a short exact sequence:

$$(1) \longrightarrow \mathcal{T}(D_1)/\mathcal{H}(D_1) \xrightarrow{\ \alpha'\ } \mathcal{C}(D) \xrightarrow{\ \phi_2'\ } \mathcal{C}(D_2) \longrightarrow (1)\,,$$

where $\mathcal{H}(D_1) \subseteq \mathcal{P}(D_1)$ is the group of principal ideals of D_1 whose contraction to D remains principal and α', ϕ_2' are the homomorphisms induced by α and ϕ_2 respectively. In a similar way, we have a short exact sequence:

$$(1) \longrightarrow \mathcal{T}(D_2)/\mathcal{H}(D_2) \xrightarrow{\beta'} \mathcal{C}(D) \xrightarrow{\phi_1'} \mathcal{C}(D_1) \longrightarrow (1),$$

where $\beta : \mathcal{T}(D_1) \longrightarrow \mathcal{T}(D)$ is the canonical injection, $\mathcal{H}(D_2) \subseteq \mathcal{P}(D_2)$ is the group of principal ideals I of D_2 such that $\beta(I)$ is principal in D and β', ϕ_1' are the homomorphisms induced by β and ϕ_1 respectively.

This implies that, if $\mathcal{C}(D)$ is isomorphic to $\mathcal{C}(D_1)$, then $\mathcal{C}(D_2)$ is trivial (because $\mathcal{T}(D_2) = \mathcal{H}(D_2) \subseteq \mathcal{P}(D_2)$). On the contrary, when $\mathcal{C}(D_2)$ is trivial, then $\mathcal{C}(D)$ is generated by the classes of the (t-invertible) t-primes of D which are in Λ_1. Domains with class group generated by the classes of the t-invertible t-primes have been characterized in [20, Section 1] (see also [14]).

If $\mathcal{H}(D_1) = \mathcal{P}(D_1)$, then we have an induced splitting short exact sequence of class groups:

$$(1) \longrightarrow \mathcal{C}(D_1) \xrightarrow{\alpha} \mathcal{C}(D) \xrightarrow{\phi_2} \mathcal{C}(D_2) \longrightarrow (1).$$

This happens in particular when each t-prime in Λ_1 is principal.

Theorem 4.5. Let D be any domain. Assume that $\Lambda_1 := \{p_\alpha D\}$ is a set of height-one principal prime ideals and that the intersection $D_1 = \cap\{D_P ; P \in \Lambda_1\}$ has finite character. Let S be the multiplicative part of D generated by $\{p_\alpha\}$ and $T := D \setminus S$. Then:

(a) $\mathcal{F}_2 \cap t(D) = \{sD, s \in S\}$.

(b) $D_1 = D_T$ is a unique factorization domain and $D_2 = D_S$.

(c) ϕ_2 is surjective.

(d) $\mathcal{C}(D)$ is isomorphic to $\mathcal{C}(D_S)$.

Proof. (a) By Theorem 1.10, we have that $\mathcal{F}_2 \cap t(D) = \{(P_1^{n_1} \dots P_k^{n_k})_v ; P_1, \dots, P_k \in \Lambda_1$ and $n_i \geq 0\} = \{(p_1^{n_1} \dots p_k^{n_k}D)_v ; p_1, \dots, p_k \in S$ and $n_i \geq 0\} = \{sD, s \in S\}$.

(b) D_1 is Krull and each t-maximal ideal of D_1 is of type $p_\alpha D_1$ by Theorem 2.9. Hence D_1 is a unique factorization domain. It is clear that $D_T \subseteq D_1$. Conversely, if $x \in D_1$, we have $x = p_1^{m_1} \dots p_k^{m_k} u$, with u invertible in D_1. Now observe that an element $a \in D$ is invertible in D_1 if and only if $a \notin \cup\{p_\alpha D_1\}$. Whence $a \in D \setminus S = T$. Thus $u = a/b$, with $a, b \in T$. It follows that $xb \in D$, for some $b \in T$. Hence $x \in D_T$ and $D_1 \subseteq D_T$. By part (a), $D_2 = D_S$.

(c) Since $D_2 = D_S$, ϕ_2 is surjective by [19, Proposition 1.10].

(d) By Proposition 4.1, we have a splitting exact sequence:

$$(1) \longrightarrow \mathcal{T}(D_T) \xrightarrow{\alpha} \mathcal{T}(D) \xrightarrow{\phi_2} \mathcal{T}(D_S) \longrightarrow (1).$$

It is immediate that $\mathrm{Im}(\alpha) \subseteq \mathcal{P}(D)$. Then we have also a splitting exact sequence of class groups:

$$(1) \longrightarrow \mathcal{C}(D_T) \xrightarrow{\alpha'} \mathcal{C}(D) \xrightarrow{\phi_2'} \mathcal{C}(D_S) \longrightarrow (1),$$

where $\mathcal{C}(D_T)$ is trivial because D_T is a unique factorization domain. Hence $\mathcal{C}(D)$ is isomorphic to $\mathcal{C}(D_S)$.

Regarding Theorem 4.5 (c), we recall that, if S is a multiplicative part of D generated by primes, then the natural homomorphism $\mathcal{C}(D) \longrightarrow \mathcal{C}(D_S)$ is injective by [1, Theorem 2.3]. Using this result, (d) follows immediately from (c).

In the terminology of [2], the multiplicative set S defined in Theorem 4.5 is a *splitting set* generated by prime elements and T is the *m*-complement of S. Hence the factoriality of D_T follows also from [2, Proposition 2.6] and the isomorphism between $\mathcal{C}(D)$ and $\mathcal{C}(D_S)$ from [2, Corollary 3.8].

Corollary 4.6. Let D be a domain with t-finite character and assume that each t-invertible t-prime ideal of D has height one. Let Λ_1 be a set of t-invertible t-prime ideals of D and $\Lambda_2 := \text{t-Max}(D)\setminus\Lambda_1$. Then the following conditions are equivalent:

(i) Each t-invertible t-prime ideal of D is principal;

(ii) $S = D\setminus\cup\{P\,;\, P \in \Lambda_2\}$ is generated by primes and $D_2 = D_S$;

(iii) $\mathcal{C}(D)$ is isomorphic to $\mathcal{C}(D_2)$.

Thus $\mathcal{C}(D)$ is trivial if and only if each t-invertible t-prime ideal of D is principal and $\mathcal{C}(D_2)$ is trivial.

Proof. (i) $\Rightarrow$ (ii). By Theorem 4.5 (b), $D_2 := \cap\{D_P\,;\, P \in \Lambda_2\} = D_S$, where, if $\Lambda_1 := \{p_\alpha D\}$, S is the multiplicative part of D generated by $\{p_\alpha\}$. Necessarily $S = D\setminus\cup\{P\,;\, P \in \Lambda_2\}$ [10, Lemma 5.5].

(ii) $\Rightarrow$ (iii). The natural homomorphism ϕ_2 is injective by [1, Theorem 2.3] and surjective by [19, Proposition 1.10].

(iii) $\Rightarrow$ (i). Let Λ_1 be the set of all t-invertible t-prime ideals of D and $P \in \Lambda_1$. Then $(PD_2)_t = P_{\mathcal{F}_2} = D_2$, because $\text{t-Max}(D_2) = \{Q_{\mathcal{F}_2}\,;\, Q \in \Lambda_2\}$ by Proposition 1.17. By (iii), P is principal.

To conclude, if $\mathcal{C}(D)$ is trivial, clearly each t-invertible t-prime ideal of D is principal. Hence $\mathcal{C}(D_2)$ is trivial by (i) $\Rightarrow$ (iii). On the contrary, if $\mathcal{C}(D_2)$ is trivial, as already observed, $\mathcal{C}(D)$ is generated by the classes of the t-invertible t-primes of D which are in Λ_1. Since these ideals are all principal, then $\mathcal{C}(D)$ is trivial.

We recall that a domain D with finitely many t-maximal ideals has trivial class group. In fact, each proper ideal of D is contained in the union of the maximal t-ideals. Hence, if t-Max(D) is finite, each maximal ideal of D is a t-ideal. It follows that D is semilocal and so each invertible ideal of D is principal. But each t-invertible t-ideal of D is invertible, because is v-finite and locally principal [12, Corollary 2.9]. Therefore, from the Corollary above, we obtain that, if D has t-finite character, each t-invertible t-prime ideal of D is principal and D has finitely many t-maximal ideals which are not t-invertible, then $\mathcal{C}(D)$ is trivial.

A Mori domain satisfies the hypoyheses of Theorem 4.4. Thus we reobtain [9, Corollary 2.12] and [10, Theorem 7.7 and Proposition 7.10].

References

[1] D.D. Anderson and D.F. Anderson, Some remarks on star operations and the class group, J. Pure Appl. Algebra 51 (1988), 27-33.

[2] D.D. Anderson, D.F. Anderson and M. Zafrullah, Splitting the t-class group, J. Pure Appl. Algebra 74 (1991), 17 - 37.

[3] D.D. Anderson, E. G. Houston and M. Zafrullah, t-linked extensions, the t-class group and Nagata's theorem, J. Pure Appl. Algebra 86 (1993), 109-124.

[4] D.D. Anderson, J. L Mott and M. Zafrullah, Finite character representation for integral domains, Boll. UMI 6-B (1992), 613-630.

[5] D.D. Anderson, J. L Mott and M. Zafrullah, Unique factorization in non-atomic integral domains, Boll. UMI, to appear.

[6] D.F. Anderson, A general theory of class groups, Comm. Algebra 16 (1988), 805-847.

[7] D. F. Anderson and A. Ryckaert, The class group of D+M, J. Pure Appl. Algebra 52 (1988), 199-212.

[8] J. T. Arnold and J. W. Brewer, On flat overrings, ideal transforms and generalized transforms of a commutative ring, J. Algebra 18 (1971), 254-263.

[9] V. Barucci and S. Gabelli, On the class group of a Mori domain, J. Algebra 108 (1987), 161-173.

[10] V. Barucci, S. Gabelli and M. Roitman, The class group of a strongly Mori domain, Comm. Algebra 22 (1994), 173-211.

[11] A. Bouvier, Le groupe des classes d'un anneau intègre, 107ème Congrès des Sociétés Savantes, Brest, 1982, IV, 85-92.

[12] A. Bouvier and M. Zafrullah, On some class groups of an integral domain, Bull. Soc. Math. Grèce 29 (1988), 45-59.

[13] M. Fontana and S. Gabelli, On the class group and the local class group of a pull-back, J. Algebra 181 (1996), 803-835.

[14] M. Fontana and S. Gabelli, Prüfer domains with class group generated by the classes of the invertible maximal ideals, Comm. Algebra 25 (1997), 3993-4008.

[15] M. Fontana, J. A. Huckaba and I. J. Papick, Prüfer Domains, M. Dekker, 1997.

[16] R. M. Fossum, The divisor class group of a Krull domain, Springer-Verlag, Berlin, 1973.

[17] S. Gabelli, Completely integrally closed domains and t-ideals, Boll. UMI 3-B (1989), 327-342.

[18] S. Gabelli, Prüfer (##)-domains and finitely generated localizing systems of ideals, Proceedings of the Third International Conference on Commutative Ring Theory, Fès 1997, Lecture Notes in Pure and Applied Mathematics, Dekker, to appear.

[19] S. Gabelli and M. Roitman, On Nagata's Theorem for the class group, J. Pure Appl. Algebra 66 (1990), 31-42.

[20] S. Gabelli and F. Tartarone, On the class group of integer-valued polynomial rings over Krull domains, J. Pure Appl. Algebra, to appear.

[21] R. Gilmer, Multiplicative ideal theory, Dekker, New York, 1972.

[22] W. J. Heinzer and J. Ohm, An essential ring which is not a v-multiplication ring, Can. J. Math. 25 (1973), 856-861.

[23] W. J. Heinzer, J. Ohm and R. L. Pendleton, On integral domains of the form $\cap D_P$, P minimal, J. Reine Angew. Math. 241 (1970), 147-159.

[24] P. Jaffard, Les systemes d'ideaux, Dunod, Paris, 1969.

[25] B. G. Kang, Prüfer v-multiplication domains and the ring $R[X]_{N_v}$, J. Algebra 123 (1989), 151-170.

[26] B. G. Kang, On the converse of a well-known fact about Krull domains, J. Algebra 124 (1989), 284-299.

[27] D. Nour El Abidine, Sur le groupe des classes d'un anneau intègre, Ann. Univ. Ferrara, Sez. VII, Sc. Mat. 36 (1990), 175-183.

[28] D. Nour El Abidine, Sur un Théorème de Nagata, Comm. Algebra 20 (1992), 2127-2138.

[29] J. Querré, Sur les anneaux reflexifs, Can. J. Math 6 (1975), 1222-1228.

[30] M. Zafrullah, The D+XD[X] construction for GCD domains, J. Pure Appl. Algebra 50 (1988), 93-107.

Catalecticant Varieties

Anthony V. Geramita
Dept of Mathematics and Statistics, Queen's University
Kingston, Ontario, Canada
&
Dip. Di Matematica, Università di Genova, Genova, Italy

For Mario:

Due to circumstances that were beyond my control, it was impossible to actually be present at the celebration for Mario. I was truly sorry for this. When Mario suggested that I might be able to still include my talk in the volume for him, I leapt at the chance to do so.

It is now over 20 years since I first met Mario at a party given by our friend, and my colleague at Queen's University, Paulo Ribenboim. Mario discovered, that evening, that I was the grandchild of immigrants from Italy and that I spoke not a word of Italian!

His (gentle) reproach about losing my heritage encouraged me to enroll in an Italian language course that year, something I had been thinking about doing for some time. That "reproach" put me on the path that would (eventually) end up with me living half my life in Italy. A path I would never have thought possible when first we met.

Mario's kindness to me (and my daughter, Katharine) during an extended sojourn in Italy in 1988 will never be forgotten either by me or by my daughter. She still keeps two of the wonderful prints that Mario gave her, on the walls of her bedroom!

Mario's kindness and energy have made him friends throughout the world, but I have to say that his modesty never permitted him to tell me of his courageous behaviour as a partisan at the end of the second world war (something which I had to learn from his friend, Giorgio Bassani, the author of "The Garden of the Finzi-Contini"). I hope to hear more about this from him in some future meeting in which we will have a chance to enjoy a glass of wine together.

Enjoy your retirement, Mario.

Introduction:

I've arranged the talk in a way that I thought would respond to certain questions from an imagined (friendly) audience.

A) What is a catalecticant variety?

B) O.K.- the definitions are clear, but are there any interesting examples?

C) Good - those are interesting examples, but do you have anything new to say about these things?

Oh yeah, by the way, where'd you pick up this word "*catalecticant*"?

Catalecticant Varieties

Let $R = k[x_1, \ldots, x_n] = \oplus_{i \geq 0} R_i$ with $k = \overline{k}$ an algebraically closed field, $char(k) = 0$. If $\alpha \in \mathbb{N}^n$, $\alpha = (a_1, \ldots, a_n)$ we write

$$x^\alpha := x_1^{a_1} \ldots x_n^{a_n} \text{ where } \mid \alpha \mid = \sum_{i=1}^{n} a_i$$

Fix integers $d,\ i,\ j\ > 0$ such that $i + j = d$ and consider the bilinear map, given by multiplication,

$$R_i \times R_j \longrightarrow R_d\ .$$

We keep track of this multiplication in a matrix whose rows are indexed by the monomials of R_i (say in lex ordering) and whose columns are indexed by the monomials of R_j. In each place in the matrix we first enter the result of multiplying the appropriate row monomial by the appropriate column monomial and then replace the product $x^\alpha x^\beta$ by a new variable, which we will denote $z_{(\alpha,\beta)}$.

We denote the resulting matrix of variables

$$Cat(i, j; n)$$

and call it the (i, j) *catalecticant matrix* for R.

Example: Let $n = 2,\ i = 1,\ j = 3.$

Then $Cat(1, 3; 2)$ is the matrix

$$\begin{array}{c} \\ x_1 \\ x_2 \end{array} \begin{array}{c} \begin{array}{cccc} x_1^3 & x_1^2x_2 & x_1x_2^2 & x_2^3 \end{array} \\ \begin{pmatrix} z_{(4,0)} & z_{(3,1)} & z_{(2,2)} & z_{(1,3)} \\ z_{(3,1)} & z_{(2,2)} & z_{(1,3)} & z_{(0,4)} \end{pmatrix} \end{array}$$

Notice that

$$Cat(i, j; n)^t = Cat(j, i; n)$$

and so we need usually only consider these matrices when $i \leq j$.

Now let i, j and d be as above and choose any integer t so that

$$t \leq \min\left\{\binom{i+n-1}{n-1}, \binom{j+n-1}{n-1}\right\}.$$

Then we will denote the ideal of $t \times t$ minors of the matrix $Cat(i, j : n)$, by

$$I_t(Cat(i, j; n)) .$$

These ideals of minors define subschemes of $\mathbb{P}^N$, where $N + 1 = \binom{d+n-1}{n-1}$ and those subschemes will be called *catalecticant varieties*. Some authors reserve the word *variety* for a reduced irreducible subscheme of $\mathbb{P}^N$, but I am being loose in the use of this word. For the moment, at least, there is no scheme structure on this object and it is simply an algebraic subset of $\mathbb{P}^N$.

What are some examples of catalecticant varieties?

Example: $n = 3, i = j = 1$.

Then $Cat(1, 1; 3)$ is the generic symmetric 3×3 matrix

$$\begin{array}{c} \\ x_1 \\ x_2 \\ x_3 \end{array} \begin{array}{c} \begin{array}{ccc} \quad x_1 \quad & \quad x_2 \quad & \quad x_3 \quad \end{array} \\ \begin{pmatrix} z_{(2,0,0)} & z_{(1,1,0)} & z_{(1,0,1)} \\ z_{(1,1,0)} & z_{(0,2,0)} & z_{(0,1,1)} \\ z_{(1,0,1)} & z_{(0,1,1)} & z_{(0,0,2)} \end{pmatrix} \end{array} \leftrightarrow \begin{pmatrix} z_1 & z_4 & z_5 \\ z_4 & z_2 & z_6 \\ z_5 & z_6 & z_3 \end{pmatrix}.$$

(where, for convenience, we have relabeled the variables).

It is well known that the ideal $I_2(Cat(1, 1; 3))$ is a prime ideal in $k[z_1, \ldots, z_6]$ which describes the Veronese embedding of $\mathbb{P}^2$ into $\mathbb{P}^5$ (or, as Hartshorne calls it, the 2-uple embedding of $\mathbb{P}^2$)

$$\nu_2 : \mathbb{P}^2 \longrightarrow \mathbb{P}^5$$

It is easy to see why this might be true: the elements of $\mathbb{P}^5$, symmetric matrices, can be identified with quadratic curves in $\mathbb{P}^2$ and those of rank 1 are identified with the equations which are the squares of a linear form. But, those are exactly the image of ν_2. So, at least up to radical, we can understand this catalecticant variety.

Now,

$$I_3(Cat(1,1;3)) = \det(Cat(1,1;3))$$

is a cubic hypersurface in $\mathbb{P}^5$. That hypersurface is precisely the secant line variety to $\nu_2(\mathbb{P}^2)$ i.e.

$$Sec_1(\nu_2(\mathbb{P}^2))$$

Again, up to radical, this makes sense since symmetric matrices of rank 2 are symmetric matrices which correspond to a sum of two squares of linear forms, i.e. $L_1^2 + L_2^2$, and we can think of that form as defining the line joining the points L_1^2 and L_2^2, which are on the Veronese variety.

Example: Now let $n = 2$ (i.e. $R = k[x_1, x_2]$) and let d be arbitrary. We can consider all the ways that $d = i + j$ ($i,\ j > 0$ of course) and so get a family of catalecticant matrices for our choices

$$(i,j) \in \{(1, d-1),\ (2, d-2),\ \dots, (r, d-r)\}$$

where, by our earlier observations, we might as well only consider the choices where $i \leq j$.

Recall that $\dim R_d = d + 1$. So, if we use $z_0, \dots, z_d$ as coordinates on $\mathbb{P}(R_d)$, we get:

$$Cat(1, d-1; 2) = \begin{pmatrix} z_0 & z_1 & \cdots & \cdots & z_{d-1} \\ z_1 & z_2 & \cdots & z_{d-1} & z_d \end{pmatrix}$$

$$Cat(2, d-2; 2) = \begin{pmatrix} z_0 & z_1 & \cdots & z_{d-2} \\ z_1 & z_2 & \cdots & z_{d-1} \\ z_2 & z_3 & \cdots & z_d \end{pmatrix}$$

$$\vdots$$

$$Cat(j, d-j; 2) = \begin{pmatrix} z_0 & z_1 & \cdots & z_{d-j} \\ z_1 & z_2 & \cdots & z_{d-j+1} \\ \vdots & \vdots & \vdots & \vdots \\ z_j & z_{j+1} & \cdots & z_d \end{pmatrix}$$

These matrices are usually referred to as *Hankel* matrices. In fact, the catalecticant varieties associated to Hankel matrices are all well known.

Proposition: (*Gruson-Peskine*) Choose t, j_1, j_2 such that

$$t \leq \min\{j_1+1, d-j_1+1\} \ , t \leq \min\{j_2+1, d-j_2+1\} \ .$$

Then,

1) $I_t(Cat(j_1+1, d-j_1+1; 2)) = I_t(Cat(j_2+1, d-j_2+1; 2))$

2) $I_t(Cat(j_1+1, d-j_1+1; 2))$ is a prime ideal.

So, in these cases the catalecticant varieties are all reduced and irreducible.

3) Moreover, the ideal

$$I_t(Cat(j_1+1, d-j_1+1; 2))$$

defines the variety

$$Sec_{t-2}(\nu_d(\mathbb{P}^1)) \subseteq \mathbb{P}^d$$

i.e. the variety of secant $\mathbb{P}^{t-2}$'s to the rational normal curve in $\mathbb{P}^d$.

These secant varieties are all *non-deficient*, i.e. have the expected dimension for a secant variety. More precisely,

$$\dim Sec_{t-2}(\nu_d(\mathbb{P}^1)) = (t-1) + (t-2) = 2t-3$$

where the $t-1$ corresponds to choosing $t-1$ points in $\mathbb{P}^1$ and the $t-2$ mirrors the fact that we have joined these $t-1$ points with a $\mathbb{P}^{t-2}$.

4) These secant varieties to the rational normal curve are arithmetically Cohen-Macaulay, i.e. they have homogeneous coordinate rings which are CM.

This follows from the well-known theorem of Eagon-Northcott describing the ideal of maximal minors of a generic matrix. For a suitable choice of j_1 the ideal I_t is an ideal of maximal minors, although not of a "generic" matrix. However of a matrix with "generic" height and that is enough to allow us to apply the Eagon-Northcott theorem.

There is another family of catalecticant matrices which have been much studied.

Proposition:(*Bertini; Pragacz-Weyman*)

The ideals

$$I_t(Cat(1,1;n)) \subseteq k[z_\alpha \mid \alpha \in \mathbb{N}^n, \mid \alpha \mid = 2]$$

are all prime.

They are the defining ideals for the varieties

$$Sec_{t-2}(\nu_2(\mathbb{P}^{n-1})) \subseteq \mathbb{P}^{\binom{2+n-1}{n-1}-1}$$

Moreover, these varieties are all arithmetically Cohen Macaulay and are **ALL** deficient.

They have dimension

$$\frac{(t-1)(2n+1)-(t-1)^2-2}{2}$$

and degree

$$\frac{\binom{n}{n-t+1}\cdots\binom{2n-(t+1)}{2}\binom{2n-t}{1}}{\binom{2n-(2t-1)}{n-t+1}\cdots\binom{3}{2}\binom{1}{1}}$$

□

These two theorems are benchmarks to which I will return frequently. But, to carry this discussion on, it will be worthwhile to see an alternate way to view catalecticant matrices.

For this point of view it will be convenient to have two copies of the polynomial ring available:

$$R = k[x_1, \ldots, x_n] \qquad S = k[y_1, \ldots, y_n]$$

I want to think of S as a (graded) R-module in two different, but related, ways. Recall that in this talk $k = \overline{k}$ and char$k = 0$. The graded action will take something in R_i and multiply something in S_j into something in S_{j-i}! Thus the action *lowers* degree, i.e.

$$R_i \times S_j \longrightarrow S_{j-i} \ .$$

1) (*Differentiation*): We think of x_i as the operator $\frac{\partial}{\partial y_i}$ and let it act on S in this way.

If $\alpha, \beta \in \mathbb{N}^n$ we can define $\binom{\alpha}{\beta} = \Pi_{i=1}^{n}\binom{\alpha_i}{\beta_i}$ and $\binom{\alpha_i}{\beta_i} = 0$ if $\alpha_i < \beta_i$. Then it's easy to see that

$$x^{\alpha} \times y^{\beta} = \binom{\beta}{\alpha} y^{\beta-\alpha}$$

2) The binomial coefficients are a bit inconvenient, so we also consider the *Contraction* action, where

$$x^{\alpha} \times y^{\beta} = \begin{cases} 0 & \text{if } \beta_i - \alpha_i < 0 \text{ for any } i \\ y^{\beta-\alpha} & \text{otherwise.} \end{cases}$$

(Note that, in char $= 0$ the actions are equivalent and I will use them interchangeably in this talk.)

With these actions, S is not a finitely generated R-module. Note also that

$$R_i \times S_i \to k \ .$$

is a non-degenerate pairing.

Definition-Proposition: If I is a homogeneous ideal in R, set

$$I^{-1} = \{F \in S \mid I \circ F = 0\} \ .$$

Then I^{-1} is called the *inverse system of I*. It is a graded R-submodule of S for which

a) $(I^{-1})_j = I_j^{\perp}$ in the pairing

$$\begin{array}{ccccc} R_j & \times & S_j & \longrightarrow & k \\ \cup & & \cup & & \\ I_j & \times & I_j^{\perp} & \longrightarrow & 0 \end{array}$$

b) $\dim_k(I^{-1})_j = \dim_k(R_j/I_j) = H(R/I, j)$, the Hilbert function of R/I.

Theorem:(*Macaulay*) I^{-1} is finitely generated $\Leftrightarrow$ R/I is artinian.

Moreover, $I^{-1} = (F)$ with $F \in S_j$, $\Leftrightarrow$ R/I is a Gorenstein artinian ring with socle degree j.

(i.e. $R/I = A = k \oplus A_1 \oplus \cdots \oplus A_j$ where $\dim A_j = 1$ and

$$ann_A(A_1 \oplus \cdots \oplus A_j) = A_j \ .$$

equivalently

$$A_t \times A_{j-t} \to A_j \simeq k \quad (\text{ multiplication })$$

is a non-degenerate pairing for every t.)

In view of the general description of I^{-1}, if $F \in S_j$ and $I^{-1} = (F)$ then

$$H(R/I, t) = \dim_k (F)_{j-t} \ .$$

So, we would like to know the size of $(F)_{j-t}$.

Let $\mathcal{F} \in S_j$ have "generic" coefficients, i.e. if $M_1, \ldots, M_{N+1}$, where $N+1 = \binom{j+n-1}{n-1}$, are the monomials of degree j in S. Then

$$\mathcal{F} = \sum_{j=1}^{N+1} z_j M_j$$

is a *generic form of degree* j. We let R_t act on $\mathcal{F}$ and then $< R_t \mathcal{F} > = < \mathcal{F} >_{j-t}$.

Now, let the action be *contraction* and so

$$< R_t \mathcal{F} > = < m_i \circ \mathcal{F} >$$

where the $m_i \in R_t$ are the monomials.

If we write the $m_i \circ \mathcal{F}$ as a sum of monomials of degree $j - t$ in S_{j-t} and display this information in a matrix, we find that this matrix is nothing more than

$$Cat(t, j-t; n) \ .$$

Example: Let $\mathcal{F} = z_1 y_1^3 + z_2 y_1^2 y_2 + z_3 y_1 y_2^2 + z_4 y_2^3$ be the generic form of degree 3 of $S = k[y_1, y_2]$. Then

$$< R_1 \mathcal{F} > = < x_1 \circ \mathcal{F}, x_2 \circ \mathcal{F} >$$
$$x_1 \circ \mathcal{F} = z_1 y_1^2 + z_2 y_1 y_2 + z_3 y_2^2$$
$$x_2 \circ \mathcal{F} = z_2 y_1^2 + z_3 y_1 y_2 + z_4 y_2^2$$

which we summarize in

$$\begin{pmatrix} z_1 & z_2 & z_3 \\ z_2 & z_3 & z_4 \end{pmatrix}$$

and this is $Cat(1, 2; 2)$, as noted.

In particular, if we specialize $\mathcal{F}$ to a specific $F \in S_j$ and place the coefficients of F in the various catalecticant matrices, denoting them by

$$Cat_F(i, j - i; n) ,$$

we find:

$$\dim_k < F >_{j-t} = rk\ Cat_F(t, j - t : n)$$

$$= H(R/I, t) = H(R/I, j - t).$$

Putting this all together we can view the various catalecticant varieties in the following way:

let $\mathbb{P}^N = \mathbb{P}(S_j)$, so $N + 1 = \binom{j+n-1}{n-1}$, then

$a)$ the points $[F] \in \mathbb{P}^N$ are in 1-1 correspondence with homogeneous ideals

$$I \subseteq k[x_1, \ldots, x_n] = R$$

such that R/I is Gorenstein with socle degree j.

$b)$ the ideal $\sqrt{I_{r+1}(Cat(t, j - t; n))}$ is the defining ideal of the set of points $[F] \in \mathbb{P}^N$ such that

$$\text{if } [F] \leftrightarrow I \text{ then } H(R/I, t) = H(R/I, j - t) \leq r$$

Example: Consider $I_2(Cat(1, j - 1 : n)) = \mathcal{I}$. From what we have just seen, $\sqrt{\mathcal{I}}$ defines the set of $[F] \in \mathbb{P}(S_j)$ such that $rk < R_1 \circ F > \leq 1$ (hence $= 1$).

If we use the differentiation action, it's easy to prove that:

$$F \in S_j \text{ and } \dim_k < \frac{\partial F}{\partial y_i} \mid i = 1, \ldots, n >= 1$$

$$\Leftrightarrow F = L^j \text{ where } L = a_1 y_1 + \cdots + a_n y_n .$$

Since $\{\ [L^j]\ |\ L$ a linear form $\}$ is the Veronese variety, we get that

$$\sqrt{\mathcal{I}} \text{ is the defining ideal of } \nu_j(\mathbb{P}^{n-1})$$

Notice further that if $rk < R_1 \circ F >= 1$ then

$$rk < R_t \circ F >= 1 \text{ for all } 1 \leq t \leq j-1 \ .$$

Conversely, if $rk < R_t \circ F >= 1$ for one t with, $1 \leq t \leq j-1$, then it is true for all. It follows that

$$\sqrt{I_2(Cat(t, j-t; n))} = \sqrt{\mathcal{I}} \text{ for any } 1 \leq t \leq j-1$$

(compare with Gruson-Peskine for the 2×2 minors in the case $n = 2$.)

Having now seen the connection with Gorenstein rings, let's return to the result of Gruson and Peskine.

Recall that in that case we were discussing catalecticant matrices associated to $R = k[x_1, x_2]$. In this case it is well known that the only Gorenstein ideals I of height 2 are the complete intersection ideals, i.e.

R/I is a graded artinian Gorenstein ring with socle degree $j \ \Leftrightarrow \ I = (F_1, F_2)$ where the F_i are forms of degree d_i $(d_1 \leq d_2)$ and $(d_1 - 1) + (d_2 - 1) = j$.

In fact, if one draws the graph of this Hilbert Function one sees that we completely know the Hilbert function of R/I as soon as we have one time where

$$H(R/I, t) < \min\{\dim_k R_t, \dim_k R_{j-t}\} \ .$$

This simple observation would completely explain the result of Gruson and Peskine **IF** we knew, *a priori* :

$$\sqrt{I_{r+1}(Cat(t, j-t; 2))} = I_{r+1}(Cat(t, j-t; 2)$$

This leads us to the following obvious question:

Q1: When is $I_{r+1}(Cat(t, j-t; n))$ a prime ideal?

Quick Answer: **Sometimes, but not always!**

The first examples of this phenomena were discovered by Iarrobino and Kanev. A recent simple example was given by Mats Boij:

Let $M = Cat_{\mathcal{F}}(2,3;4)$. This is a 10×20 matrix. Then $I_7(M)$ is not prime, in fact, it has at least two irreducible components.

On the positive side we have the results of Gruson and Peskine and that of Pragacz and Weyman. More recently we have:

Theorem: *(Pucci)* The ideal

$$\mathcal{I} = I_2(Cat(1, j-1; n)) \text{ is prime.}$$

Moreover,

$$\mathcal{I} = I_2(Cat(t, j-t; n)) \text{ for all } t,\ \ 1 \le t \le j-1\ .$$

These ideals all define the Veronese variety in $\mathbb{P}^N$, where $N+1 = \binom{j+n-1}{n-1}$.

Aside: It is well known that the Veronese variety is an arithmetically Cohen-Macaulay variety but I have not ever seen a description of the resolution of those ideals. I would be interested in knowing if there is such a description.

Q2: Is $I_{r+1}(Cat(t, j-t; n))$ a radical ideal?

This problem is completely open. Apart from the times we can prove the ideals in question are prime, and some calculations by computer, there are no general results in this direction. If this were true it would explain a great deal about the interrelation of catalecticant varieties.

Since there is so little we can say about *Q2* I have tried to think of several things which might test that hypothesis. The weakest thing I could think of along those lines was the following question.

Q3: Is $I_{r+1}(Cat(t, j-t; n))$ a saturated ideal?

Theorem: *(Deery)* The ideals $I_{r+1}(Cat(t, j-t; n))$ are saturated ideals for every r, t, j and n.

Deery's argument is very clever and uses an interesting way to grade the polynomial ring. This will be part of Deery's Ph.D. thesis at Queen's University.

The next question I want to mention is less easy to state precisely, so let me state it imprecisely and then give an example to illustrate what I mean.

Q4: Is Macaulay's Theorem on the growth of Hilbert functions reflected in catalecticant ideals?

Consider the following: Let $R = k[x_1, \ldots, x_4]$, then the Hilbert function of R is:

$$1 \quad 4 \quad 10 \quad 20 \quad 35 \quad 56 \quad \cdots$$

Now suppose that I is a homogeneous ideal in R for which $A = R/I$ is a Gorenstein Artinian algebra of socle degree j (which is at least 8) and for which $H(A,3) \leq 11$.

Since the 3-binomial expansion of 11 is $\binom{5}{3} + \binom{2}{2}$ Macaulay's Theorem on the growth of Hilbert functions says that $H(A,4) \leq 16$.

Let's consider the two relevant catalecticant matrices,

$$Cat(3, j-3; 4) \quad \text{and} \quad Cat(4, j-4; 4)$$

which are matrices of size, respectively,

$$20 \times \binom{(j-3)+3}{3} \quad \text{and} \quad 35 \times \binom{(j-4)+3}{3}$$

Macaulay's Theorem tells us:

$$\sqrt{I_{17}(Cat(4, j-4; 4))} \subseteq \sqrt{I_{12}(Cat(3, j-3; r))}$$

The question I am posing in *Q4* is, essentially, if this inequality is true without the radicals.

A simple test case for this question is the following:

Q5 a)

Is it true that

$$I_3(Cat(2, j-2; n)) = I_3(Cat(t, j-t; n))$$

for all t with $2 \leq t \leq j-2$ and

Q5 b)

$$I_3(Cat(1, j-1; n)) \supsetneq I_3(Cat(2, j-2; n)) \ ?$$

We had conjectured (by analogy with the result of Pucci) that the ideals

$$I_3(Cat(t, j-t; n) \text{ for } 2 \leq t \leq j-2 \ ,$$

should be the ideals defining the secant line varieties to the Veronese varieties.

****

Oh yes! Where does the word *catalecticant* come from? If F is a quadratic form, i.e. $F \in S_2$, the only interesting catalecticant to consider is the square matrix

$$Cat_F(1, 1; n)$$

(roughly, the symmetric matrix associated to the form F). We know that the vanishing of the determinant of this matrix (the *discriminant* of the quadratic form) implies some degeneracy of the form F.

Sylvester considered an arbitrary form F of even degree in S_{2j} and looked at the matrix I've called

$$Cat_F(j, j; n)$$

He called the determinant of that matrix the *catalecticant of the form* F, since if it vanished there was something "missing" from F – and the word *catalecticant* is derived from the Greek for *incomplete*.

As Bruce Reznick notes in his book, Sylvester knew poetry well and a line of poetry which is missing an "iamb" is called a *catalectic* line of verse.

Apparently, Sylvester was not completely happy with his choice of this name. Again, Reznick quotes (with a straight face) Sylvester's unhappiness:

"*Meicatalecticizant* would more completely express the meaning of that which, for the sake of brevity, I denominate the *catalecticant*."

References

1. Derry, T. S., *Unpublished Notes*, 1998.

2. Geramita, A. V., *Inverse Systems of Fat Points : Waring's Problem, Secant Varieties of Veronese and Parameter Spaces for Gorenstein Ideals*, The Curves Seminar at Queen's, Vol 102, Queen's Papers in Pure and Applied Mathematics, 1996.

3. Gruson, L., Peskine, C., *Courbes de l'espace projectif : Variétés de secantes*, in :P. Le Barz and Y. Hervier, editors, Enumerative Geometry and Classical Algebraic Geometry, Vol. 24, Progress in Mathematics, 1982.

4. Iarrobino, A., Kanev, V., *The Length of a Homogeneous Form, Determinantal Loci of Catalecticants and Gorenstein Algebras*, Preprint, May 1996.

5. Jozefiak, T., Pragacz, P., Weyman, J., *Resolutions of Determinantal Varieties*, Asterique 87-88, 1981, 109-190.

6. Macaulay, F. H. S., *Some Problems of Enumeration in the Theory of Modular Systems*, Proc. Lond. Math. Soc., 26, (1927), 531-555.

7. Pucci, M., *The Veronese Variety and Catalecticant Matrices*, J. of Alg., Vol. 202, 1998, 72-95.

Bounds for the Betti Numbers of Shellable Simplicial Complexes and Polytopes

JÜRGEN HERZOG
FB 6 MATHEMATIK UND INFORMATIK
UNIVERSITÄT-GHS-ESSEN
POSTFACH 103764
ESSEN 45117, GERMANY
MAT300@UNI-ESSEN.DE

AND

ENZO MARIA LI MARZI
DIPARTIMENTO DI MATEMATICA
UNIVERSITA' DI MESSINA
CONTRADA PAPARDO, SALITA SPERONE 31
98166 SANT'AGATA - MESSINA, ITALY
LIMARZI@RISCME.UNIME.IT

Introduction

In this paper we give upper bounds (Theorem 2.1 and Proposition 3.4) for the Betti numbers of shellable simplicial complexes with a given number of vertices and facets, and of the boundary complex of certain classes of simplicial polytopes. For shellable simplicial complexes our bound is attained when these complexes are $(d-1)$-trees, and for the class of polytopes which we are considering the given bound is attained when the polytope is stacked, that is, when it admits a triangulation which is a $(d-1)$-tree.

Recall that a $(d-1)$- dimensional shellable simplicial complex Δ is called a $(d-1)$-tree if in the shelling of Δ each facet intersects the previous facets in only one subfacet. It follows at once that the h-vector of a $(d-1)$-tree is of the form $(1, h, 0, \ldots, 0)$.

The Betti numbers of stacked simplicial polytopes have first been computed by Hibi and Terai [7]. In this paper we give a different proof of their result, see 3.3. It has been shown by Terai [9] that for 3-polytopes with a given number of vertices the boundary complex of a stacked polytope has the maximal Betti numbers. One could hope that this is true in all dimensions. With the methods developed in this paper we can only give an upper bound for the Betti numbers of a d-polytope, if this polytope admits a proper triangulation, that is, a shellable triangulation with no interior vertices.

1. Basic concepts

We first recall some basic definitions on simplicial complexes. The reader is referred to [2], [6] or [8] for further details.

A simplicial complex on a vertex set $V = \{v_1, \dots, v_n\}$ is a collection Δ of subsets of V such that

(1) $\{v_i\} \in \Delta$;
(2) if $F, G \subset V$ with $F \in \Delta$ and $G \subset F$, then $G \in \Delta$.

The elements of Δ are called the *faces* of Δ. The *dimension* of a face F of Δ, denoted by $\dim F$, is the number $|F| - 1$. Faces of dimension 0 are called *vertices*, those of dimension 1 *edges* of Δ. The maximal faces under inclusion are called *facets*. The dimension of Δ is defined to be

$$\dim \Delta = \max\{\dim F \colon F \in \Delta\}.$$

A simplicial complex is called *pure* if all facets have the same dimension.

Let Δ be a $(d-1)$-dimensional simplicial complex. We denote by f_i the number of i-dimensional faces of Δ, and set $f_{-1} = 1$. The vector of integers $(f_{-1}, f_0, \dots, f_{d-1})$ is called the f-vector of Δ.

Given a collection $F_1, \dots, F_m$ of subsets of a vertex set $V = \{v_1, \dots, v_n\}$, there exists a unique smallest simplicial complex $\Delta = \langle F_1, \dots, F_m \rangle$ on the vertex set $\bigcup_{i=1}^m F_i$ containing all F_i as faces. Indeed,

$$\Delta = \{G \subset V \colon G \subset F_i \text{ for some } i, \dots, m\}.$$

We say that Δ is *spanned by* $F_1, \dots, F_m$.

Recall that a $(d-1)$-dimensional simplicial complex is called *shellable*, if Δ is pure, and if there exists an order of the facets of Δ, say, $F_1, \dots, F_m$, such that

$$\langle F_1, \dots, F_{i-1} \rangle \cap \langle F_i \rangle$$

is spanned by $(d-2)$-simplices. For $i = 2, \dots, m$ we denote by k_i the number of the $(d-2)$-simplices spanning these intersections, and set $k_1 = 0$. We call $(k_1, \dots, k_m)$ the k-vector of the shelling.

Definition 1.1. A $(d-1)$-dimensional shellable simplicial complex Δ with shelling $F_1, \dots, F_m$ is called a $(d-1)$-tree, if $k_i = 1$ for $i = 2, \dots, m$.

Let K be a field, and Δ a simplicial complex on the vertex set $V = \{v_1, \dots, v_n\}$. The *Stanley-Reisner ring* of Δ over K is the factor ring of the polynomial ring

$$K[\Delta] = K[x_1, \dots, x_n]/I_\Delta,$$

where I_Δ is the ideal generated by all monomials

$$x_F = \prod_{v_i \in F} x_i \quad \text{with} \quad F \notin \Delta.$$

The polynomial ring $K[x_1, \dots, x_n]$ is multigraded. The homogeneous elements are the terms λx^a with $\lambda \in K$ and $a \in \mathbb{Z}^n$. The multidegree of such a term is $a \in \mathbb{Z}^n$. Since the defining ideal of a simplicial complex Δ is defined by monomials, the Stanley-Reisner ring $K[\Delta]$ inherits a multigraded structure.

We set

$$H_{K[\Delta]}(\mathbf{t}) = \sum_{a \in \mathbb{Z}} \dim_K K[\Delta]_a \mathbf{t}^a,$$

and call it the *multigraded Hilbert function of* $K[\Delta]$. Here $\mathbf{t} = (t_1, \ldots, t_n)$ and $\mathbf{t}^a = t_1^{a_1} \ldots t_n^{a_n}$ for $a \in \mathbb{Z}^n$.

One has

(1) $$H_{K[\Delta]}(\mathbf{t}) = \sum_{F \in \Delta} \prod_{v_i \in F} \frac{t_i}{1 - t_i}.$$

Of course, since I_Δ is a graded ideal in the ordinary sense, too, the algebra $K[\Delta]$ is homogeneous, that is, $K[\Delta]$ is a finitely generated K-algebra which is generated over K by elements of degree 1.

Recall that an arbitrary homogeneous K-algebra R has a Hilbert function of the form

(2) $$H_R(t) = \frac{Q(t)}{(1-t)^d},$$

where $d = \dim R$ and $Q(t) = \sum_{i=0}^m h_i t^i$ is a polynomial with $Q(1) \neq 0$; see [2, Lemma 4.1.7(b)]. The vector $(h_0, \ldots, h_m)$ is called the h-vector of R, and $Q(1) = \sum_{i=0}^m h_i$ is called the *multiplicity of* R. We denote the multiplicity of R by $e(R)$.

The Hilbert function of $K[\Delta]$ is obtained from (1) by replacing all t_i by t. Therefore,

(3) $$H_{K[\Delta]}(t) = \sum_{i=-1}^{d-1} \frac{f_i t^{i+1}}{(1-t)^{i+1}}.$$

A comparison of (1) and (2) yields the identity

(4) $$\sum_i h_i t^i = \sum_{i=0}^{d} f_{i-1} t^i (1-t)^{d-i}$$

which shows that the h-vector of a $(d-1)$-dimensional simplicial complex has length at most d. Moreover, using (4), one can compute the h-vector from the f-vector, and vice versa.

For a shellable simplicial complex Δ there is a result of McMullen (cf. [2, Corollary 5.1.14]) which gives the h-vector from the shelling.

Theorem 1.2. *Let Δ be a shellable $(d-1)$-dimensional simplicial complex with a shelling whose k-vector is $(k_1, \ldots, k_m)$. Then the h-vector $(h_0, \ldots, h_d)$ of Δ is given by*

$$h_j = |\{i : k_i = j\}| \quad \textit{for} \quad j = 0, \ldots, d.$$

As an immediate consequence of the McMullen formulas one obtains

Corollary 1.3. *Let Δ be a shellable $(d-1)$-dimensional simplicial complex. The following conditions are equivalent:*

(a) *Δ is a $(d-1)$-tree;*
(b) *$h_i = 0$ for $i > 1$*

It follows from (4) that $h_1 = f_0 - d$. We also conclude from (4) that the multiplicity of $K[\Delta]$ equals

$$e(K[\Delta]) = \sum_i h_i = f_{d-1}.$$

In other words, $e(K[\Delta])$ is equal to the number of facets of Δ. It follows from 1.3 that for a $(d-1)$-tree one has $e(K[\Delta]) = f_{d-1} = f_0 - d + 1$.

For an arbitrary homogeneous K-algebra R with h-vector $(h_0, \dots, h_m)$ it is known that $h_1 = \operatorname{emb}\dim R - \dim R$, and that $h_i \geq 0$ if R is Cohen-Macaulay; see [2, Proposition 4.3.1]. Here $\operatorname{emb}\dim R$ denotes the embedding dimension of R. Therefore

$$e(R) \geq \operatorname{emb}\dim R - \dim R + 1.$$

if R is Cohen-Macaulay. This is the inequality of Abhyankar. If equality holds, then R is said to have *minimal multiplicity*. It is clear that R has minimal multiplicity if and only if $h_i = 0$ for $i \geq 2$.

Since shellable simplicial complexes are always Cohen-Macaulay (see [2]), it follows that for a shellable simplicial complex $f_{d-1} \geq f_0 - d + 1$, with equality if and only if it is a $(d-1)$-tree. Of course, the inequality follows also directly by induction on the number of facets. More precisely, one has: if $(k_1, \dots, k_m)$ is the k-vector of the shelling, then

$$f_{d-1} - f_0 + d - 1 = |\{i : k_i > 1\}|.$$

2. Upper Bound for the Betti Numbers of a Shellable Simplicial Complex

In this section we will study the minimal free resolution $\mathbb{F}$ over the polynomial ring $P = K[x_1, \dots, x_n]$ of the Stanley-Reisner ring $K[\Delta] = P/I_\Delta$ of a simplicial complex Δ. Since $K[\Delta]$ is multigraded, the resolution is multigraded as well, that is,

$$\mathbb{F}: \quad 0 \to F_p \to F_{p-1} \to \dots \to F_1 \to F_0 \to K[\Delta] \to 0,$$

where each $F_i = \bigoplus_{a \in \mathbb{Z}^n} P(-a)^{\beta_{ia}}$, and where all maps in this complex are homogeneous (of degree 0). Here we denote, as usual, for any $b \in \mathbb{Z}^n$ by $P(b)$ the shifted rank 1 free graded P-module with $P(b)_a = P_{a+b}$ for all $b \in \mathbb{Z}^n$. The numbers β_{ia} are called the *multigraded Betti numbers* of $K[\Delta]$. For each i there are only finitely many $\beta_{ia} \neq 0$. The *graded Betti numbers* of $K[\Delta]$ are defined to be $\beta_{ij}(K[\Delta]) = \sum_{a \in \mathbb{Z}^n, |a| = j} \beta_{ia}$, where for $a = (a_1, \dots, a_n) \in \mathbb{Z}^n$ we set $|a| = \sum_i a_i$, and the numbers $\beta_i(K[\Delta]) = \sum_{j \in \mathbb{Z}} \beta_{ij}$ are simply called the *Betti numbers of* $K[\Delta]$. Finally, the series

$$\operatorname{Poin}(\mathbf{t}, s) = \sum_{i \in \mathbb{Z}, a \in \mathbb{Z}^n} \beta_{ia} \mathbf{t}^a s^i$$

is called the *multigraded Poincaré series of* $K[\Delta]$. Similarly one defines the (graded) Poincaré series.

We will prove the following

Theorem 2.1. *Let Δ be a shellable $(d-1)$-dimensional simplicial complex with n vertices and m facets. Then*

$$\beta_i(K[\Delta]) \leq \sum_{j=1}^{m-1} j \binom{j-1}{i-1} - \sum_{k=1}^{i} \binom{n-d}{i-k} \binom{m+d-1-n}{k}.$$

Moreover, the bound is reached if and only if Δ is a $(d-1)$-tree.

We first show that for a $(d-1)$-tree Δ the given bound is attained. As observed in the previous section, one has $n = m + d - 1$ when Δ is a $(d-1)$-tree. Hence the second sum in this upper bound is 0, and we have to show that

$$\beta_i(K[\Delta]) = \sum_{j=1}^{m-1} j\binom{j-1}{i-1}. \tag{5}$$

More generally we note the following well-known fact, whose proof we outline for the convenience of the reader.

Lemma 2.2. *Let R be a Cohen-Macaulay ring with minimal multiplicity of embedding dimension n and dimension d. Then $\beta_{ii+j}(R) = \sum_{k=1}^{n-d} k\binom{k-1}{i-1}$ if $j = 1$, and 0 otherwise.*

Proof. We may assume that K is infinite, because otherwise we may choose an infinite base field extension without changing the Betti numbers. Then there exists (c.f. [2, Proposition 1.5.12]) a regular sequence $f_1, \ldots, f_d$ of forms of degree 1. If $\mathbb{F}$ is a minimal graded free $P = K[x_1, \ldots, x_n]$ resolution of R, then $\mathbb{F}/(f_1, \ldots, f_d)\mathbb{F}$ is a graded minimal free $\bar{P} = P/(f_1, \ldots, f_d)P$ resolution of $\bar{R} = R/(f_1, \ldots, f_d)R$.

It follows that

$$\beta_{ij}^{P}(R) = \beta_{ij}^{\bar{P}}(\bar{R}).$$

Since $f_1, \ldots, f_d$ is a regular sequence of 1-forms, $\bar{P}$ is isomorphic to a polynomial ring in $n-d$ variables, say $\bar{P} = K[y_1, \ldots, y_{n-d}]$, and $\bar{R}$ is a quotient ring of $\bar{P}$ with Hilbert function $1 + h_1 t = 1 + (n-d)t$.

Therefore,

$$\bar{R} \cong \bar{P}/(y_1, \ldots, y_{n-d})^2.$$

The ideal $(y_1, \ldots, y_{n-d})^2$ is strongly stable, hence its resolution over $\bar{P}$ is given by Eliahou-Kervaire ([4]). The explicit formulas for the graded Betti numbers in [4] (see also [1]) applied to this situation yield the desired result.

(Alternatively one may view $(y_1, \ldots, y_{n-d})^2$ as an ideal of maximal minors of the matrix

$$\begin{pmatrix} y_1 & y_2 & \cdots & y_{n-d} & 0 \\ 0 & y_1 & y_2 & \cdots & y_{n-d} \end{pmatrix},$$

whose resolution is a special case of the Eagon-Northcott resolution; see [3]). □

Proof. [*Proof of Theorem* 2.1] We prove the asserted inequalities for an arbitrary shellable simplicial complex Δ of dimension $d-1$ and with n vertices by induction on the number m of facets. We choose a $(d-1)$-tree Γ with m facets, and let $V = \{v_1, \ldots, v_r\}$ be the vertex set of Γ. Then $n \leq r = m + d - 1$. Hence may assume that the subset $W = \{v_1, \ldots, v_n\}$ of V is the vertex set of Δ. For any integer $i \leq r$ we set $P_i = K[x_1, \ldots, x_i]$. Then $\beta_i(K[\Delta]) = \beta_i^{P_n}(K[\Delta])$.

We may view $K[\Delta]$ also as a P_r-module, since

$$K[\Delta] = P_r/(I_\Delta, x_{n+1}, \ldots, x_r).$$

Let $\mathbb{F}$ be a minimal free P_n-resolution of $K[\Delta]$. Since $x_{n+1}, \ldots, x_r$ is a regular sequence on P_r/I_Δ, we see that

$$(\mathbb{F} \otimes_{P_n} P_r) \otimes_{P_r} K(x_{n+1}, \ldots, x_r; P_r)$$

is a minimal free P_r-resolution of $K[\Delta]$. Here $K(x_{n+1}, \ldots, x_r; P_r)$ is the Koszul complex of the sequence $x_{n+1}, \ldots, x_r$.

From this we deduce the following equations

$$\beta_i^{P_r}(K[\Delta]) = \sum_{j=0}^{i} \beta_j^{P_n}(K[\Delta]) \binom{r-n}{i-j}.$$

So

$$\beta_i^{P_n}(K[\Delta]) = \beta_i^{P_r}(K[\Delta]) - \sum_{j=0}^{i-1} \beta_j^{P_n}(K[\Delta]) \binom{r-n}{i-j}.$$

The Betti numbers $\beta_i(K[\Delta])$ are bounded below by $\binom{\text{height}\, I_\Delta}{j}$; see for example [5]. Therefore, since height $I_\Delta = n-d$, and $r = m+d-1$, we get

$$\beta_i^{P_n}(K[\Delta]) \leq \beta_i^{P_r}(K[\Delta]) - \sum_{j=0}^{i-1} \binom{n-d}{j} \binom{m+d-1-n}{i-j}.$$

Thus if we can show that

$$\beta_i^{P_{m+d-1}}(K[\Delta]) \leq \beta_i^{P_{m+d-1}}(K[\Gamma])$$

for all i, then the theorem follows from (5).

We prove this inequality by induction on m. For $m = 1$ the assertion is trivial. Now we assume that the assertion is proved for $m \geq 1$, and prove it for $m+1$.

Let

$$\Delta' = \Delta \cup \langle F_{m+1} \rangle \quad \text{and} \quad \Sigma = \Delta \cap \langle F_{m+1} \rangle.$$

In order to simplify notation we set

$$T_i^{m+d}(M) = \operatorname{Tor}_i^{P_{m+d}}(K, M)$$

for any P_{m+d}-module M, and if $M = K[\Pi]$, we set $T_i^{m+d}(\Pi) = T_i^{m+d}(K[\Pi])$. Recall that $\beta_i(M) = \dim_K \operatorname{Tor}_i^{P_{m+d}}(K, M)$.

There is a short exact sequence of P_{m+d}-modules (c.f. [2, Sequence (3), page 210])

$$0 \to K[\Delta'] \to K[\Delta] \oplus K[\langle F_{m+1} \rangle] \to K[\Sigma] \to 0$$

which gives rise to the long exact homology sequence

$$\begin{aligned} \ldots \quad & \to T_i^{m+d}(\Delta') \to T_i^{m+d}(\Delta) \oplus T_i^{m+d}(\langle F_{m+1} \rangle) \to T_i^{m+d}(\Sigma) \\ & \to T_{i-1}^{m+d}(\Delta') \to \ldots \end{aligned}$$

We first show that for all i the map

$$T_i^{m+d}(\langle F_{m+d} \rangle) \to T_i^{m+d}(\Sigma)$$

is injective.

Since Δ is shellable, Σ is the union of $(d-2)$-simplices, say, of the simplices $\langle F_{m+1} \setminus \{v_{r_l}\} \rangle$, $l = 1, \ldots, k$. Then we have

$$K[\langle F_{m+1} \rangle] = P_{m+d}/(x_i \colon v_i \notin F_{m+1}),$$

and

$$K[\Sigma] = P_{m+d}/(x_i \colon v_i \notin F_{m+1}) + (\prod_{l=1}^{k} x_{r_l}).$$

In both cases the defining ideals are generated by regular sequences, and hence the Koszul complexes

$$K(x_i; P_{m+d})_{v_i \notin F_{m+1}} \quad \text{and} \quad K(x_i, \prod_{l=1}^{k} x_{r_l}; P_{m+d})_{v_i \notin F_{m+1}}$$

are the multigraded free P_{m+d}-resolutions of $K[\langle F_{m+1}\rangle]$ and $K[\Sigma]$, respectively.

The natural inclusion map

$$(6) \qquad K(x_i; P_{m+d})_{v_i \notin F_{m+1}} \xrightarrow{\alpha} K(x_i, \prod_{l=1}^{k} x_{r_l}; P_{m+d})_{v_i \notin F_{m+1}}$$

is a lifting of the epimorphism

$$K[\langle F_{m+1}\rangle] = P_{m+d}/(x_i \colon v_i \notin F_{m+1}) \to P_{m+d}/(x_i \colon v_i \notin F_{m+1}) + (\prod_{l=1}^{k} x_{r_l}) = K[\Sigma].$$

Since α is split exact, it induces for all i an injective map $T_i^{m+d}(\langle F_{m+1}\rangle) \to T_i^{m+d}(\Sigma)$.

For all i we now set $M_i = T_i^{m+d}(\Sigma)/T_i^{m+d}(\langle F_{m+1}\rangle)$. Then we get the exact sequence

$$\dots \xrightarrow{\alpha_{i+1}} M_{i+1} \to T_i^{m+d}(\Delta') \to T_i^{m+d}(\Delta) \xrightarrow{\alpha_i} M_i \to T_{i-1}^{m+d}(\Delta') \to \dots$$

From this we deduce

$$(7) \qquad \begin{aligned} \dim_K T_i^{m+d}(\Delta') &= \dim_K T_i^{m+d}(\Delta) + \dim_K M_{i+1} \\ &- \dim_K \operatorname{Im} \alpha_i - \dim_K \operatorname{Im} \alpha_{i+1}. \end{aligned}$$

Similarly for Γ we get

$$(8) \qquad \begin{aligned} \dim_K T_i^{m+d}(\Gamma') &= \dim_K T_i^{m+d}(\Gamma) + \dim_K M'_{i+1} \\ &- \dim_K \operatorname{Im} \beta_i - \dim_K \operatorname{Im} \beta_{i+1}, \end{aligned}$$

where the M'_i and β_i are defined for Γ as the M_i and α_i for Δ.

We note that M_i is $\mathbb{Z}^r$-graded, set $M = \oplus M_i$ and

$$H_M(t_1, \dots, t_r, s) = \sum_i H_{M_i}(t_1, \dots, t_r) s^i.$$

Then it follows from the definition of M and (6) that

$$H_M(t_1, \dots, t_r, s) = \prod_{v_i \notin F_{m+1}} (1 + t_i s)(\prod_{l=1}^{k} t_{r_l}) s.$$

Similarly, if $\Gamma' = \Gamma \cup \langle F'_{m+1}\rangle$, with $\Gamma \cap \langle F'_{m+1}\rangle = F'_{m+1} \setminus \{v_{m+d}\}$ (since we assume that Γ' is a $(d-1)$-tree) we get for $M' = \oplus M'_i$ the formula

$$H_{M'}(t_1, \dots, t_r, s) = \prod_{v_i \notin F'_{m+1}} (1 + t_i s) t_{m+d} s.$$

It follows that

$$\dim_K M_{i+1} = \binom{r-d}{i} = \dim_K M'_{i+1}$$

for all i. Therefore, the equations (7) and (8) together with our induction hypothesis imply that $\beta_i^{P_{m+d}}(\Delta') \leq \beta_i^{P_{m+d}}(\Gamma')$ if

(9) $$\dim_K \operatorname{Im} \alpha_i + \dim_K \operatorname{Im} \alpha_{i+1} \geq \dim_K \operatorname{Im} \beta_i + \dim_K \operatorname{Im} \beta_{i+1}$$

for all i.

We claim that $\beta_i = 0$ for $i > 1$. Indeed, the Poincaré series $\operatorname{Poin}_{m+d}(\mathbf{t}, s)$ of $K[\Gamma]$ as a P_{m+d}-module, and Poincaré series $\operatorname{Poin}_{m+d}(\mathbf{t}, s)$ of $K[\Gamma]$ as a P_{m+d-1}-module, are related by the equation

$$\operatorname{Poin}_{m+d}(\mathbf{t}, s) = \operatorname{Poin}_{m+d-1}(\mathbf{t}, s)(1 + t_{m+d}s).$$

It follows from 2.2 that

$$\operatorname{Poin}_{m+d-1}(\mathbf{t}, s) = 1 + \sum_{i\geq 1} P_i(\mathbf{t})s^i$$

where $P_i(\mathbf{t}) \in K[t_1, \ldots, t_{d+m-1}]$ is homogeneous of degree $i + 1$.

Therefore,

$$\operatorname{Poin}_{m+d}(\mathbf{t}, s) = 1 + t_{m+d}s + P_1(\mathbf{t})s + \sum_{i\geq 2}(P_i(\mathbf{t}) + P_{i-1}(\mathbf{t})t_{m+d})s^i.$$

On the other hand,

$$\begin{aligned} H_{M'}(\mathbf{t}) &= \prod_{v_j \notin F_{m+1}} (1 + t_j s)t_{d+m}s \\ &= t_{d+m}s + \sum_{i\geq 2} Q_{i-1}(\mathbf{t})t_{m+d}s^i \end{aligned}$$

where $Q_{i-1}(\mathbf{t})$ is a homogeneous polynomial of degree $i - 1$ over K. Assume that $\beta_i \neq 0$ for $i \geq 2$, then $P_i(\mathbf{t}) + P_{i-1}(\mathbf{t})t_{m+d}$ and $Q_i(\mathbf{t})t_{m+d}$ must have a common monomial in the variables t_j, $v_j \notin F'_{m+1}$. But this is impossible by degree reasons.

Since we now know that $\beta_i = 0$ for $i > 1$, it follows that $\dim_K \operatorname{Im} \alpha_i \geq \dim \operatorname{Im} \beta_i$ for $i > 1$.

We finally show that $\dim_K \operatorname{Im} \alpha_1 = \dim_K \operatorname{Im} \beta_1 = 1$, then the desired inequalities (9) follow for all i, and the theorem is proved.

Notice that $\dim_K M_1 = \dim_K M'_1 = 1$. Therefore since

$$K[\Delta \cap \langle F_{m+1} \rangle] = P_{m+d}/(x_i \colon v_i \notin F_{m+1}) + (\prod_{l=1}^{k} x_{r_l})$$

and

$$K[\Gamma \cap \langle F'_{m+1} \rangle] = P_{m+d}/(x_i \colon v_i \notin F'_{m+1}) + (x_{m+d}),$$

we see from the definition of M_1 and M'_1, that the generator of M_1 corresponds to $\prod_{l=1}^{k} x_{r_l}$ and the generator of M'_1 to x_{m+d}. Thus in order to prove that $\dim_K \operatorname{Im} \alpha_1 = \dim_K \operatorname{Im} \beta_1$, we must show that the defining ideal (I_Γ, x_{m+d}) of the P_{m+d}-module $K[\Gamma]$ contains the element x_{m+d} as a minimal generator (which is trivial), and that the defining ideal $(I_\Delta, x_{n+1}, \ldots, x_{m+d})$ of the P_{m+d}-module $K[\Delta]$ contains the element $\prod_{l=1}^{k} x_{r_l}$ as a minimal generator. The second assertion is seen by noting that F_{m+1} is a non-face of Δ (which we add in order to obtain Δ'). However, it is not a minimal non-face of Δ. The minimal non-face of Δ which is contained in F_{m+1} is the face $G = \{v_{r_1} \ldots, v_{r_k})$. Since the minimal non-faces of Δ correspond to minimal generators of I_Δ, we conclude that $\prod_{v_i \in G} x_i = \prod_{l=1}^{k} x_{r_l}$ is a minimal generator of I_Δ and hence of $(I_\Delta, x_{n+1}, \ldots, x_{m+d})$, as desired. □

3. Betti numbers of the boundary complex of polytopes

In this section we give a short proof of a result of Hibi and Terai [7] who computed the Betti numbers of the boundary complex of a stacked polytope, and give an upper bound for the Betti numbers of the boundary complex of a polytope with a proper triangulation.

Recall that a *triangulation of a simplicial d-polytope P* is a d-dimensional simplicial complex Γ whose geometric realization is P. A simplicial d-polytope P is called *stacked* if it admits a triangulation which is a $(d-1)$-tree. In other words, starting with a d-simplex, one adds new vertices by building shallow pyramides over facets to obtain P. Let $P(n,d)$ be a such stacked d-polytope with n vertices. We denote by $\Delta P(n,d)$ the boundary complex of $P(n,d)$, that is, the simplicial complex whose facets are the boundary faces of $P(n,d)$. Note that $\dim \Delta P(n,d) = d-1$.

It is well-known that the boundary complex $\Delta(P)$ of any simplicial d-polytope is Gorenstein; see [2, Corollary 5.5.6]. In other words, for any field K the Stanley-Reisner ring $K[\Delta(P)]$ is Gorenstein. In particular, P has a symmetric h-vector. More precisely, one has $h_i = h_{d-i}$ for $0 \le i \le d$. These are the famous Sommerville equations. (Here we follow the common convention to define the h-vector of P to be the h-vector of $\Delta(P)$.) For stacked polytopes the Sommerville equations are more special.

Proposition 3.1. *The h-vector of $P(n,d)$ is the vector $(1, n-d, n-d, \ldots, n-d, 1)$ of length d.*

The proof could be easily done by induction on the number of vertices. Instead we will use the following theorem of Hochster (see [2, Theorem 5.6.2]), since this theorem will be crucial for other arguments of this section as well.

We denote by $\omega_{K[\Delta]}$ the canonical module of a Cohen-Macaulay simplicial complex Δ over a field K.

Theorem 3.2 (Hochster). *Let K be a field, and Γ a Cohen–Macaulay complex of dimension d over K whose geometric realization $X = |\Gamma|$ is a manifold with a non-empty boundary ∂X. Further let Δ be the subcomplex of Γ whose geometric realization is ∂X, and J the ideal in $K[\Gamma]$ generated by the monomials $x^F = \prod_{v_i \in F} x_i$, $F \in \Gamma \setminus \Delta$. Then the following conditions are equivalent:*

(a) *$\omega_{K[\Gamma]} \cong J$ as a $\mathbb{Z}^n$-graded $K[\Gamma]$-module;*
(b) *Δ is a Gorenstein complex over K;*

We apply this theorem to a d-polytope P with triangulation Γ and boundary complex Δ. Since $\omega_{K[\Gamma]} \cong J$ it clear that

$$K[\Gamma]/\omega_{K[\Gamma]} \cong K[\Delta]. \tag{10}$$

It follows then from [2, 3.3.18(b)] that Δ is Gorenstein. Moreover, if

$$H_{K[\Gamma]}(t) = \frac{\sum_{i=0}^{d+1} h_i t^i}{(1-t)^{d+1}}$$

is the Hilbert function of $K[\Gamma]$, then, by a result of Stanley [8],

$$H_{\omega_{K[\Gamma]}}(t) = \frac{\sum_{i=0}^{d+1} h_{d-i}t^i}{(1-t)^{d+1}}$$

is the Hilbert function of $\omega_{K[\Gamma]}$; see [2, Corollary 4.3.8]. Therefore (10) yields

$$\begin{aligned} H_{K[\Delta]}(t) &= H_{K[\Gamma]}(t) - H_{\omega_{K[\Gamma]}}(t) \\ &= \frac{\sum_{i=0}^{d+1} h_i t^i - \sum_{i=0}^{d+1} h_{d-i}t^i}{(1-t)^{d+1}} \\ &= \frac{\sum_{i=0}^{d} g_i t^i}{(1-t)^d}, \end{aligned}$$

where

$$g_i = \sum_{j=0}^{i}(h_j - h_{d+1-j}) \quad \text{for} \quad j = 0, \ldots, i. \tag{11}$$

Now let us apply this to prove Proposition 3.1: Let Γ be the stacked shelling of $P(n,d)$. We know from 1.3 that the h-vector of Γ is $(1, n-d-1, 0, \ldots, 0)$, so that from (11) we obtain for $\Delta(P(n,d))$ the h-vector, as asserted in 3.1.

Next we give a short proof for the Hibi-Terai formulas [7].

Theorem 3.3 (Hibi-Terai). *Let $P(n,d)$ be a stacked simplicial d-polytope with n vertices, and $\Delta P(n,d)$ its boundary complex. Then the minimal graded free resolution of the Stanley-Reisner ring has only a 2-linear and a d-linear strand. More precisely, one has*

$$\beta_{ii+j}(K[\Delta P(n,d)]) = \begin{cases} \sum_{k=i}^{n-d-1} k\binom{k-1}{i-1} & \text{for } 0 \le i \le n-d-1 \text{ and } j = 1, \\ \sum_{k=n-d-i}^{n-d-1} k\binom{k-1}{n-d-i-2} & \text{for } 1 \le i \le n-d \text{ and } j = d-1. \end{cases}$$

Proof. We apply Hochster's theorem 3.2 to $P(n,d)$. Let $\Gamma(n,d)$ be the stacked shelling of $P(n,d)$. By (10) and 3.2 we have

$$K[\Delta P(n,d)] = K[\Gamma(n,d)]/\omega_{K[\Gamma(n,d)]}.$$

Furthermore, $\omega_{K[\Gamma(n,d)]}$ is generated by all monomials $x^F \in K[\Gamma(n,d)]$ where $F \in \Gamma(n,d) \setminus \Delta P(n,d)$. Since the shelling of $P(n,d)$ is stacked all such faces are of dimension $d-1$. Therefore all generators of $\omega_{K[\Gamma(n,d)]}$ are of degree d.

By 2.2, $K[\Gamma(n,d)]$ has the graded minimal free $S = K[x_1, \ldots, x_n]$-resolution

$$\mathbb{F}: 0 \to F_{n-d-1} \to \ldots \to F_1 \to F_0 \to K[\Gamma(n,d)] \to 0,$$

where $F_i = S(-i-1)^{b_i}$ for $i = 1, \ldots, n-d-1$ with $b_i = \sum_{k=i}^{n-d-1} k\binom{k-1}{i-1}$.

Let $\mathbb{G}$ be the minimal free graded S-resolution of $\omega_{K[\Gamma(n,d)]}$. The resolution $\mathbb{G}$ is obtained from $\mathbb{F}$ by dualizing $\mathbb{F}$ into S and shifting it suitably, see [2, Exercise 3.3.25]. Since in our case all generators of $\omega_{K[\Gamma(n,d)]}$ are of degree d, we see that $G_i \cong F^*_{n-d-1-i}(-n)$ for $i = 0, \ldots, n-d-1$.

The inclusion $\omega_{K[\Gamma(n,d)]} \to K[\Gamma(n,d)]$ lifts to a map of complexes $\alpha : \mathbb{G} \to \mathbb{F}$. The mapping cone of α gives the resolution of $K[\Delta P(n,d)]$. It has the form

$$\begin{aligned} 0 \to F_0^*(-n) \to \quad & F_1^*(-n) \oplus F_{n-d-1} \to \dots \\ & \to F_{n-d-1}^*(-n) \oplus F_1 \to F_0 \to K[\Delta P(n,d)] \to 0. \end{aligned}$$

This yields the assertion of the theorem. □

The Sommerville equations for a 3-polytope P with $n = h + 3$ vertices imply that P has the h-vector $(1, h, h, 1)$. Here we see that the combitorial type of $\Delta(P)$ is not reflected by the h-vector since it only depends on the number of vertices of P. In particular, a stacked 3-polytope with the same number of vertices as P has the same h-vector. Nevertheless the stacked 3-polytope is distinguished by the fact that it has the largest Betti numbers among all 3-polytopes with the same number of vertices. This is the theorem of Terai [9].

Let Γ be a triangulation of a d-polytope P. The boundary complex ΔP of P is generated by all $d-1$-dimensional faces F of Γ which belong to exactly one facet of Γ. The faces of Γ which do not belong to ΔP are called the interior faces of Γ. We say that Γ is a *proper triangulation of* P if Γ is shellable and has no interior vertices.

We conclude this paper with the following

Proposition 3.4. *Let P be a simplicial d-polytope admitting a proper triangulation with m facets. Then*

$$\beta_i(K[\Delta P]) \le b_i + b_{n-d-i}$$

for all $i = 0, \dots, n-d$, where $b_i = \sum_{j=1}^{m-1} j\binom{j-1}{i-1}$.

Proof. Let Γ be the proper triangulation of P. Since Γ has no interior vertices it follows that $\omega_{K[\Gamma]}$ is contained in the square of the graded maximal ideal of $K[\Gamma]$. Therefore $K[\Gamma]$ and $K[\Delta P] = K[\Gamma]/\omega_{K[\Gamma]}$ are defined over the same polynomial ring S, and a free S-resolution of $K[\Delta P]$ is obtained as a mapping cone of the S-resolution of $K[\Gamma]$ and the S-resolution of $\omega_{K[\Gamma]}$, cf. the proof of 3.3. This resolution need not to be minimal. Thus, if β_i denotes the i-th Betti number of $K[\Gamma]$ we get $\beta_i(K[\Delta P]) \le \beta_i + \beta_{n-d-i}$. Hence together with 2.1 the desired inequality follows. □

References

1. A. Aramova - J. Herzog. Koszul cycles and and Eliahou-Kervaire type resolutions. J. Algebra **183**, 347-370 (1996).
2. W. Bruns - J. Herzog. *Cohen-Macaulay rings. (Revised edition).* Cambridge studies in advanced mathematics **39**. Cambridge University Press, 1998.
3. W. Bruns - U. Vetter. *Determinantal rings.* LNM **1327**, Springer, 1988.
4. S. Eliahou - M. Kervaire. Minimal resolutions of some monomial ideals. J. of Alg. **129**, 1-25 (1990).
5. J. Herzog - M. Kühl. On the Betti numbers of pure and linear resolutions. Comm. in Alg. **12 (13)**, 1627-1646 (1984).
6. T. Hibi. *Algebraic Combinatorics on Convex Polytopes.* Carslaw Publications, 1992.
7. T. Hibi - N. Terai. Computation of Betti numbers of monomial ideals associated with stacked polytopes. manuscripta math. bf 92, 447-453.
8. R. Stanley. *Combinatorics and Commutative algebra.* 2nd Edition, Birkhäuser, 1996.
9. N. Terai. Upper bound for Betti numbers of 3-dimensional Gorenstein Stanley-Reisner rings. Preprint 1996.

Generators for the Generic Rational Space Curve: Low Degree Cases

Monica Idà

Dipartimento di Matematica, Università di Bologna, Piazza di Porta S. Donato, 5, 40127 Bologna, Italia. e-mail: ida@dm.unibo.it

INTRODUCTION.

Equations defining projective variety is a classical topic (Noether, Enriques, Petri) which has developed in many directions: vanishing theorems, regularity questions (after Castelnuovo, Mumford, Gruson-Lazarsfeld-Peskine), connections with linear systems (Green's conjecture), just to quote a few.

Recently attention has been attracted by problems of "minimal resolution conjecture". If X is a closed subscheme in $\mathbb{P}^n$, I_X denotes its saturated ideal, v is the minimum degree of a hypersurface containing X, and $\mathcal{I}_X$ is (v+1)-regular (i.e. $H^i(\mathcal{I}_X(v+1-i)) = 0$ for each $i \geq 1$), then X is of maximal rank (we recall that X is of maximal rank if $\forall k \geq 0$ the natural map $\rho_k: H^0(\mathcal{O}_{\mathbb{P}^n}(k)) \to H^0(\mathcal{O}_X(k))$ is of maximal rank) and the minimal free resolution looks like :

$$0 \to \mathcal{O}^{\oplus\alpha_n}(-v-n+1) \oplus \mathcal{O}^{\oplus\beta_n}(-v-n) \to \ldots \to \mathcal{O}^{\oplus\alpha_2}(-v-1) \oplus \mathcal{O}^{\oplus\beta_2}(-v-2) \to$$
$$\to \mathcal{O}^{\oplus\alpha_1}(-v) \oplus \mathcal{O}^{\oplus\beta_1}(-v-1) \to \mathcal{I}_X \to 0.$$

If the numbers α_i, β_i are "as small as possible" (this means that I_X, as well as the kernels of the arrows in the sequence, is "minimally generated": see below) we can say that X is "minimally resolved".

Minimal resolution conjectures state that if d is big enough and if X, of degree d, is sufficiently general in its Hilbert scheme, then X (i.e. I_X) is minimally resolved.

This circle of ideas has motivated most of the work on zero-dimensional scheme in the last few years and we will just mention two recent results, referring the reader to those works for more bibliographical informations: on one hand, Hirschowitz and Simpson ([H-S]) proved, for d>>0, the conjecture for d points in $\mathbb{P}^n$, on the other hand, Eisenbud and Popescu ([E-P]) have shown the existence of counterexamples for d points in $\mathbb{P}^n$ with d low.

For what concerns curves, very little is known. This paper deals with this problem for rational curves of low degree in $\mathbb{P}^3$. We work over an algebraically closed field of characteristic zero.

Let C be a curve of $\mathbb{P}^3$; if C is non special and of maximal rank, its postulation $\{h^0(\mathcal{I}_C(k))\}_{k\geq 0}$ is known, since $h^0(\mathcal{I}_C(k)) = \dim \ker \rho_k =$ $= \max\{0, h^0(\mathcal{O}_{\mathbb{P}^3}(k))\text{-}h^0(\mathcal{O}_C(k))\}$, and $h^0(\mathcal{O}_C(k))$ is given by Riemann-Roch.

This maximal rank property is shared by "many" generic curves of $\mathbb{P}^3$: for each d, the generic union of d lines ([H-H]), the generic rational curve of degree d ([H]), a "generic" curve of degree d and genus g, with $d \geq 3/4g+3$, which is of general moduli ([B-E], [B-E,2]).

Now let C be a curve in $\mathbb{P}^3$ with known postulation, let $I_C := \oplus_{k\geq 0} H^0(\mathcal{I}_C(k))$ be its saturated homogeneous ideal, and look at the natural maps:

σ_k: $H^0(\mathcal{I}_C(k)) \otimes H^0(\mathcal{O}_{\mathbb{P}^3}(1)) \to H^0(\mathcal{I}_C(k+1))$.

A minimal system of generators for I_C can be obtained taking a basis for the first non zero $H^0(\mathcal{I}_C(k))$, and adding a basis for a supplementary of $\mathrm{Im}(\sigma_k)$ in $H^0(\mathcal{I}_C(k+1))$ for all k for which σ_k is not onto; hence if we know $\{\dim \mathrm{coker}(\sigma_k)\}_{k\geq 0}$ we know the number, for each degree, of a minimal system of generators for I_C.

This happens in particular if all the σ_k's are of maximal rank; we say in this case that C, or I_C, is minimally generated.

A maximal rank, minimally generated curve C is said to be naturally generated (these two properties are independent: see [I,3]). We recall that if C is a non special naturally generated curve in $\mathbb{P}^3$, the minimal free resolution of $\mathcal{I}_C$ looks like (here $\mathcal{O} = \mathcal{O}_{\mathbb{P}^3}$):

$0 \to \mathcal{O}^{\oplus\alpha_3}(-v-2) \oplus \mathcal{O}^{\oplus\beta_3}(-v-3) \to \mathcal{O}^{\oplus\alpha_2}(-v-1) \oplus \mathcal{O}^{\oplus\beta_2}(-v-2) \to$
$\to \mathcal{O}^{\oplus\alpha_1}(-v) \oplus \mathcal{O}^{\oplus\beta_1}(-v-1) \to \mathcal{I}_C \to 0,$

where $v(C) := \min\{k \geq 1 \mid h^1(\mathcal{I}_C(k)) = 0\}$, $\alpha_1 = h^0(\mathcal{I}_C(v))$,
$\beta_1 = \max\{0, h^0(\mathcal{I}_C(v+1)\text{-}4h^0(\mathcal{I}_C(v))\}$, $\beta_3 = h^1(\mathcal{I}_C(v-1))$.

Moreover, if σ_v is injective (that is, if $\beta_1 \neq 0$ or if σ_v is bijective) the resolution is completely determined, since $\alpha_2=0$ and $\alpha_3=0$, and $\beta_2=\beta_3+\alpha_1+\beta_1-1$ (see [I] sec.7).

The hope is to prove that also the minimal generation property is true for many generic curves; in fact, there is the following conjecture, due to A.Hirschowitz:

For any g ≥ 0 there exists d(g) such that for d$\geq$d(g) the generic curve of some irriducible component of $H_{d,g}$ is naturally generated.

Although not concerned by the conjecture, the generic union of d lines in $\mathbb{P}^3$, d $\neq 4$, goes in the sense of the conjecture, since it is of maximal rank (as already said), and minimally generated (see [I]). In this paper we prove the following:

Theorem 1. The generic rational curve of degree d in $\mathbb{P}^3$ is minimally generated for d$\neq$ 5, d$\leq$73.

The generic quintic rational curve has a 4-secant and hence it is not minimally generated, because it can't be generated by cubics while numerically it should be (see [G-L-P]).

Hence, the first condition on the degree is obligatory, but the second one, i.e. d$\leq$73, is not! In fact, the author would like very much to take it away, but unfortunately she finds some difficulties with the construction in the general case. Notice however that the number 73 is not significant, in the sense that it does not exist a particular obstacle for the next cases; it was chosen because it corresponds to the initial 18 cases of a possible inductive proof for the general case.

The main tecniques used to prove theorem 1 are "la méthode d'Horace" (see [H,2]) combined with the use of elementary transformations (see [M]), which allow to "break up" points of $\mathbb{P}^3$; this is very useful especially for the low degrees.

Elementary transformations where used also in [I], but the ones introduced here (section 1) are more efficient. Section 2 contains some lemmas which are used in the proofs of section 4.

In section 3 we show that, in order to prove theorem 1, it is enough to prove, for each k$\leq$18, the statement R(k); that is, the restriction map: $H^0(\mathcal{O}_{\mathbb{P}(\Omega_3)}(1)\otimes\pi^*\mathcal{O}_{\mathbb{P}^3}(k+1)) \to H^0(\mathcal{O}_{\mathbb{P}(\Omega_3)}(1)\otimes\pi^*\mathcal{O}_{\mathbb{P}^3}(k+1)\,|_Z)$ is bijective for a particular configuration Z pull back of a rational curve and lines, together with points of $\mathbb{P}(\Omega_{\mathbb{P}^3})$.

In section 4 we prove R(k) for $k \leq 18$; this is done with the support of pictures.

ACKNOWLEDGMENTS.

I wish to thank A.Hirschowitz, who introduced me to these problems and tecniques and with whom I have had so many interesting talks; I also wish to thank A.Vistoli and R.Hartshorne for the useful discussions.

NOTATIONS AND PRELIMINARIES.

i) With **Q** we shall always denote a smooth quadric in $\mathbb{P}^3$. Denoting by p_1, p_2 the two canonical projections:

$$\mathbb{P}^1 \xleftarrow{p_2} Q \xrightarrow{p_1} \mathbb{P}^1$$

we set $\mathcal{O}_Q(a,b) := p_1{}^*\mathcal{O}_{\mathbb{P}^1}(a) \oplus p_2{}^*\mathcal{O}_{\mathbb{P}^1}(b)$.

Recall that $\Omega_Q \cong p_1{}^*\Omega_{\mathbb{P}^1} \oplus p_2{}^*\Omega_{\mathbb{P}^1}$, and that if L is a line of type (1,0), and M is a line of type (0,1), $p_1{}^*\Omega_{\mathbb{P}^1}|_M \cong \Omega_M$, $p_2{}^*\Omega_{\mathbb{P}^1}|_L \cong \Omega_L$.

ii) We set $\Omega_3 := \Omega_{\mathbb{P}^3}$, and $\Lambda := \Omega_3|_Q$.

iii) If Z is a closed subscheme of a projective scheme S, and L is a locally free coherent sheaf on S, we say that Z is L-settled if the restriction map: $H^0(L) \to H^0(L|_Z)$ is bijective.

iv) If f: $Y \to S$ is a morphism of schemes, and Z is a closed subscheme of S, $f^{-1}(Z)$ denotes the inverse image scheme, that is, the closed subscheme $Z \times_S Y$ of Y, having as ideal sheaf the inverse image ideal sheaf of I_Z.

v) We set: $\mathbf{X} := \mathbb{P}(\Omega_3)$; $\pi: \mathbf{X} \to \mathbb{P}^3$ the canonical projection; $\mathbf{L_k} := \mathcal{O}_X(1) \otimes \pi^*\mathcal{O}_{\mathbb{P}^3}(k+1)$.

vi) If A is a given subset of X, "**R(A,n)**" denotes the statement "$H^0(I_A \otimes L_n) = 0$".

vii) Let E be a rank ≤ 3 vector bundle on a smooth variety S, and let q: $\mathbb{P}(E) \to S$ the canonical projection; with **s-point**, **d-point**, **t-point**, we mean respectively : a point of $\mathbb{P}(E)$, two points of $\mathbb{P}(E)$ lying in the same fiber $q^{-1}(x)$, three points of $\mathbb{P}(E)$ lying in the same fiber $q^{-1}(x)$ but not on a line of $q^{-1}(x)$. Let L be an invertible sheaf on S; if we are concerned with the zeros of the sections of the sheaf $\mathcal{O}_{\mathbb{P}(E)}(1) \otimes q^*L$, which are linear on the fibers of q, a d-point can be identified with a line of $q^{-1}(x)$, and a t-point with $q^{-1}(x)$.

When we say that a t-point, resp. a d-point, resp. an s-point lies on a subscheme A of S, we mean that it is in $q^{-1}(x)$ with $x \in A$.

viii) If Y and Z are closed subschemes of a projective variety, the union $Y \cup Z$ (resp. the intersection $Y \cap Z$) denotes the schematic union (resp. intesection) of Y and Z.

ix) We shall also make use of the following notations:

$\mathbf{C_d}$:=the generic rational curve of degree d;

$\chi(\mathbf{p})$:= the first infinitesimal neighbourhood of the point p in $\mathbb{P}^3$;

a grille **G(n,m)**:= a set of nm points on Q obtained as intersection of n lines of type (1,0) with m lines of type (0,1);

$\Gamma(\mathbf{n,m})$:= the union of n lines $L_1,\dots,L_n$ on Q of type (1,0), of m lines $M_1,\dots,M_m$ on Q of type (0,1), and of $\chi(L_i \cap M_j)$, i=2,...,n, j=2,...m; in other words, $\Gamma(n,m)$ is a specialization of a rational curve of degree n+m on Q, and $res_Q\Gamma(n,m)$ is a grille of points G(n-1,m-1);

$\Gamma(\mathbf{n,m})_{\mathbf{red}}$:= the union of n lines on Q of type (1,0) and of m lines on Q of type (0,1);

r(m,n,p):= the generic union of m s-points, n d-points, and p t-points, lying on the same line $r \subset \mathbb{P}^3$;

$\mathbf{F_{a,b}}$:=$\mathcal{O}_X(1)|_Q \otimes (\pi|_Q)^* \mathcal{O}_Q(a,b)$; recall that $h^0(F_{a,b})$=3ab-a-b-1.

Some more notations are given in section 4.

SECTION 1: TWO ELEMENTARY TRANSFORMATIONS.

1.1 Proposition. There are two elementary transformations over $\mathbb{P}^3$ of the form:

$$\text{(e1)} \qquad 0 \to F \to \Omega_3 \xrightarrow{\alpha} \mathcal{O}_Q(-2,0) \to 0$$

$$\text{(e2)} \qquad 0 \to \Omega_3(-2) \to F \xrightarrow{\beta} \mathcal{O}_Q(-1,-2)^{\oplus 2} \to 0,$$

where α is the natural map $\Omega_3 \to \mathcal{O}_Q(-2,0) \cong p_1{}^*\Omega_{\mathbb{P}^1}$, and F denotes the rank 3 vector bundle ker α.

Proof. We have the exact sequences:

$$\text{(I)} \qquad 0 \to \mathcal{I}_{Q|Q} \to \Lambda \xrightarrow{\gamma} \Omega_Q \to 0,$$

$$\text{(II)} \qquad 0 \to p_2{}^*\Omega_{\mathbb{P}^1} \to \Omega_Q \overset{\delta}{\to} p_1{}^*\Omega_{\mathbb{P}^1} \to 0,$$

hence there is a canonical map ε: $\Lambda \to p_1{}^*\Omega_{\mathbb{P}1} \cong O_Q(-2,0)$, where $\varepsilon = \delta \circ \gamma$. Now we set K:=Ker ε. Denoting by F the suitable rank 3 vector bundle on $\mathbb{P}^3$, we have a commutative diagram:

$$\begin{array}{ccccccccc}
 & & 0 & & 0 & & & & \\
 & & \uparrow & & \uparrow & & & & \\
 & & & & & \varepsilon & & & \\
0 & \to & K & \to & \Lambda & \to & p_1{}^*\Omega_{\mathbb{P}1} & \to & 0 \\
 & & \uparrow & & \uparrow & & \| & & \\
0 & \to & F & \to & \Omega_3 & \to & p_1{}^*\Omega_{\mathbb{P}1} & \to & 0 \\
 & & \uparrow & & \uparrow & & & & \\
 & & \Omega_3(-2) & = & \Omega_3(-2) & & & & \\
 & & \uparrow & & \uparrow & & & & \\
 & & 0 & & 0 & & & &
\end{array}$$

and its first column and second row are the elementary transformations we were looking for, once proved $K \cong O_Q(-1,-2)^{\oplus 2}$. Now consider p_{2*} of the first row of the diagram twisted by 1:

$$0 \to p_{2*}K(1) \overset{\cong}{\to} p_{2*}\Lambda(1) \to 0$$

since $(p_1{}^*\Omega_{\mathbb{P}1})(1)|_{p_2{}^{-1}(x)} \cong O_{\mathbb{P}1}(-1)$.

The pull down by p_{2*} of sequence (I) twisted by 1 gives:

$$0 \to p_{2*}O_Q(-1,-1) \to p_{2*}\Lambda(1) \to p_{2*}(O_Q(-1,1) \oplus O_Q(1,-1)) \to R^1p_{2*}O_Q(-1,-1)$$

and since $p_{2*}O_Q(-1,-1) = 0 = R^1p_{2*}O_Q(-1,-1)$, while

$$p_{2*}(O_Q(-1,1) \oplus O_Q(1,-1)) \cong p_{2*}(p_1{}^*O_{\mathbb{P}1}(1) \otimes p_2{}^*O_{\mathbb{P}1}(-1)) \cong$$
$$\cong p_{2*}(p_1{}^*O_{\mathbb{P}1}(1)) \otimes O_{\mathbb{P}1}(-1) \cong O_{\mathbb{P}1}(-1)^{\oplus 2}$$

(in fact $p_1{}^*O_{\mathbb{P}1}(1) \cong O_{\mathbb{P}(O_{\mathbb{P}1}{}^{\oplus 2})}(1)$ relatively to: $\mathbb{P}(O_{\mathbb{P}1}{}^{\oplus 2}) \to^{p_2} \mathbb{P}^1$), we conclude $p_{2*}K(1) \cong p_{2*}\Lambda(1) \cong O_{\mathbb{P}1}(-1)^{\oplus 2}$.

Now we have an injection of vector bundles of the same rank: $0 \to p_2{}^*p_{2*}K(1) \to$ $\to K(1)$ which is hence an isomorphism, so we conclude $K \cong O_Q(-1,-2)^{\oplus 2}$.

1.2 Remark. Putting in a diagram sequences (I), (II) and

$$0 \to K \to \Lambda \to p_1{}^*\Omega_{\mathbb{P}1} \to 0 \qquad \text{we get:}$$

$$\begin{array}{ccccccccc} & & & & 0 & & & & \\ & & & & \uparrow & & & & \\ 0 & \to & p_2{}^*\Omega_{\mathbb{P}^1} & \to & \Omega_Q & \to & p_1{}^*\Omega_{\mathbb{P}^1} & \to & 0 \\ & & \uparrow & & \uparrow & & \| & & \\ 0 & \to & K & \to & \Lambda & \to & p_1{}^*\Omega_{\mathbb{P}^1} & \to & 0 \\ & & & & \uparrow & & & & \\ & & & & I_{Q|Q} & & & & \\ & & & & \uparrow & & & & \\ & & & & 0 & & & & \end{array}$$

hence we have $0 \to I_{Q|Q} \to K \to p_2{}^*\Omega_{\mathbb{P}^1} \to 0.$

Twisting by (0,2) we get: $0 \to O_Q(-2,0) \to O_Q(-1,0)^{\oplus 2} \to O_Q \to 0$, that is, the pull-back through $p_1{}^*$ of Euler sequence. Hence K is the non trivial extension in $\mathrm{Ext}^1(p_2{}^*\Omega_{\mathbb{P}^1}, I_{Q|Q})$, and $\mathbb{P}(p_2{}^*\Omega_{\mathbb{P}^1}) \subset \mathbb{P}(K)$ in a natural way.

SECTION 2: SOME LEMMAS.

2.1 Lemma. Let S be a smooth variety, A an invertible sheaf on S, E a vector bundle on S, q: $\mathbb{P}(E) \to S$ the canonical projection, Z a closed subscheme of S, and let $T := Z \times_S \mathbb{P}(E)$. Then:

$q_*((O_{\mathbb{P}(E)}(1) \otimes q^*A)|_T) \cong (E \otimes A)|_Z$, and $q_*(O_{\mathbb{P}(E)}(1) \otimes q^*A \otimes I_T) \cong E \otimes A \otimes I_Z$.

Proof. (the proof is the same as [I], 2.10, where the lemma is proved for $S=\mathbb{P}^3$, $E=\Omega_3$, $A=O_{\mathbb{P}^3}(k)$).

Using projection formula we see that it is enough to prove $q_*(O_{\mathbb{P}(E)}(1)|_T) \cong E|_Z$, and $q_*(O_{\mathbb{P}(E)}(1) \otimes I_T) \cong E \otimes I_Z$. Moreover, the second statement follows from the first, since the exact sequence: $0 \to O_{\mathbb{P}(E)}(1) \otimes I_T \to O_{\mathbb{P}(E)}(1) \xrightarrow{\rho} O_{\mathbb{P}(E)}(1)|_T \to 0$, together with the first statement, give the exact sequence

$0 \to q_*(O_{\mathbb{P}(E)}(1) \otimes I_T) \to E \xrightarrow{q_*\rho} E|_Z \to \ldots$, where $q_*\rho$ is the restriction map, hence $q_*(O_{\mathbb{P}(E)}(1) \otimes I_T) \cong E \otimes I_Z$.

So we now prove $q_*(O_{\mathbb{P}(E)}(1)|_T) \cong E|_Z$. Denote by $i: Z \hookrightarrow S$ the closed embedding. By [EGA I] 9.7.9 the exact sequence $E \to E|_Z \to 0$ gives a closed embedding $j: \mathbb{P}(E|_Z) \hookrightarrow \mathbb{P}(E)$; by [EGA I] 9.7.6 we have $\mathbb{P}(E|_Z) \cong Z \times_S \mathbb{P}(E)$ canonically with j the second projection:

$$\begin{array}{ccc} \mathbb{P}(E|_Z) & \stackrel{j}{\hookrightarrow} & \mathbb{P}(E) \\ \phi\downarrow & & \downarrow q \\ Z & \stackrel{i}{\hookrightarrow} & S \end{array}$$

and still by [EGA I] 9.7.6, $j^*(\mathcal{O}_{\mathbb{P}(E)}(1)) \cong \mathcal{O}_{\mathbb{P}(i^*E)}(1)$ (that is, $\mathcal{O}_{\mathbb{P}(E)}(1)|_T \cong$ $\cong \mathcal{O}_{\mathbb{P}(E|_Z)}(1)$).

Hence $q_*(j_*(\mathcal{O}_{\mathbb{P}(E)}(1)|_T)) \cong i_*(\phi_*(\mathcal{O}_{\mathbb{P}(E)}(1)|_T)) \cong i_*(\phi_*(\mathcal{O}_{\mathbb{P}(E|_Z)}(1))) = i_*(E|_Z)$.

2.2 Remark. Let S, Z, E, A, T be as in lemma 2.1. Then, $H^0(E\otimes A\otimes \mathcal{I}_Z) = 0$ if and only if $H^0(\mathcal{O}_{\mathbb{P}(E)}(1)\otimes q^*A\otimes \mathcal{I}_T) = 0$; so again by 2.1, Z is $E\otimes A$-settled if and only if T is $\mathcal{O}_{\mathbb{P}(E)}(1)\otimes q^*A$-settled.

2.3 Lemma. Let V be a smooth variety, $W \subset V$ a smooth subvariety and $f: Y \to V$ the blowing- up of V along W. Let Z be a closed subscheme of V and let $T := Z \times_V Y$. If Z has nor proper neither embedded components contained in W, then $f_*\mathcal{I}_T = \mathcal{I}_Z$.

Proof. Since V is normal and integral, and Y is integral, $\mathcal{O}_V = f_*\mathcal{O}_Y$.

The natural map $\Phi: \mathcal{O}_Z \to f_*\mathcal{O}_T$ induces an injective map $\vartheta: \mathcal{I}_Z \to f_*\mathcal{I}_T$ which makes the following diagram commute:

$$\begin{array}{ccccccccc} 0 \to & f_*\mathcal{I}_T & \to & f_*\mathcal{O}_Y & \to & f_*\mathcal{O}_T & \to & R^1f_*\mathcal{I}_T & \to \dots \\ & \uparrow\vartheta & & \| & & \uparrow\Phi & & & \\ 0 \to & \mathcal{I}_Z & \to & \mathcal{O}_V & \to & \mathcal{O}_Z & \to & 0 & \end{array}$$

If Φ is injective, then ϑ is surjective, and thesis follows. Now set $E := \ker \Phi$; then SuppE is a closed subscheme of Z contained in $Z \cap W$, and the assumption on the components of Z says that $E=0$.

2.4 Lemma. Let V be a smooth variety, $W \subset V$ a smooth subvariety and $f: Y \to V$ the blowing-up of V along W, with exceptional divisor D; set $\lambda := f|_D : D \to W$.

Let Z be a closed subscheme of W and let $T := Z \times_V Y$. Then $\lambda_*\mathcal{I}_{\lambda^{-1}(Z),D} = \mathcal{I}_{Z,W}$, and $f_*\mathcal{I}_T = \mathcal{I}_Z$.

Proof. By [A-H] prop.2.3.3, $\lambda_* O_T = f_* O_T = O_Z$. Moreover, $\lambda_* O_D = O_W$, and $f_* O_Y = O_V$ (since in particular V and W are normal and integral, and Y and D are integral); hence the exact sequence: $0 \to I_T \to O_Y \xrightarrow{\rho} O_T \to 0$ gives $0 \to f_* I_T \to O_V \xrightarrow{f_*\rho} O_Z \to \ldots$, where $f_*\rho$ is the restriction map, hence $f_* I_T = I_Z$, and analougously for the other equality.

2.5 Lemma. Let V be a smooth variety, $W \subset V$ a smooth subvariety and f: $Y \to V$ the blowing- up of V along W. Let Z be a closed subscheme of V and let $T := Z \times_V Y$. If Z is reduced and can be written as union of two closed reduced subschemes K and J, such that K is contained in W, and none of the components of J is contained in W, then $f_* I_T = I_Z$.

Proof. Exactly as in proof of lemma 2.3, we have a commutative diagram:

$$\begin{array}{ccccc} 0 \to & f_* I_T & \to & f_* O_Y & \to \\ & \uparrow \vartheta & & \| & \\ 0 \to & I_Z & \to & O_V & \to \\ & \uparrow & & & \\ & 0 & & & \end{array}$$

Hence $f_* I_T$ is an ideal sheaf of O_V, defining a closed subscheme R; moreover, the commutativity of this diagram gives $R \subset Z$.

On any open subset U (resp.U') not meeting K (resp. J), $f_* I_T|_U = I_J|_U$ by lemma 2.3 (resp. $f_* I_T|_{U'} = I_K|_{U'}$ by lemma 2.4); hence, Z being reduced, $Z \subset R$, and so we conclude Z=R, that is, $I_Z = I_R = f_* I_T$.

2.6 Corollary. Let S be a smooth variety, E a vector bundle on S, and F a quotient sheaf of E, with F vector bundle on a smooth divisor of S, and rank(F) < rank(E); let q: $\mathbb{P}(E) \to S$ be the canonical projection, and f: $Y \to \mathbb{P}(E)$ be the blowing-up of $\mathbb{P}(E)$ along $\mathbb{P}(F)$.

Let Z be a closed subscheme of $\mathbb{P}(E)$, disjoint union of four closed subschemes M, R, P, and $q^{-1}(N)$, where: $M \subset \mathbb{P}(F)$; $N \subset S$; $P \subset \mathbb{P}(E)$, and P does not meet $\mathbb{P}(F)$; R is reduced and can be written as union of two closed reduced subschemes K and J, such that K is contained in $\mathbb{P}(F)$, and none of the components of J is contained in $\mathbb{P}(F)$.

Then, $f_* I_{f^{-1}(Z)} = I_Z$.

Proof. The question is local on $\mathbb{P}(E)$, hence we conclude by 2.3, 2.4, and 2.5 ($q^{-1}(N)$ satisfies the assumptions of lemma 2.3, since rank F < rank E).

SECTION 3: REDUCTION TO THE STATEMENTS R(K).

Let C be a closed subscheme of $\mathbb{P}^3$. Tensoring the Euler sequence in $\mathbb{P}^3$ by $I_C(k)$, and taking cohomology, we have:

$$0 \to H^0(I_C \otimes \Omega_3(k+1)) \to H^0(I_C(k)) \otimes H^0(O_{\mathbb{P}^3}(1)) \xrightarrow{\sigma_k} H^0(I_C(k+1)) \to$$
$$\to H^1(I_C \otimes \Omega_3(k+1)) \to \dots$$

hence $\ker(\sigma_k) = H^0(I_C \otimes \Omega_3(k+1))$; in the following we shall therefore study these vector spaces for particular choices of C.

3.1 Definition. For $d \geq 1$, we set $v(d) := \min\{k \geq 1 \mid \binom{k+3}{3} - (dk+1) \geq 0\}$

and $w(d) := \min\{k \geq 1 \mid \frac{3k}{k+1} \cdot \binom{k+3}{3} - 3 \geq d(3k-1)\}$.

3.2 Remark. Let C be the generic rational curve of degree d in $\mathbb{P}^3$; then, $v(C) := v(d)$ is the critical value for the postulation (see [B-E]), and $w(C) := w(d)$ is the critical value for the generation, that is, the least integer k for which $h^0(I_C(k)) \neq 0$ and the map σ_k can be surjective, since

$w(d) = \min\{k \mid 4\, h^0(I_C(k)) \geq h^0(I_C(k+1)) > 0\}$.

In fact, since C is of maximal rank and not special, we have, if $h^0(I_C(k)) > 0$:

$$4h^0(I_C(k)) - h^0(I_C(k+1)) = 4\binom{k+3}{3} - 4(dk+1) - \binom{k+4}{3} + (d(k+1)+1) =$$
$$= \frac{3k}{k+1} \cdot \binom{k+3}{3} - 3 - d(3k-1).$$

Hence, if k is such that $4h^0(I_C(k)) \geq h^0(I_C(k+1)) > 0$, then k satisfies $\frac{3k}{k+1} \cdot \binom{k+3}{3} - 3 \geq d(3k-1)$.

Viceversa, if k is an integer such that $\frac{3k}{k+1} \cdot \binom{k+3}{3} - 3 \geq d(3k-1)$, then

$\binom{k+3}{3} \geq \frac{k+1}{3k}(d(3k-1)+3) > dk+1$, the last inequality being true as soon as $k \geq 1$.

The curve C being of maximal rank, this implies $h^0(I_C(k)) > 0$ and hence k is such that $4\, h^0(I_C(k)) \geq h^0(I_C(k+1)) > 0$.

3.3 Definition. For every $k \geq 1$, let $t(k)$, $s(k)$ be the integers such that:

$$\frac{3k}{k+1} \cdot \binom{k+3}{3} - 3 = t(k)(3k-1) + s(k), \text{ with } 0 \leq s(k) \leq 3k-2.$$

3.4 Remark. Setting $k = 18n+j$, $0 \leq j \leq 17$, definition 3.3 gives the following values for $t(k)$, $s(k)$, for any $k \geq 1$:

$k = 18n$	$t(k) = 54n^2 + 16n + 1$	$s(k) = 16n-2$	
$k = 18n+1$, $n>0$	$t(k) = 54n^2 + 22n + 2$	$s(k) = 19n-1$;	$t(1)=1$, $s(1)=1$;
$k = 18n+2$	$t(k) = 54n^2 + 28n + 3$	$s(k) = 40n+2$	
$k = 18n+3$	$t(k) = 54n^2 + 34n + 5$	$s(k) = 25n+2$	
$k = 18n+4$	$t(k) = 54n^2 + 40n + 7$	$s(k) = 28n+4$	
$k = 18n+5$	$t(k) = 54n^2 + 46n + 9$	$s(k) = 49n+11$	
$k = 18n+6$	$t(k) = 54n^2 + 52n + 12$	$s(k) = 34n+9$	
$k = 18n+7$	$t(k) = 54n^2 + 58n + 15$	$s(k) = 37n+12$	
$k = 18n+8$	$t(k) = 54n^2 + 64n + 19$	$s(k) = 4n$	
$k = 18n+9$	$t(k) = 54n^2 + 70n + 22$	$s(k) = 43n+19$	
$k = 18n+10$	$t(k) = 54n^2 + 76n + 26$	$s(k) = 46n+23$	
$k = 18n+11$	$t(k) = 54n^2 + 82n + 31$	$s(k) = 13n+6$	
$k = 18n+12$	$t(k) = 54n^2 + 88n + 35$	$s(k) = 52n+32$	
$k = 18n+13$, $n>0$	$t(k) = 54n^2 + 94n + 41$	$s(k) = n-1$;	$t(13)=40$, $s(13)=37$;
$k = 18n+14$	$t(k) = 54n^2 + 100n + 46$	$s(k) = 22n+15$	
$k = 18n+15$	$t(k) = 54n^2 + 106n + 52$	$s(k) = 7n+4$	
$k = 18n+16$	$t(k) = 54n^2 + 112n + 58$	$s(k) = 10n+7$	
$k = 18n+17$	$t(k) = 54n^2 + 118n + 64$	$s(k) = 31n+27$	

3.5 Remark. For every d such that $t(k-1) < d \leq t(k)$, we have $w(d) = k$.

3.6 Definition. For each $k \geq 1$, with Y_k^+ we denote the union in $\mathbb{P}^3$ of the generic rational curve C_d of degree d with m lines $r_1, \ldots, r_m$ such that

i) $d+m = t(k)$;

ii) $m \geq t(k)-t(k-2)$ (we set $t(-1)=t(0)=0$);

iii) it is possible to give an order to the m lines so that r_i meets (transversally) $C_d \cup r_1 \cup \ldots \cup r_{i-1}$ exactly at one smooth point, for $i=1,\ldots,m$.

Hence $\deg(Y_k^+)=t(k)$, $p_a(Y_k^+)=0$, and Y_k^+ is a specialization of the generic rational curve of degree $t(k)$.

For each $k \geq 1$, with $\mathbf{Y_k}$ we denote a scheme generic union in X of $\pi^{-1}(Y_k^+)$ with n(k) t-points, δ(k) d-points and ε(k) s-points lying on a line r of $\mathbb{P}^3$, such that $3n(k) + 2\delta(k) + \varepsilon(k) = s(k)$, and with r meeting the curve Y_k^+ at one smooth point. For $k\geq 19$ let n(k) be the integral part of s(k)/3, and $0\leq \delta(k) + \varepsilon(k) \leq 1$; for $k\leq 18$ the values of the functions n(k), δ(k), ε(k) are given in paragraphs 4.5 (k odd) and 4.6 (k even).

In the following, if $k \geq 1$, "**R(k)**" denotes the statement "there exists a scheme Y_k such that $H^0(L_k \otimes I_{Y_k})=0$".

3.7 Lemma. For $k \geq 1$ the statement R(k) is equivalent to "there exists a scheme Y_k which is L_k-settled".

Proof. If C is a smooth rational curve of degree d, $c_1(\Omega_3(k+1)|_C) = d(3k-1)$; $\Omega_3(k+1)$ being generated by global section for $k \geq 2$, the same is true for $\Omega_3(k+1)|_C$, hence if $k \geq 2$, $h^0(\Omega_3(k+1)|_C) = d(3k-1)+3$.

Let C be any curve and R a line meeting C only transversally in a point P; then, the cohomology of the exact sequence

$0 \to O_{C\cup R}\otimes\Omega_3(k+1) \to (O_C \oplus O_R) \otimes \Omega_3(k+1) \to^{\rho} O_P\otimes\Omega_3(k+1) \to 0$ gives, if $h^0\rho$ is onto, $h^0(\Omega_3(k+1)|_{C\cup R}) = h^0(\Omega_3(k+1)|_C) + h^0(\Omega_3(k+1)|_R) -3$; but $h^0\rho$ is onto since $h^0\rho'$ is onto, where

$0 \to O_R(-1)\otimes\Omega_3(k+1) \to O_R \otimes \Omega_3(k+1) \to^{\rho'} O_P\otimes\Omega_3(k+1) \to 0$

$(h^1(O_R(-1)\otimes\Omega_3(k+1)) = h^1(O_R(k-2)\oplus O_R(k-1)^{\oplus 2})= 0$ since $k>0)$.

Recall that for $k\geq 3$ Y_k^+ is the union of a rational curve C_d of degree d union with $m \geq t(k)-t(k-2)$ lines r_i; if k=1, Y_k^+ is a line, and if k=2, Y_k^+ is union of three lines. We hence conclude that, if $k \geq 1$, $h^0(L_k|_{\pi^{-1}(Y_k^+)}) = h^0(\pi_*(L_k|_{\pi^{-1}(Y_k^+)})) =$ $=h^0(\Omega_3(k+1)|_{C_d\cup r_1\cup\ldots\cup r_m})=h^0(\Omega_3(k+1)|_{C_d})+m(h^0(\Omega_3(k+1)|_{r_i}-3)= t(k)(3k-1)+3$ (in the 2° equality use lemma 2.1).

Now look at the exact sequence $0 \to H^0(L_k\otimes I_{Y_k}) \to H^0(L_k) \to H^0(L_k|_{Y_k}) \to\ldots$ and notice that $h^0(L_k)=h^0(L_k|_{Y_k})$, since $h^0(L_k) = h^0(\pi_*L_k) = h^0(\Omega_3(k+1)) =$ $= \frac{3k}{k+1}\cdot\binom{k+3}{3}$, and $h^0(L_k|_{Y_k}) = h^0(L_k|_{\pi^{-1}(Y_k^+)}) + s(k) = t(k)(3k-1)+s(k)+3$.

3.8 Proposition. Let C be the generic rational curve of degree d; then, $v(C) \le w(C) \le v(C) + 1$. Moreover, all the σ_k are of maximal rank if and only if $\sigma_{v(C)}$ is injective when $v(C) = w(C) - 1$, surjective when $v(C) = w(C)$.

Proof. The same proof as in [I], 2.6.

3.9 Proposition. Let q, k be positive integer such that $t(k-1)<q\le t(k)$, and let C_q be the generic rational curve of degree q. Then if R(k-1) is true, σ_{k-1} is injective for C_q; if R(k) is true, σ_k is surjective for C_q.

Proof. Almost the same proof as in [I], 2.8; just be careful that:

- when proving injectivity, we add to our scheme Y_{k-1} the inverse image through π of $(q-t(k-1))$ lines $s_1,\dots, s_{q-t(k-1)}$ of $\mathbb{P}^3$, so that: $s_1 = r$, and s_i meets $Y_{k-1} \cup s_1 \cup \dots \cup s_{i-1}$ transversally exactly at one smooth point for $i=1,\dots,q-t(k-1)$ (this is done in order to have a curve specialization of C_q);
- when proving surjectivity, we have to construct a curve $C^\sim$, specialization of C_q, such that $h^0(I_{C^\sim}\otimes\Omega_3(k+1)) \le (3k-1)(t(k)-q)+s(k)$. We proceed as in [I], 2.8 taking away from Y_k all the points and the inverse image of $t(k)-q$ lines; this can be done; in fact, recall that $Y_k = \pi^{-1}(Y_k^+) \cup$ points, and $Y_k^+ = C_d \cup r_1 \cup \dots \cup r_m$; moreover, $t(k)-q<t(k) - t(k-1)< m$.

Taking only points away from Y_k we get $h^0(I_{Y_k^+}\otimes\Omega_3(k+1)) \le s(k)$; now if we take the line r_m away from Y_k^+, we are left with a curve C s.t. $h^0(I_C\otimes\Omega_3(k+1)) \le$ $\le s(k)+(3k-1)$ (from the exact sequence $0 \to I_{Y_k^+}\otimes\Omega_3(k+1) \to I_C\otimes\Omega_3(k+1) \to$ $\to I_{C,Y_k^+}\otimes\Omega_3(k+1) \to 0$, since $I_{C,Y_k^+} \cong O_{r_m}(-1)$), and repeating the procedure conclusion follows.

3.10 Corollary. If R(k) is true for every $1 \le k \le 18$, $k \ne 3$, theorem 1 is proved for the generic rational curve of degree $d \le 73$, $d \ne 5$.

Proof. If $d \ne 1,4,5,6,7$, $d \le 71$, there exists k such that $t(k-1) < d \le t(k)$ with $1 \le k \le 18$, and R(k-1), R(k), are true, hence by 3.5, 3.8 and 3.9 we conclude. If d= 72,73, $v(d) = w(d)-1=18$, hence we conclude again by 3.8 and 3.9. Now denote by C_4,C_6,C_7, the generic rational quartic, sextic, septic. We have:

$v(C_4)=2$, $h^0(I_{C_4}(2))=1$, $h^0(I_{C_4}(3))=7$; $w(C_4)=3$

$v(C_6)=3$, $h^0(I_{C_6}(3))=1$, $h^0(I_{C_6}(4))=10$; $w(C_6)=4$

$v(C_7)=4$, $h^0(I_{C_7}(4))=6$, $h^0(I_{C_7}(5))=20$; $w(C_7)=4$.

Hence σ_v is of maximal rank in the three cases, since it is trivially injective for C_4 and C_6, and surjective for C_7 since R(4) is true (see 3.9). We conclude by 3.8.

SECTION 4: PROOF OF R(k) FOR $k \leq 18$, $k \neq 3$.

4.0 La méthode d'Horace. We recall briefly la méthode d'Horace (see [H,2]), which is used in each proof of this section and of sections 5 and 6:
Let U be a quasiprojective variety, D a Cartier divisor on U, L a locally free coherent sheaf on U, and Z a specialization of a subscheme Y of U; then the following sequence is exact:

$$0 \to L\otimes I_{res_D Z}(-D) \to L\otimes I_Z \to L\otimes I_{Z\cap D,D} \to 0.$$

Hence, in order to prove $H^0(L\otimes I_Y) = 0$, it is enough to prove the following claim on the residue (r-claim for short): $H^0(L\otimes I_{res_D Z}(-D)) = 0$, and the following claim on the intersection (i-claim): $H^0(L\otimes I_{Z\cap D,D}) = 0$.
Following the notations of [H,2], we say we have exploited D.

We use this in various contexts; the principal ones are:

a) $U=X$, $L=L_k$, $D = \pi^{-1}(Q)$.

b) If E is a rank 3 vector bundle on a projective variety S, T a Cartier divisor of S, E' is a rank 1 or 2 vector bundle supported on T, and E' is a quotient of E, then we can apply la méthode d'Horace to the following situation: U= the blow up of $\mathbb{P}(E)$ along $\mathbb{P}(E')$, L=the pull-back of $\mathcal{O}_{\mathbb{P}(E)}(1)$, D = the exceptional divisor.

We use this applied to the quotient sheaves given by the two elementary transformations (e1) and (e2) of section 1, namely:

- $S= \mathbb{P}^3$, $T = Q$, $E = \Omega_3(k+1)$, and $E'= \mathcal{O}_Q(k-1,k+1)$;
- $S= \mathbb{P}^3$, $T = Q$, $E = F(k+1)$, and $E'= \mathcal{O}_Q(k,k-1)^{\oplus 2}$.

4.1 Sketch of the proof. Our aim is to prove R(k) for $1\leq k \leq 18$, $k \neq 3$. We make the following conventions (for the notation R(A,n) see preliminaries vi)):
"**R(T,k-2) $\Rightarrow$ R(B,k), proof of type a**" means we are using la méthode d'Horace in context a. We have an assumption R(T,k-2) where T is specified by the context, and we build a scheme $Z \subset X$ such that Z is a specialization of B and such

that, setting $r(Z):=res_{\pi^{-1}(Q)}Z$ and $i(Z):=Z\cap\pi^{-1}Q$, r(Z) is a generalization of T in X, and i(Z) verifies the i-claim
$H^0(L_{k|\pi^{-1}(Q)} \otimes I_{i(Z),\pi^{-1}(Q)}) = H^0(F_{k+1,k+1} \otimes I_{i(Z),\pi^{-1}(Q)}) = 0$. So we conclude that R(B,k) is true.

"**R(T,k-2) ⇒ R(B,k), proof of type b**" means we are using la méthode d'Horace in context b, that is, in the context of the elementary transformations, and we need two steps. In this case we are working with (prop.1.1):

$$(e1)\otimes O_{\mathbb{P}^3}(k+1) \qquad 0 \to F(k+1) \to \Omega_3(k+1) \xrightarrow{\alpha} O_Q(k-1,k+1) \to 0$$

$$(e2)\otimes O_{\mathbb{P}^3}(k+1) \qquad 0 \to \Omega_3(k-1) \to F(k+1) \xrightarrow{\beta} O_Q(k,k-1)^{\oplus 2} \to 0;$$

we set $V:= \mathbb{P}(\Omega_3(k+1))$, $\pi: V \to \mathbb{P}^3$ the canonical projection;
$V':=\mathbb{P}(F(k+1))$, $p: V' \to \mathbb{P}^3$ the canonical projection;
$W:= \mathbb{P}(O_Q(k-1,k+1))$;
$W':= \mathbb{P}(O_Q(k,k-1)^{\oplus 2})$;
$f: V^{\sim} \to V$ the blow up of V along W with exceptional divisor G.

We recall that there is a birational morphism $g: V^{\sim} \to V'$ such that g is the blow up of V' along W', and, denoting by G' its exceptional divisor, the following hold (see [M],§1):
$g^*O_{V'}(1) \cong f^*O_V(1)\otimes O_{V^{\sim}}(-G)$, and
$f^*(O_V(1)\otimes\pi^*O_{\mathbb{P}^3}(-2)) \cong g^*O_{V'}(1)\otimes O_{V^{\sim}}(-G')$.
In [M] ,§1, and in [I], first part of 1.9.3, the reader can find a geometric description of this situation .

The proof goes on like that:
we have the assumption $H^0(L_{k-2}\otimes I_T) = 0$ with T given subscheme of X;

1st step: We exploit G'; we work with $V^{\sim} \xrightarrow{g} V' \xrightarrow{p} \mathbb{P}^3$, that is, with $(e2)\otimes O_{\mathbb{P}^3}(k+1)$.

We exhibit a scheme $K'\subset V'$ such that, setting $r(K'):= res_{G'}(g^{-1}(K'))$, $i(K'):= g^{-1}(K')\cap G'$, then

(i) $f_*I_{r(K')} = I_{T^g}$, where T^g is a generalization of T (in X, or, which is the same, in V, since $V \cong X$ canonically). This condition is realized building K' such that $r(K') = f^{-1}(T^g)$, with T^g satisfying the assumptions of corollary 2.6.
(ii) $K' \cap W' = (p_{|W'})^{-1}(P')$, with $P' \subset \mathbb{P}^3$ satisfying $H^0(O_Q(k,k-1)^{\oplus 2} \otimes I_{P'}) = 0$; lemma 2.1 hence gives $H^0(O_{W'}(1) \otimes I_{K' \cap W'}) = 0$.
(iii) K' satisfies assumptions of cor.2.6, hence $g_*I_{g^{-1}(K')} \cong I_{K'}$.

We consider the residual sequence:
$$0 \to g^*O_{V'}(1)(-G') \otimes I_{r(K')} \to g^*O_{V'}(1) \otimes I_{g^{-1}(K')} \to g^*O_{V'}(1)_{|G'} \otimes I_{i(K')} \to 0.$$

The <u>r-claim</u> is proved in the following way:
$$H^0(g^*O_{V'}(1)(-G') \otimes I_{r(K')}) \cong H^0(f^*(O_V(1) \otimes \pi^*O_{\mathbb{P}^3}(-2)) \otimes I_{r(K')}) \cong$$
$$\cong^{(\text{use } f_*)} H^0(O_V(1) \otimes \pi^*O_{\mathbb{P}^3}(-2) \otimes f_*I_{r(K')})) \cong H^0(L_{k-2} \otimes I_{T^g,X}) = 0$$

The <u>i-claim</u> is proved in the following way:
set $\lambda := g_{|G'}: G' \to W'$, and consider the commutative diagram

$$\begin{array}{ccc} G' & \overset{j}{\hookrightarrow} & V^{\sim} \\ \lambda \downarrow & & \downarrow g \\ W' & \overset{i}{\hookrightarrow} & V' \end{array}$$

We have $g^*O_{V'}(1)_{|G'} \cong \lambda^* i^* O_{V'}(1) \cong \lambda^* O_{W'}(1)$.

Moreover, observe that $i(K') = \lambda^{-1}(K' \cap W')$, hence lemma 2.4 gives $\lambda_* I_{i(K'),G} \cong I_{K' \cap W',W'}$. So we conclude:
$$H^0(g^*O_{V'}(1)_{|G'} \otimes I_{i(K')}) \cong H^0(\lambda^* O_{W'}(1) \otimes I_{i(K')}) \cong^{(\text{use } \lambda_*)} H^0(O_{W'}(1) \otimes I_{K' \cap W'}) = 0.$$

The i-claim and the r-claim together give $H^0(g^*O_{V'}(1) \otimes I_{g^{-1}(K')}) = 0$.
Using g_* and (iii) we get:
$$0 = H^0(O_{V'}(1) \otimes g_*I_{g^{-1}(K')}) \cong H^0(O_{V'}(1) \otimes I_{K'}).$$

2nd step: We exploit G; we work with $V^{\sim} \xrightarrow{f} V \xrightarrow{\pi} \mathbb{P}^3$, that is, with $(e1) \otimes O_{\mathbb{P}^3}(k+1)$.

We exhibit a scheme $K \subset V$ (or,which is the same, $K \subset X$, since $V \cong X$ canonically) such that, setting $r(K) := res_G(f^{-1}(K))$, $i(K) := f^{-1}(K) \cap G$, then

(i) K is a specialization of B;

(ii) $f_*I_{f^{-1}(K)} = I_K$ (this condition is realized building K so that cor. 2.6 applies to K).

(iii) $K \cap W = (\pi_{|W})^{-1}(P)$, with $P \subset \mathbb{P}^3$ satisfying $H^0(O_Q(k-1,k+1) \otimes I_P) = 0$; lemma 2.1 hence gives $H^0(O_W(1) \otimes I_{K \cap W}) = 0$.

(iv) there exists a generalization K'^g of K' in V' such that $r(K) = g^{-1}(K'^g)$, and such that K'^g satisfies the assumptions of cor. 2.6; hence, we have $g_*I_{r(K)} \cong I_{K'g}$.

We consider the residual sequence:

$0 \to f^*O_V(1)(-G) \otimes I_{r(K)} \to f^*O_V(1) \otimes I_{f^{-1}(K)} \to f^*O_V(1)_{|G} \otimes I_{i(K)} \to 0.$

The r-claim is proved in the following way:

we have $H^0(f^*O_V(1)(-G) \otimes I_{r(K)}) \cong H^0(g^*O_{V'}(1) \otimes I_{r(K)})$;

using g_* and (iv) we get $H^0(g^*O_{V'}(1) \otimes I_{r(K)}) \cong H^0(O_{V'}(1) \otimes g_*I_{r(K)}) \cong$ $\cong H^0(O_{V'}(1) \otimes I_{K'g})$; the last cohomology group is 0 by step 1 using semicontinuity, so we are done.

The i-claim $H^0(f^*O_V(1)_{|G} \otimes I_{i(K)}) = 0$ is proved as in step 1.

So we conclude:

$$H^0(f^*O_V(1) \otimes I_{f^{-1}(K)}) \cong^{\text{(use } f_*)} H^0(O_V(1) \otimes f_*I_{f^{-1}(K)}) \cong H^0(O_V(1) \otimes I_K) = 0,$$

and by semicontinuity R(B,k) holds.

There is a variant to this proof of type b, namely:

in step 1 we work with a scheme $K^\sim \subset V^\sim$ which takes the place of $g^{-1}(K')$; condition (i) is the same, condition (ii) becames: $K^\sim \cap G'$ contains $g^{-1}(p_{|W'})^{-1}(P')$, with $P' \subset \mathbb{P}^3$ satisfying $H^0(O_Q(k,k-1)^{\oplus 2} \otimes I_{P'}) = 0$, and condition (iii) disappears; the conclusion of step 1 is $H^0(g^*O_{V'}(1) \otimes I_{K^\sim}) = 0$.

In step 2 we give the scheme K as above, except that condition (iv) becames: there is a flat family in $V^\sim$ with generic fiber r(K) and special fiber $K^\sim$; the first step hence gives by semicontinuity $H^0(g^*O_{V'}(1) \otimes I_{r(K)}) = 0$, and we conclude as above.

4.2 In order to describe the flat families in $V^\sim$ and their special fibers $K^\sim$ used in the variant of type b proofs, we need some lemmas and definitions:

4.2.1 Lemma. We denote by m a line in Q, by O a point in m, and by B_0 an s-point in $\pi^{-1}(O)\subset V$, $B_0 \not\in W$. We set $n := g^{-1}(p^{-1}(m)\cap W')$.

Then:

a) There is a flat family $\mathcal{F}$ in V~ with generic fiber $g^{-1}(p^{-1}(P))$, where P is a point in $\mathbb{P}^3$ not in Q, and special fiber $g^{-1}p^{-1}(O)$.

b) There is a flat family $\mathcal{G}$ in V~ such that:

the generic fiber is the union of n with $g^{-1}(p^{-1}(P))$, where P is a point in $\mathbb{P}^3$ not in Q; the special fiber $\mathcal{G}_0$ is the union of n with $g^{-1}p^{-1}(O)$ and with a nilpotent $\alpha(O)$ supported on $g^{-1}(p^{-1}(O)\cap W')$; $res_{G'}(\mathcal{G}_0)$ is $g^{-1}p^{-1}(O)$.

c)There is a flat family $\mathcal{L}$ in V~ such that:

the generic fiber is the union of n with $g^{-1}(B)$, where B is an s-point in V', lying out of Q; the special fiber $\mathcal{L}_0$ is the union of n with a nilpotent $\beta(B_0)$, supported on the point $f^{-1}(B_0)$; $res_{G'}(\mathcal{L}_0) = f^{-1}(B_0)$.

Proof.

a) It is possible to take local coordinates x,y,z on $\mathbb{P}^3$ so that the point O becomes the origin; so that locally on $\mathbb{P}^3$, V' is $Proj(k[x,y,z])[t_0:t_1:t_2]$, and restricting ourselves to the open subet $U=\{t_0\neq 0\}$, and setting $u:=t_1/t_0$, $v:=t_2/t_0$, V' is $Spec(k[x,y,z,u,v])$; and finally so that the quadric Q has equation z=0, and $I_{W'}= (z,u)$. Hence the blow up V~ of V' along W' is locally $Proj((k[x,y,z,u,v])[x_0:x_1] / (x_0u - x_1z))$ and on $\{x_1\neq 0\}$, setting $t:=x_0/x_1$, we have that the blow up g: V~ $\rightarrow$ V' corresponds to the morphism:

δ: $k[x,y,z,u,v] \rightarrow k[x,y,z,u,v,t]/(tu - z)$, with $\delta(x)=x$, $\delta(y)=y$, $\delta(z)=tu$, $\delta(u)=u$, $\delta(v)=v$; hence $I_{G'}=(u)$. Now take in $Spec(k[x,y,z,u,v,t]/(tu - z)) \times Spec(k[\lambda])$ the family $\mathcal{F}$ (with base space $Spec(k[\lambda])$ corresponding to the ideal $I= (x,y,z-\lambda)$; the total space of the family is reduced and irreducible, since this ideal is prime. Hence its closure in V~ gives a flat family by [Ha], prop.III.9.7.

b) and c) Let $\mathcal{S}$ be a flat family with generic fiber $g^{-1}(B)$, where B is an s-point in V' lying out of Q, and special fiber a point $P_0=f^{-1}(B_0)$, where B_0 is an s-point in $\pi^{-1}(O)\subset V$, $B_0\not\in W$, that is, $P_0\in G'$, but $P_0\not\in G'\cap G$; for example, in the previous choice of coordinates, choose $I_{\mathcal{S}}=(x,y,z-\lambda,t+1,v)$.

Let $\mathcal{H}$ be the constant family $n \times Spec(k[\lambda]) \rightarrow Spec(k[\lambda])$; lemma 4.2.2 applied to $\mathcal{H}$ and to the family constructed in a) (respectively $\mathcal{S}$) gives a family as wanted in b) (resp.c)); the assertions on the special fibers are easily proved taking local coordinates (in each cart of an affine covering of V~) and computing.

4.2.2 Lemma. Let Y be a smooth projective variety, $Z:=Y\times \mathrm{Spec}(k[\lambda])$, and let $\mathcal{F}\to \mathrm{Spec}(k[\lambda])$, $\mathcal{H}\to \mathrm{Spec}(k[\lambda])$ be two flat families in Z. Then, the schematic union of $\mathcal{F}$ and $\mathcal{H}$ in Z is flat over $\mathrm{Spec}(k[\lambda])$.

Proof. The associated points of $\mathcal{F}\cup\mathcal{H}$ are associated points of $\mathcal{F}$ or of $\mathcal{H}$; in fact, let η be an associated point of $\mathcal{F}\cup\mathcal{H}$ and let U be an open affine subset of Z containing η; then, if $\mathfrak{A}$ is the ideal of $\mathcal{F}$ in U with minimal primary decomposition $\mathfrak{A} = \mathfrak{P}_1\cap \dots \cap \mathfrak{P}_n$, and $\mathfrak{B}$ is the ideal of $\mathcal{H}$ in U with minimal primary decomposition $\mathfrak{B}= \mathfrak{q}_1\cap \dots \cap\mathfrak{q}_t$, the ideal of $\mathcal{F}\cup\mathcal{H}$ in U is $\mathfrak{P}_1\cap\dots\cap\mathfrak{P}_n\cap\mathfrak{q}_1\cap\dots$ $\dots \cap\mathfrak{q}_t$, and this is a primary decomposition (eventually not minimal); if $\mathfrak{P}$ is the associated prime corresponding to η, it is hence the radical of some $\mathfrak{P}_i$ or $\mathfrak{q}_j$. Denoting by i: $\mathcal{F}\cup\mathcal{H}\to$ Z the natural immersion, and with p: $Y\mathrm{x}\mathrm{Spec}(k[\lambda]) \to$ $\to \mathrm{Spec}(k[\lambda])$ the canonical projection, we conclude that $p\circ i$: $\mathcal{F}\cup\mathcal{H}\to \mathrm{Spec}(k[\lambda])$ is flat by [Ha], III 9.7.

4.2.3 Lemma. Let L, resp. M be a line of type (1,0), resp.(0,1), on Q, and let $x=L\cap M$; the subscheme $T:=\mathrm{res}_G(f^{-1}\pi^{-1}(L\cup M\cup\chi(x)))$ of $V^\sim$, which is the union of $g^{-1}(p^{-1}(L\cup M)\cap W')$ with a nilpotent which will be denoted by $\nu(x)$, and whoose support is contained in $g^{-1}p^{-1}(x)$, verifies: $\mathrm{res}_{G'}(T)=f^{-1}\pi^{-1}(x)$.

Proof. For the first statement, observe that T is a closed subscheme which contains $g^{-1}(p^{-1}(L\cup M)\cap W')$ away from x (see 4.3 d). For the claim on the residue, recall that, since $f^{-1}\pi^{-1}(Q)=G\cup G'$, for a subscheme Y of $V^\sim$ one has: $\mathrm{res}_{G'}(\mathrm{res}_G(Y))=\mathrm{res}_{f^{-1}\pi^{-1}(Q)}(Y)$; in fact, let $\mathfrak{A}$, (g), (g') be the ideals of Y, G, G' in an affine open subset; then, $\mathrm{res}_G Y$ has ideal $(\mathfrak{A} : g)$; moreover, $((\mathfrak{A} : g) : g') = (\mathfrak{A} : gg')$, and $(gg')=(g)\cap(g')$ since G, G' have no common component. Now it is immediate to conclude computing in local coordinates.

4.2.4 Lemma. Let L, M be lines meeting Q transversally at the common point $x=L\cap M$; the subscheme $T:=\mathrm{res}_G(f^{-1}\pi^{-1}(L\cup M\cup\chi(x)))$ of $V^\sim$, which is the union of $g^{-1}(p^{-1}(L\cup M))$ with a nilpotent which will be denoted by $\mu(x)$, and whoose support is contained in $g^{-1}p^{-1}(x)$, verifies: $\mathrm{res}_{G'}(T)=f^{-1}\pi^{-1}(L\cup M)$, and $T\cap G'$ contains $g^{-1}(p^{-1}(\eta(x))\cap W')$, $\eta(x)$ denoting the first infinitesimal neighbourhood of x on Q.

Proof. It is enough to compute using local coordinates, taking into account the remark in proof of 4.2.3.

4.2.5 Definition. Let $L_1,\ldots,L_n$, resp. $M_1,\ldots,M_m$ be lines of type (1,0), resp. (0,1), on Q. With $\Gamma^\sim(n,m)$ we shall denote the union (in $V^\sim$) of $g^{-1}(p^{-1}(L_1\cup\ldots\cup L_n\cup M_1\cup\ldots\cup M_m)\cap W')$ with $v(L_i\cap M_j)$, i=2,...,n, j=2,...m. Hence for m=1 it is a reduced scheme, and for m≥2, $res_{G'}\Gamma^\sim(n,m)$ is the inverse image of a grille of points G(n-1,m-1). Moreover, $res_G f^{-1}\pi^{-1}(\Gamma(n,m)) = \Gamma^\sim(n,m)$ by lemma 4.2.3.

4.3 We shall use freely the following facts:

a) The generic rational curve C_d has generic intersection with the quadric if d≥3 ([P], th.5.12);

b) Given d, δ, 0<δ<d, C_d can be specialized to the union of C_δ with d-δ lines $r_1,\ldots,r_{d-\delta}$ such that r_i meets (transversally) $C_\delta \cup r_1 \cup\ldots\cup r_{i-1}$ exactly at one smooth point, for i=1,...,d-δ.

c) Given a line ℓ in Q, it is possible to build a flat family in X such that the generic fiber is the disjoint union of $\pi^{-1}(\ell)$ with a t-point, resp. a d-point, resp. an s-point, and the special fiber is $\pi^{-1}(\ell)$ with a nilpotent supported on a t-point, resp. a d-point, resp. an s-point lying on ℓ, and giving as residual scheme with respect to $\pi^{-1}(Q)$ a t-point, resp. a d-point, resp. an s-point. This is used in type a proofs.

d) In type b proofs, to understand which are the residual schemes at each step, if no nilpotents are involved, it is enough to keep in mind what happens when elementary transformation operate ([M],§1, [I],1.9.3, [H,3]); for example, let P be a point of the quadric; then, the residue with respect to G' of $g^{-1}p^{-1}(P)$ is the pull-back through f of the s-point $\pi^{-1}(P)\cap W$, while the residue with respect to G of $f^{-1}\pi^{-1}(P)$ is the pull-back through g of the d-point $p^{-1}(P)\cap W'$.

4.4 Graphic notations. Since the proofs are given essentially by pictures, we now fix some conventions.

In each proof we build a subscheme of V, or of V', or of $V^\sim$, which has a component J_1 whose projection to $\mathbb{P}^3$ is 1-dimensional, and a component J_0 whose projection to $\mathbb{P}^3$ is 0-dimensional (while J_1 itself may have irreducible components of different dimensions). We just draw the projection of J_1 to $\mathbb{P}^3$, while for J_0 we use the following notations:

a t-point is denoted in the pictures with "o" ; a d-point is denoted by " +"; an s-point is denoted by "/ ". The same symbol, but "bolded" ("● ", " **+** " ," **/** "), denotes a nilpotent giving in the residual scheme (residual respect to the divisor we are using in that particular proof) a t-point, resp. a d-point, resp. an s-point. Sometimes, the symbol "● " is also used to represent certain nilpotents not giving as residual scheme a t-point, but this is made clear by the context.
The line r in the definition of Y_k is drawn by a dashed line "- - - - - - - -".

4.5 The odd cases. Both here and in 4.6 (the even cases), in the proof of each statement R(-,k), we describe by the picture (and if necessary by words) the scheme Z if we are in a proof of type a, respectively the schemes K', K if we are in a proof of type b, respectively K~, K if we are in the variant of type b proof. Then, the proof goes on as explained in 4.1.

The type b proof is used in R(U,5) ⇒ R(I,7) inside R(13), in R(4) and in R(6), while the variant of type b proof is used in R(9), R(11), in R(9) ⇒ R(D,11) inside R(15), and in R(4) ⇒ R(N,6) inside R(8).

At the beginning of each proof R(k) we give the values, for $k \leq 18$, of the functions n(k), δ(k), ε(k) in the definition of Y_k, and for the reader's convenience we also recall the values for t(k-2) (maximum degree for the generic rational curve contained in Y_k^+) and t(k)-t(k-2) (minimum number of lines contained in Y_k^+).

R(1) **t(-1)=0, t(1)-t(-1)=1, n(1)=0, δ(1)=0, ε(1)=1.**
Y_1 is the union of a generic line L with any s-point A. Moving the line L if necessary, we can assume that if H is the plane of $\mathbb{P}^3$ containing Y_1, then $A \notin \mathbb{P}(\Omega_H)$. We conclude as in [I], 5.2.

R(3) is false, since, if it were true, the generic quintic rational curve would have a σ_3 surjective, which is not possible since it has a 4-secant line (see [G-L-P]).

R(5) **t(3)=5, t(5)-t(3)=4, n(5)=3, δ(5)=0, ε(5)=2.**
R(1) ⇒ R(A,3), where $A = \pi^{-1}(\Gamma(2,1) \cup \text{a line } \ell) \cup s(2,0,1)$; the s-point, x, is chosen no matter how in the fiber $X_{\pi(x)}$.

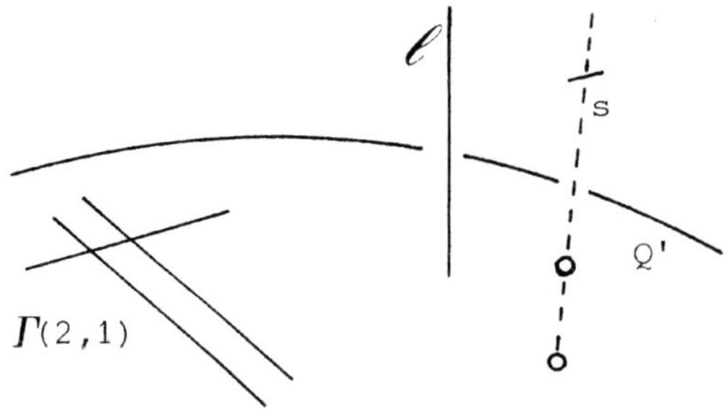

proof of type a

Choose Z = A; the two t-points are on Q' and the s-point is outside. Then, r(Z) is Y_1, while the i-claim follows by [I], lemma 3.3.1 (for f=1, h=1, m=0, u=4, n=δ=ε=0).

R(A,3) ⇒ R(5)

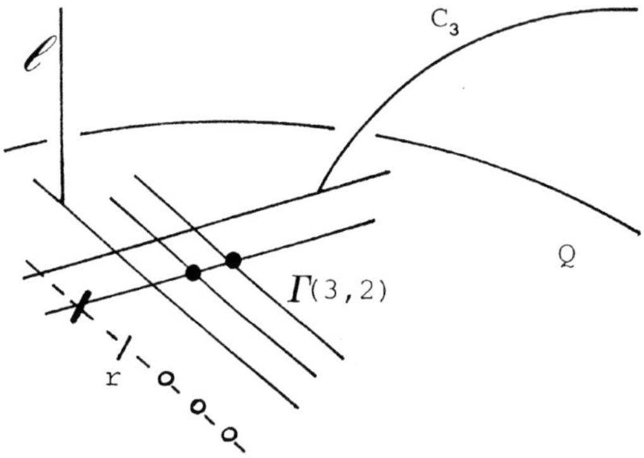

proof of type a

We change the quadric. Z is the union of $\pi^{-1}(\Gamma(3,2) \cup C_3 \cup \ell)$, of r(3,0,1), and of a nilpotent supported on an s-point of $\Gamma(3,2) \cap r$, giving an s-point as residual scheme (see picture); the points are on Q, the line ℓ and the curve C_3 both meet Γ(3,2) at a smooth point.

Then, r(Z) is a generalization of A, while the i-claim follows by the forthcoming lemma 4.5.2.

4.5.1 Lemma. Let $\mathcal{Y}$ be the generic union in $\mathbb{P}(\Lambda) = \pi^{-1}(Q)$ of 1 t-point with 1 t-point and 1 s-point lying on a line r of type (1,0) of Q (in particular, the s-point is not in $\mathbb{P}(\Omega_r)$); then $H^0(I_{\mathcal{Y}} \otimes F_{2,2})=0$.

Proof. Let m be a line of type (0,1) on Q, and M the point $r \cap m$; first notice that a d-point not intersecting $\mathbb{P}(\Omega_m)$ is $F_{1,2}$-settled (see [I] 3.2.1).

We have $\Lambda(2,2)|_r \cong \Omega_r(2) \oplus E \oplus E'$ with $E \cong E' \cong \mathcal{O}_r(1)$, and it is possible to choose this splitting so that in $\mathbb{P}(\Lambda(2,2))$ $\mathbb{P}(E')_M = \mathbb{P}(\Omega_m)_M$.

Setting $\mathcal{L} := \ker(\Lambda(2,2) \to \Omega_r(2) \oplus E)$ we have the following two elementary transformations:

$$1°) \quad 0 \to \Lambda(1,2) \to \mathcal{L} \to E' \to 0$$

$$2°) \quad 0 \to \mathcal{L} \to \Lambda(2,2) \to \mathcal{O}_r \oplus \mathcal{O}_r(1) \to 0.$$

We now use la méthode d'Horace in the context of these two elementary transformations, analogously to the higher dimensional case; in the 1° step we see that in $\mathbb{P}(\mathscr{L})$ the union $\mathcal{A}$ of 1 s-point with 1 t-point on r satisfies $H^0(I_{\mathcal{A}}\otimes O_{\mathbb{P}(\mathscr{L})}(1))=0$ (the residue is a d-point not intersecting $\mathbb{P}(\Omega_m)$, the intersection is a couple of s-points). In step 2, we see that $H^0(I_{\mathcal{Y}}\otimes O_{\mathbb{P}(\Lambda(2,2))}(1))=0$, since the residue is a generalization of $\mathcal{A}$, namely a s-point on r union with a t-point generic on Q; the intersection is an s-point union with a d-point, which takes care of $O_r\oplus O_r(1)$.

4.5.2 Lemma. Let $\mathcal{T}$ be the generic union in $\mathbb{P}(\Lambda)=\pi^{-1}(Q)$ of 6 t-points with 3 t-points and 1 s-point lying on a line r of type (1,0) of Q (in particular, the s-point is not in $\mathbb{P}(\Omega_r)$); then $H^0(I_{\mathcal{T}}\otimes F_{3,4})=0$.

Proof. We exploit the pull back of a smooth curve C of type (1,2) on Q. Let $\mathcal{T}_0$ be obtained from $\mathcal{T}$ specializing 2 of the t-points lying on r and 5 of the general t-points, on 7 t-points lying on C:

$$0\to I_{res_{\pi^{-1}(C)}\mathcal{T}_0}\otimes F_{2,2}\to I_{\mathcal{T}_0}\otimes F_{3,4}\to I_{\mathcal{T}_0\cap\pi^{-1}(C),\pi^{-1}(C)}\otimes F_{3,4}\,|_{\pi^{-1}(C)}\to 0$$

The r-claim is lemma 4.5.1, the i-claim follows from $\Lambda(3,4)|_C\cong O_{\mathbb{P}^1}(6)^{\oplus 3}$ (see [I] 3.2.2), using lemma 2.1.

R(7) **t(5)=9, t(7)-t(5)=6, n(7)=3, δ(7)=0, ε(7)=3.**

R(5) ⇒ R(7)

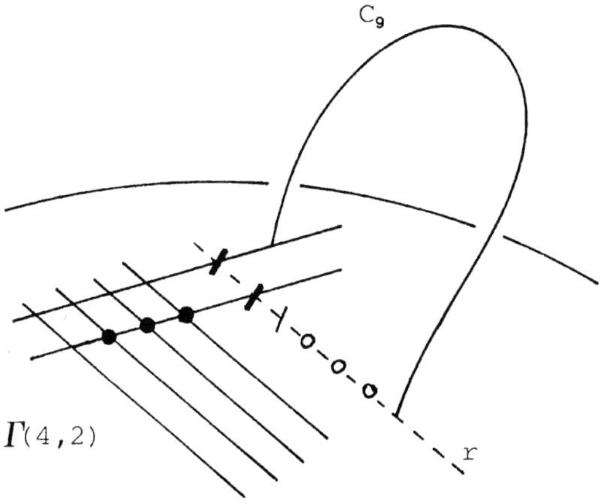

$\Gamma(4,2)$

proof of type a

Z is the union of $\pi^{-1}(\Gamma(4,2)\cup C_9)$, of r(3,0,1), and of two nilpotents supported on 2 s-points y,z of $\Gamma(4,2)\cap r$; the s-point x inside Q and y are chosen no matter how inside their fiber $X_{\pi(x)}$, resp. $X_{\pi(y)}$ (while their projections are general in r). Then, r(Z) is a generalization of Y_5, while the i-claim follows by [I], lemma 3.3.1 (for f=2, h=2, m=0, u=16, n=4, δ=0, ε=1.

Notice that the s-point in [I] 3.3.1 is general, but in fact the proof holds for any s-point; it is enough, in the proof of the initial case h=0 and f=2, to choose the plane H of the smooth conic D not containing the tangent direction corresponding to the s-point, since $\Omega_3(2)|_D \cong \Omega_H(2)|_D \oplus O_H(1)|_D$ with $\Omega_H(2)|_D \cong O_{\mathbb{P}^1}(1)^{\oplus 2}$ and $O_H(1)|_D \cong O_{\mathbb{P}^1}(2)$).

R(9) **t(7)=15, t(9)-t(7)=7, n(9)=6, δ(9)=0, ε(9)=1.**

R(7) ⇒ R(9)

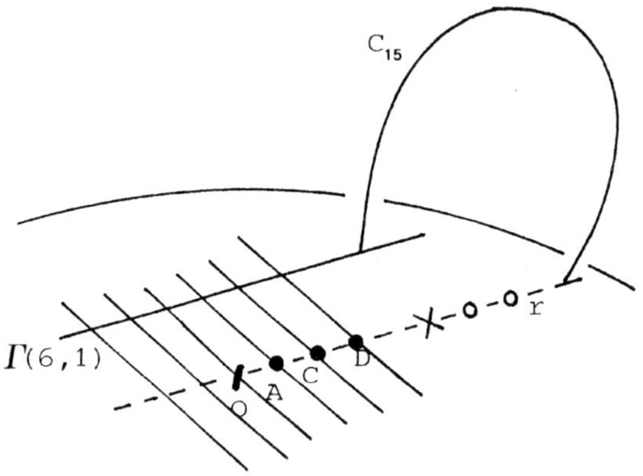

proof of type b

1st step: here we denote by A, C, D, O four of the points of $\Gamma(6,1) \cap r$ (see picture).

$K^\sim$ is the union in $V^\sim$ of $\Gamma^\sim(6,1) \cup g^{-1}(p^{-1}(C_{15}) \cup r(2,1,0))$, of $g^{-1}p^{-1}(A \cup C \cup D)$, and of the nilpotents $\alpha(A)$, $\alpha(C)$, $\alpha(D)$, and $\beta(B_0)$, B_0 denoting an s-point in $\pi^{-1}(O)$, $B_0 \notin W$; for the definitions and the properties of these nilpotents, see lemma 4.2.1 b) and c).

Then (see also 4.2.5) T^g is $\pi^{-1}(C_{15})$ union 3 t-points and 3 s-points, 2 of which in $\mathbb{P}(\Omega_r)$, and $T=Y_7$. The scheme P' is the generic union of Γ(6,1) with 4 points on a line (0,1) and 28 points on Q.

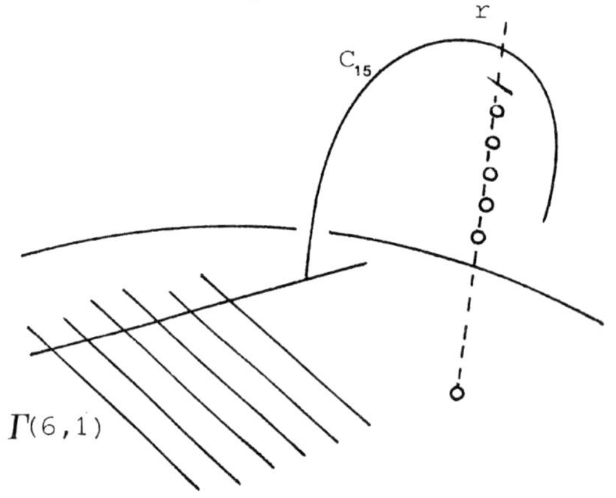

2nd step:

$K = \pi^{-1}(\Gamma(6,1) \cup C_{15}) \cup r(6,0,1)$;

the points of r lie out of Q, except 1 t-point. The flat family in $V^\sim$ with generic fiber r(K) and special fiber $K^\sim$ exists (lemma 4.2.1 b) and c)). The scheme P is the generic union of Γ(6,1) with 30 points on Q.

R(11) **t(9)=22, t(11)-t(9)=9, n(11)=2, δ(11)=0, ε(11)=0.**

R(9) ⇒ R(11)

proof of type b:

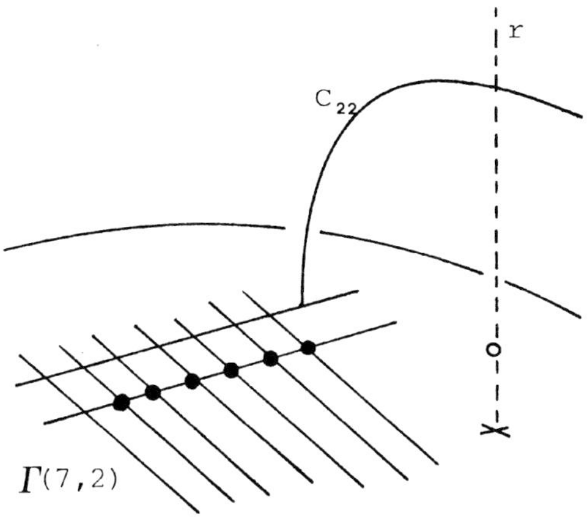

1st step:
$K^{\sim}=\Gamma^{\sim}(7,2)\cup g^{-1}\big(p^{-1}(C_{22})\cup r(1,1,0)\big)$ (the points of r lie on Q). Then (see also 4.2.5) T^g is $\pi^{-1}(C_{22})$ union 6 t-points and 1 s-point in $\mathbb{P}(p_1{}^*\Omega_{\mathbb{P}^1})$, and $T=Y_9$; it is in fact possible to specialize (in $\mathbb{P}(\Omega_3)$) all the points on a line t so that the s-point is generic (in particular, it is not in $\mathbb{P}(\Omega_t)$). The scheme P' is the generic union of $\Gamma(7,2)_{red}$ with 45 points on Q.

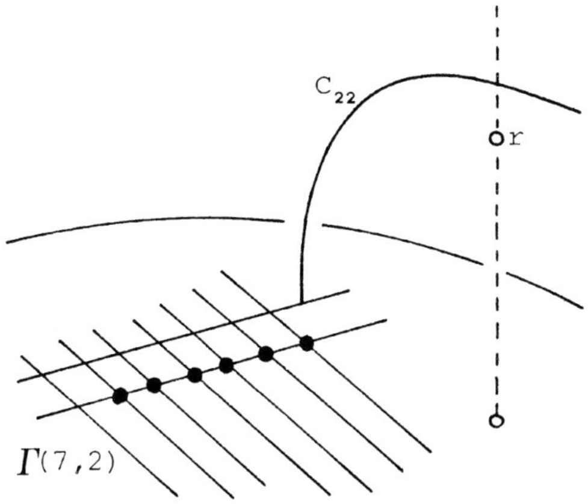

2nd step:
$K=\pi^{-1}(\Gamma(7,2)\cup C_{22})\cup r(2,0,0)$; 1 t-point of r lies out of Q. The flat family in $V^{\sim}$ with generic fiber r(K) and special fiber $K^{\sim}$ exists (lemma 4.2.1 a)). The scheme P is the generic union of $\Gamma(7,2)_{red}$ with 44 points on Q.

R(13) **t(11)=31, t(13)-t(11)=9, n(13)=11, δ(13)=2, ε(13)=0.**

$R(A,3)\Rightarrow R(U,5)$, where A was defined in R(5), and $U=\pi^{-1}\big(C_8\cup \text{a line } s\big)\cup$ $\cup$ 2 t-points and 1s-point lying on Q', and 1 s-point lying out of Q'.

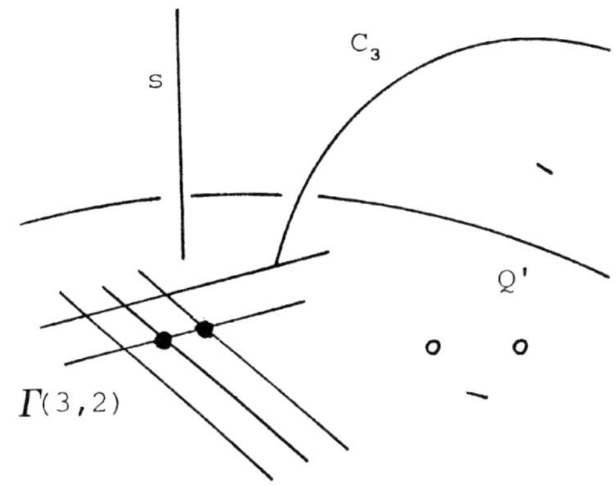

proof of type a
$Z=\pi^{-1}\big(\Gamma(3,2)\cup C_3\cup s\big)\cup$ 2 t-points and 1 s-point lying on Q', and 1 s-point lying out of Q'.
Then, r(Z) is a generalization of A, while the i-claim follows by [I], 3.3.1 (for f=2, h=1, u=9, m=n=δ=0, ε=1).

$R(U,5) \Rightarrow R(I,7)$, where $I = \pi^{-1}(C_{15}) \cup 4$ t-points

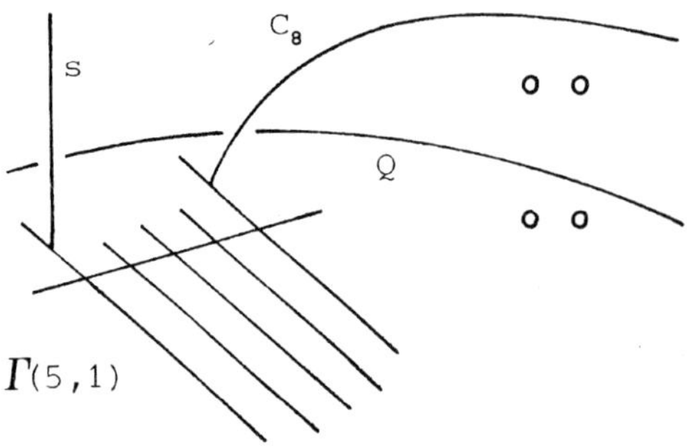

proof of type b

1st step:

$K' = (p^{-1}(\Gamma(5,1)) \cap W') \cup p^{-1}(C_8 \cup s) \cup 2$ t-points lying on Q and 2 t-points lying out of Q; hence $T = T^g = \pi^{-1}(C_8 \cup s) \cup \cup 2$ t-points $\cup$ 2 s-points $=U$.

The scheme P' is the generic union of $\Gamma(5,1)$ with 18 points on Q.

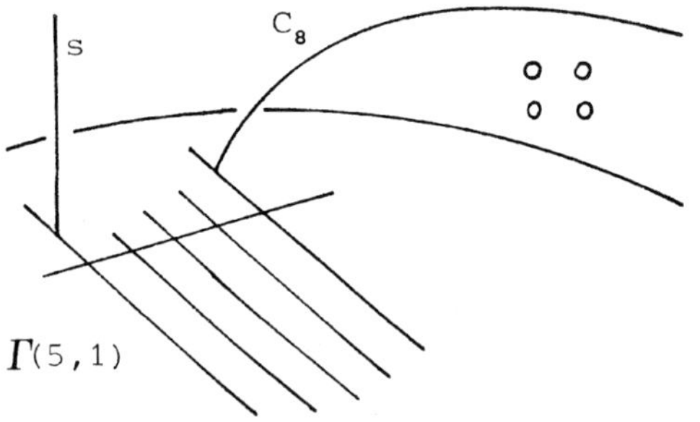

2nd step:

$K = \pi^{-1}(\Gamma(5,1) \cup C_8 \cup s)) \cup 4$ t-points lying out of Q. The scheme P is the generic union of $\Gamma(5,1)$ with 16 points on Q.

$R(I,7) \Rightarrow R(J,9)$, where $J = \pi^{-1}(C_{21}) \cup 15$ t-points

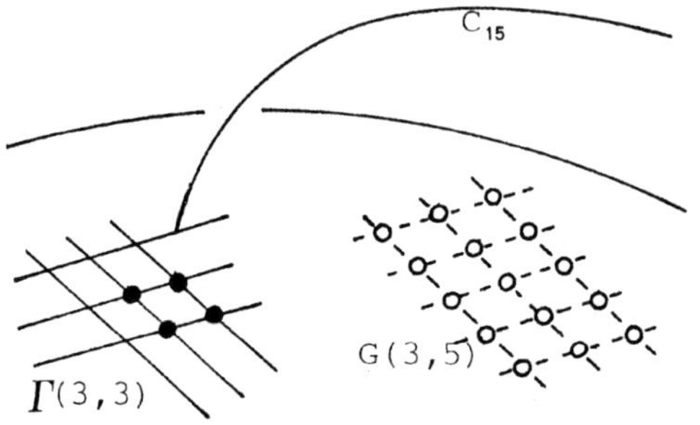

proof of type a

$Z = \pi^{-1}\big(\Gamma(3,3) \cup C_{15} \cup G(3,5)\big)$.

Then, $r(Z)=I$, while the i-claim follows by the following:

4.5.3 Lemma. Let A denote the union in $\mathbb{P}(\Lambda)$ of 29 general t-points and 15 t-points lying on a grille G(3,5); then, $H^0(F_{7,7} \otimes \mathcal{I}_A) = 0$.

Proof. Denote by r_i, resp. m_j, the 3 lines of type (1,0), resp.the 5 lines of type (0,1) intersecting at the points of the grille, and set $a_{ij} := r_i \cap m_j$. It is possible to choose four smooth cubics, C and C' of type (1,2), D and D' of type (2,1), such that $a_{11}, a_{22}, a_{33}, a_{42}, a_{51}$ are on C, $a_{13}, a_{23}, a_{32}, a_{41}, a_{52}$ are on C', a_{12}, a_{21}, a_{43} are on D, and a_{31}, a_{53} are on D'. Now the thesis follows exploiting successively $\pi^{-1}(D')$, $\pi^{-1}(C')$, $\pi^{-1}(D)$, and $\pi^{-1}(C)$, as in the general case (see proof of [I], 3.3.1).

$R(J,9) \Rightarrow R(E,11)$, where $E= \pi^{-1}(C_{27} \cup 3$ lines $a,b,c) \cup 13$ t-points and 1 d-point lying on a $G(7,2)$; the configuration C_{27} union the three lines a,b,c is described in the picture of the next step: $R(E,11) \Rightarrow R(13)$, and it is is a generalization of $\Gamma \cup C_{21}$, Γ being defined as follows:

Let $\ell_1,\ldots,\ell_7$ and m_1, m_2 be lines of type (1,0), resp. (0,1) on Q; set $a_{ij}:=\ell_i \cap m_j$, and $\Gamma:=\ell_1 \cup \ldots \cup \ell_7 \cup m_1 \cup m_2 \cup (\bigcup_{i=1,\ldots,5} \chi(a_{i2}))$.

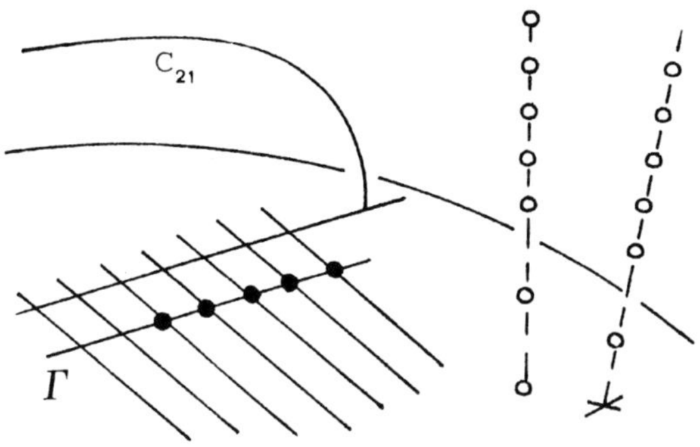

proof of type a

$Z=\pi^{-1}(\Gamma \cup C_{21}) \cup 10$ t-points out of Q and 3 t-points and 1 d-point lying on Q, as in picture. Then, $r(Z)=J$, while the i-claim follows by [I], 3.3.1 (for f=1, h=3, m=n=δ= =ε=0, u=24) and [I], 3.3.8 (for a=4, b=7, h=1, e=g=0, and f=20).

$R(E,11) \Rightarrow R(13)$

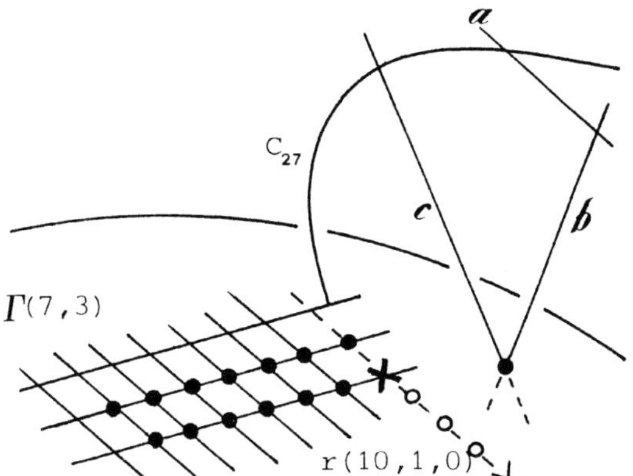

proof of type a

$Z=\pi^{-1}(\Gamma(7,3) \cup C_{27} \cup 3$ lines $a,b,c \cup \chi(b \cap c)) \cup r(10,1,0) \cup$ nilpotents supported on 1 t-point and 1 d-point lying on $\Gamma(7,3) \cap r$. Then, $r(Z)=E$, while the i-claim follows by the following:

4.5.4 Lemma. In the previous notations, there exists a subscheme B of $\pi^{-1}(Q)$ such that B is specialization of $Z \cap \pi^{-1}(Q)$, and moreover $H^0(F_{14,14} \otimes I_B)=0$.

Proof. The scheme $Z \cap \pi^{-1}(Q)$ is the union of $\pi^{-1}(\Gamma(7,3)_{red})$ with a scheme A' whoose projection to $\mathbb{P}^3$ is zero-dimensional, so we prove $H^0(F_{7,11} \otimes I_{A'})=0$; since r is a line of type (1,0) on Q, and 10 t-points and 1 d-point are on r, by [I], 3.2.1 $H^0(F_{7,11}|_r \otimes I_{A' \cap r})=0$; hence, denoting by C the curve $C_{27} \cup$(lines $a,b,c) \cup \chi(b \cap c)$ (see previous notations), using also lemma 2.1 we see that it is enough to prove that a specialization A of $C \cap Q$ satisfies $H^0(I_A \otimes \Lambda(6,11))=0$.

Fix five smooth cubics of type (1,2) on Q, D_i, i=1,...,5, such that $D_i \cap D_j \cap D_k$ is empty if $i \neq j \neq k \neq i$.
Set $a \cap Q = \{a, a'\}$, $b \cap Q = \{b, d\}$, $c \cap Q = \{c, d\}$, and denote by $\eta(d)$ the first infinitesimal neighbourhood of d on Q. It is possible to specialize the configuration of lines a, b, c so that $d \in D_1 \cap D_2$, $d \notin D_i$ for $i \geq 3$; $b \in D_3$, $b \notin D_i$ for $i \geq 4$; $c \in D_4$, $c \notin D_5$; $a, a' \in D_5$; moreover, the union of 53 points among $C_{27} \cap Q$ specializes to the union Y of 18 points on D_5, 15 points on D_4 and not on D_5, 11 points on D_3 and not on $D_4 \cup D_5$, 6 points on D_2 and not on $D_3 \cup D_4 \cup D_5$, 3 points on D_1 and not on $D_2 \cup D_3 \cup D_4 \cup D_5$. Now set: A_0=the empty set,
$A_1 = (Y \cap D_1) \cup \{d\}$, $A_2 = (Y \cap (D_1 \cup D_2)) \cup \eta(d)$, $A_3 = (Y \cap (D_1 \cup D_2 \cup D_3)) \cup \{b\} \cup \eta(d)$, $A_4 = (Y \cap (D_1 \cup \ldots \cup D_4)) \cup \{b, c\} \cup \eta(d)$, $A_5 = Y \cup \{b, c, a, a'\} \cup \eta(d)$.
We apply 5 times la méthode d'Horace on Q, exploiting D_h for h=1,...,5:

$$0 \to I_{res_{D_h} A_h} \otimes \Lambda(1+(h-1), 1+2(h-1)) \to I_{A_h} \otimes \Lambda(1+h, 1+2h) \to$$
$$\to I_{A_h \cap D_h} \otimes \Lambda(1+h, 1+2h)_{|D_h} \to 0.$$

Since $\Lambda(1+h, 1+2h)_{|D_h} \cong \mathcal{O}_{\mathbb{P}^1}(4h-1)^{\oplus 3}$, and $res_{D_h} A_h = A_{h-1}$ (notice that $\eta(d) \cap D_2$ is a double point, and the residue is the simple point $d \in D_1$), choosing $A = A_5$ we get $H^0(I_A \otimes \Lambda(6,11)) = 0$.

R(15) **t(13)=40, t(15)-t(13)=12, n(15)=1, δ(15)=0, ε(15)=1.**
$R(9) \Rightarrow R(D,11)$, where $D = \pi^{-1}(C_{30} \cup \text{a line } n) \cup 1$ t-point; C_{30} and n do not meet.

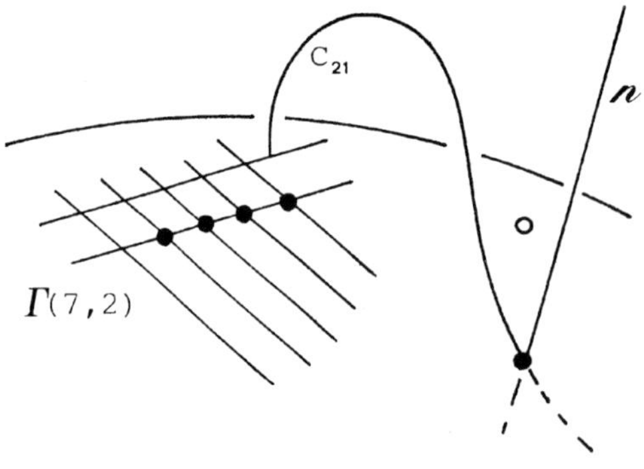

proof of type b
1st step: $K^{\sim}$ is the union in $V^{\sim}$ of $\Gamma^{\sim}(7,2)$, of $g^{-1}(p^{-1}(C_{21} \cup \text{a line } n) \cup 1$ t-point lying on Q), and of the nilpotent $\mu(x)$ where x denotes the common point of C_{21} and n on Q (for the definition of $\mu(x)$ see lemma 4.2.4).
Then (see also 4.2.5) T^g is the union of $\pi^{-1}(C_{21} \cup n)$ and 6 t-points and 1 s-point in $\mathbb{P}(p_1{}^*\Omega_{\mathbb{P}^1})$; $T = Y_9$.

The scheme P' is the generic union of $\Gamma(7,2)_{red}$ with 42 points and the triple point $\eta(x)$ on Q, hence P' satisfies condition (ii) (see [H-H] 2.3, with q=42, q"=0, d=0, t=1, n=8, r=4).

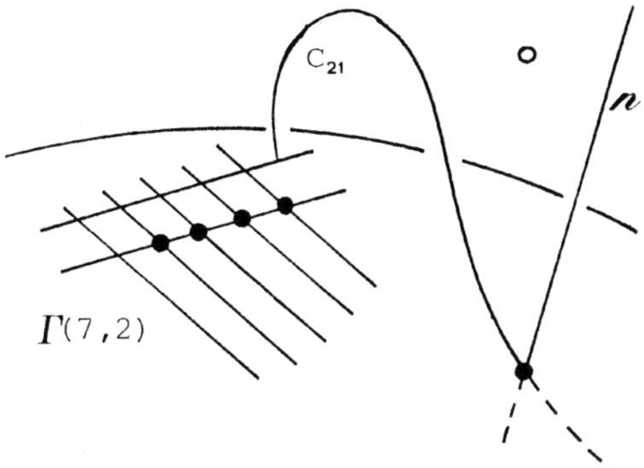

2nd step:

$K = \pi^{-1}(\Gamma(7,2) \cup C_{21} \cup n \cup \chi(x)) \cup 1$ t-point lying out of Q. The flat family in $V^{\sim}$ with generic fiber r(K) and special fiber $K^{\sim}$ exists (lemma 4.2.1 a)). The scheme P is the generic union of $\Gamma(7,2)_{red}$ with 41 points and the triple point $\eta(x)$ on Q, hence P satisfies condition iii) (see [H-H], lemme 2.3, with q=41, q"=0, d=0, t=1, n=10, r=7).

R(D,11) ⇒ R(M,13), where $M = \pi^{-1}(C_{39}) \cup G(3,8) \cup 1$ t-point on a line of the grille (see picture);

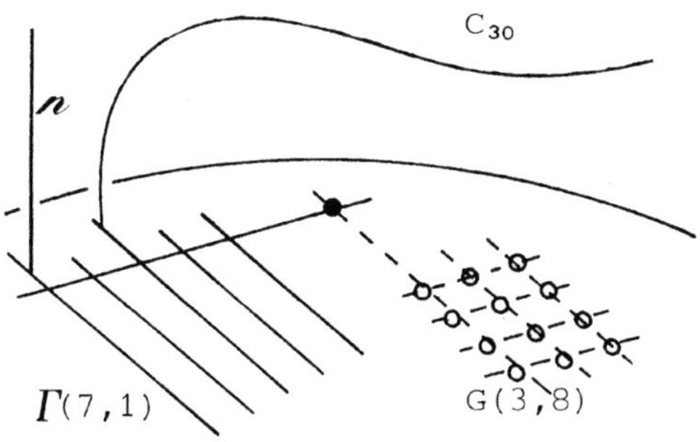

proof of type a

$Z = \pi^{-1}(\Gamma(7,1) \cup C_{30} \cup$ a line $n \cup \cup G(3,8)) \cup$ a nilpotent supported on a t-point lying on $\Gamma(7,1)$.

Then, r(Z) = D, while the i-claim follows by [I], lemma 3.3.4 (for a=1, h=6, g=24, p=3, q=8, n=60, c=0).

R(M,13) ⇒ R(15)

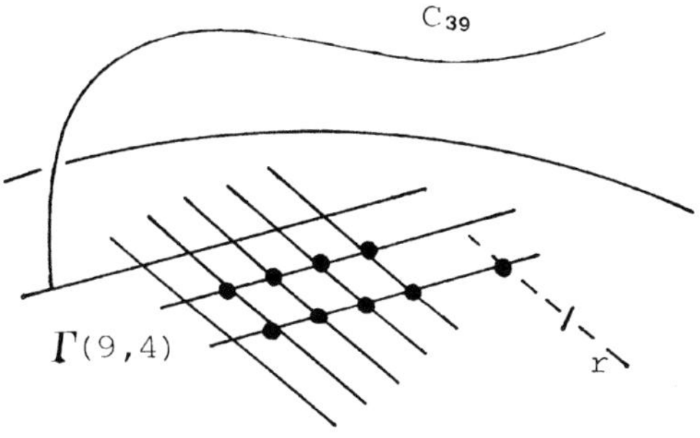

proof of type a

$Z = \pi^{-1}(\Gamma(9,4) \cup C_{39}) \cup r(0,0,1)$ union with a nilpotent supported on a t-point lying on $r \cap \Gamma(9,4)$.

Then, r(Z) = M, while the i-claim follows by [I] 3.3.1 (for f=2, h=5, m=0, u=77, n=δ=0, ε=1).

R(17) **t(15)=52, t(17)-t(15)=12, n(17)=6, δ(17)=4, ε(17)=1.**

R(M,13) ⇒ R(N,15), where M is defined in R(15), and N denotes the disjoint union of $\pi^{-1}(C_{51})$, of an s-point, and of $\pi^{-1}(m)$ where m is a line.

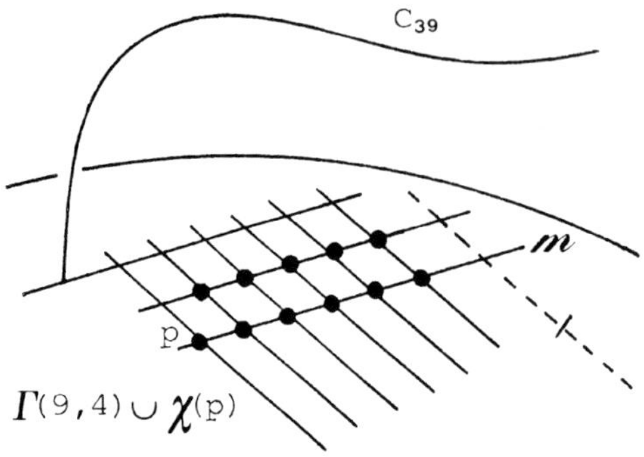

proof of type a
$Z=\pi^{-1}(\Gamma(9,4)\cup\chi(p)\cup C_{39})\cup$ 1 s-point lying on Q (the curve $\Gamma(9,4)\cup\chi(p)$ is the result of a specialization on the quadric of a C_{12} and of a line not meeting this C_{12}). Then, r(Z) is a generalization of M, while the i-claim follows by [I] 3.3.1 (for f=2, h=5, m=0, u=77, n=0, δ=0, ε=1).

R(N,15) ⇒ R(17)

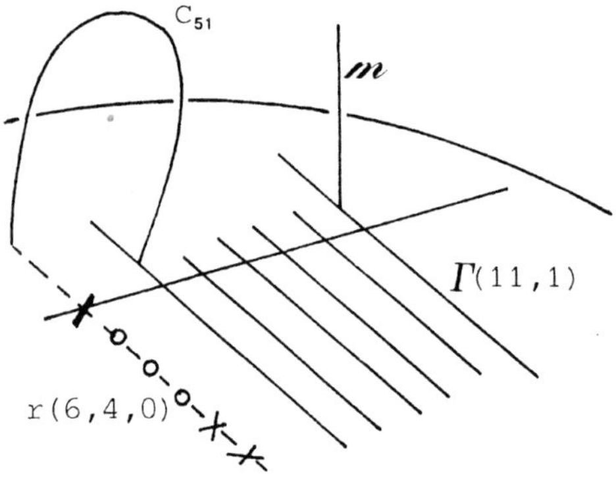

proof of type a
$Z=\pi^{-1}(\Gamma(11,1)\cup C_{51}\cup m)\cup r(6,4,0)$ union with a nilpotent supported on an s-point lying on $r\cap\Gamma(11,1)$; m is a line as in picture. Then, r(Z) = N, while the i-claim follows by [I] 3.3.8 (for a=3,b=5, h=4, f=89, e=0, g=7) and [I] 3.3.1 (for f=1, h=2, m=0, u=12, n=δ=ε=0).

4.6 The even cases.

R(2) **t(0)=0, t(2)-t(0)=3, n(2)=0, δ(2)=0, ε(2)=2.**

R(1) ⇒ R(2): this is the unique case of this section where we use la méthode d'Horace for: U=X, $L=L_2$, $D = \pi^{-1}(H)$, where H denotes a plane of $\mathbb{P}^3$.

We fix a splitting $\Omega_3|_H \cong \Omega_H \oplus E$, with $E \cong \mathcal{O}_H(-1)$, we chose Z specialization of Y_2 such that Z is the inverse image through π of the union of a singular reduced conic C on H, a line s outside H meeting C, and two s-points on r, one in $\mathbb{P}(E)$ and the other outside H; hence the r-claim is R(1), while the i-claim $H^0(L_2|_{\pi^{-1}(H)}\otimes\mathcal{I}_{Z\cap\pi^{-1}(H)})=0$ is proved as in [I], 3.2.1.

R(4) **$t(2)=3$, $t(4)-t(2)=4$, $n(4)=1$, $\delta(4)=0$, $\varepsilon(4)=1$.**

$R(2) \Rightarrow R(4)$

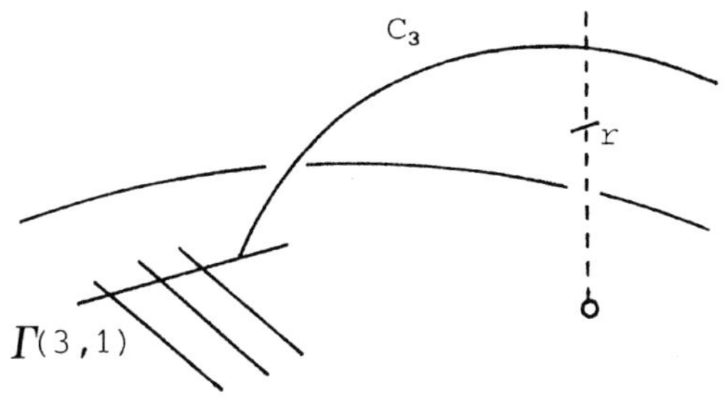

proof of type b

1st step: $K'= (p^{-1}(\Gamma(3,1))\cap W') \cup \cup p^{-1}(C_3) \cup r(1,0,1)$; the t-point lies on Q, and the s-point outside; $T^g=\pi^{-1}(C_3)$ union 2 s-points, and $T=Y_2$. The scheme P' is the generic union of $\Gamma(3,1)$ and of 6 points on Q.

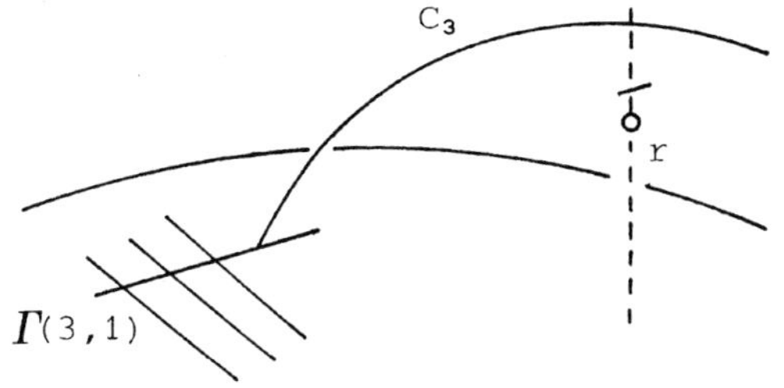

2nd step: $K = \pi^{-1}(\Gamma(3,1) \cup C_3)) \cup \cup r(1,0,1)$; the t-point and the s-point lie out of Q.The scheme P is the generic union of $\Gamma(3,1)$ and of 5 points on Q.

R(6) **$t(4)=7$, $t(6)-t(4)=5$, $n(6)=3$, $\delta(6)=0$, $\varepsilon(6)=0$.**

$R(4) \Rightarrow R(6)$

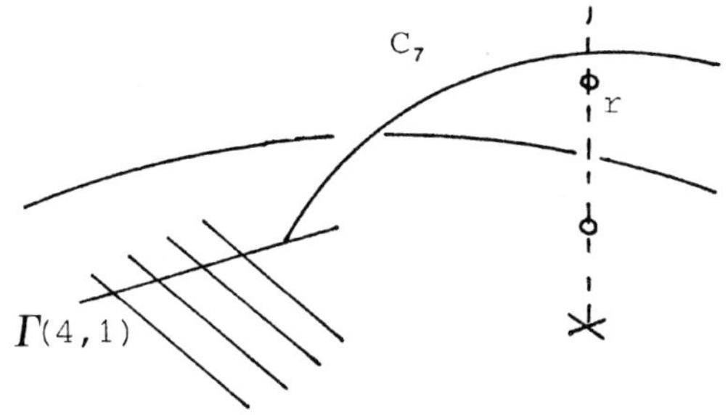

proof of type b

1st step: $K' = (p^{-1}(\Gamma(4,1))\cap W') \cup \cup p^{-1}(C_7) \cup r(2,1,0)$; one t-point lies on Q, and the d-point is in W'; $T^g=\pi^{-1}(C_7)\cup r(1,0,1)$, and $T=Y_4$. The scheme P' is the generic union of $\Gamma(4,1)$ and of 15 points on Q as in picture (the zero-dimensional part in P' is a generalization of n=4, r=2, q=12, q''=3, d=t=0 in lemme 2.3, [H-H]).

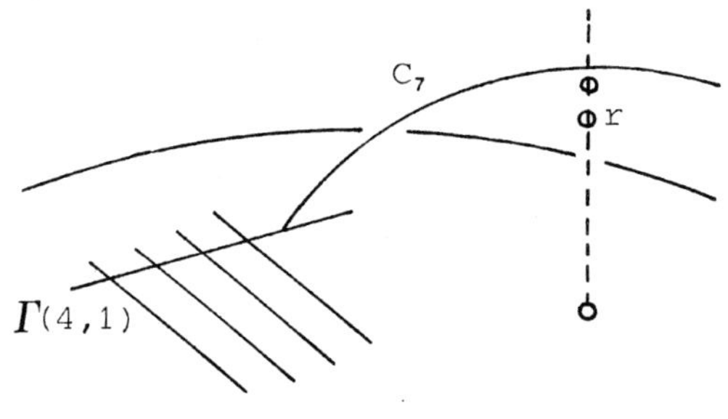

2nd step: $K = \pi^{-1}(\Gamma(4,1) \cup C_7) \cup \cup r(3,0,0)$; one t-point lies on Q. The scheme P is the generic union of $\Gamma(4,1)$ and of 14 points on Q.

R(8) **$t(6)=12$, $t(8)-t(6)=7$, $n(8)=0$, $\delta(8)=0$, $\varepsilon(8)=0$.**

$R(4) \Rightarrow R(N,6)$, where $N = \pi^{-1}(\Gamma(4,1) \cup C_6 \cup \ell) \cup r(4,0,0)$; ℓ is a line meeting C_6 in one point, and $\Gamma(4,1)$ in two points; r does not meet the other components.

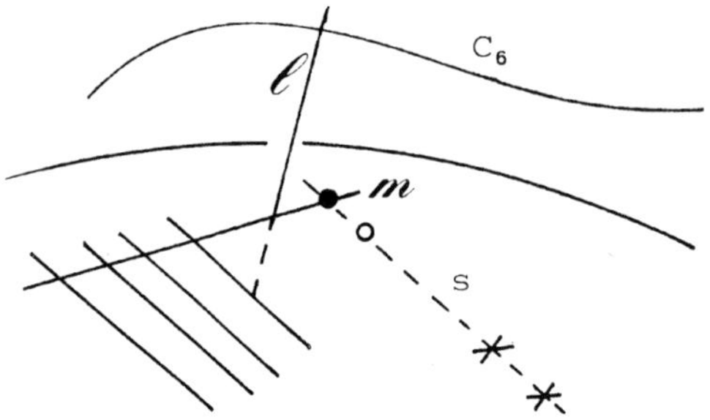

proof of type b

1st step:

$K^{\sim}$ is the union in $V^{\sim}$ of $\Gamma^{\sim}(4,1) \cup \cup g^{-1}(p^{-1}(C_6 \cup \ell) \cup s(1,2,0))$, of $g^{-1}p^{-1}(O)$, and of the nilpotent $\alpha(O)$, O denoting the point $s \cap m$; for the definitions and the properties of this nilpotent, see lemma 4.2.1 b).

Then (see also 4.2.5) T^g is $\pi^{-1}(C_6 \cup \ell) \cup s(1,0,1)$, which is a generalization of Y_4; the scheme P' is the generic union of $\Gamma(4,1)$ and of 15 points on Q as in picture (see [H-H] 2.3 for $n=4, r=2, q=12, q''=3, d=t=0$).

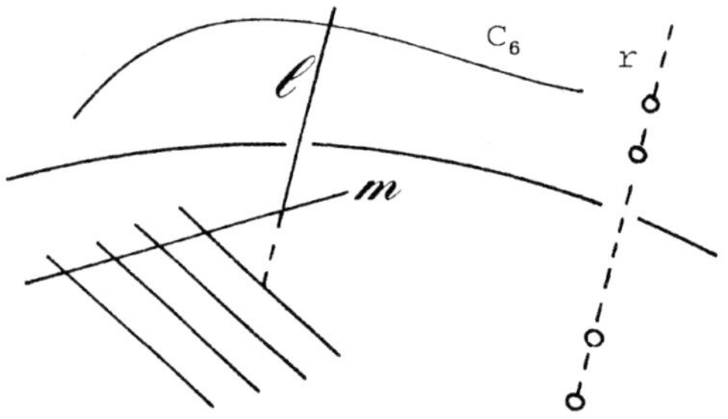

2nd step:

$K = \pi^{-1}(\Gamma(4,1) \cup C_6 \cup \ell) \cup r(4,0,0)$; 2 t-points of r lie on Q. The flat family in $V^{\sim}$ with generic fiber r(K) and special fiber $K^{\sim}$ exists (lemma 4.2.1 b)). The scheme P is the generic union of $\Gamma(4,1)$ and of 14 points on Q.

$R(N, 6) \Rightarrow R(8)$

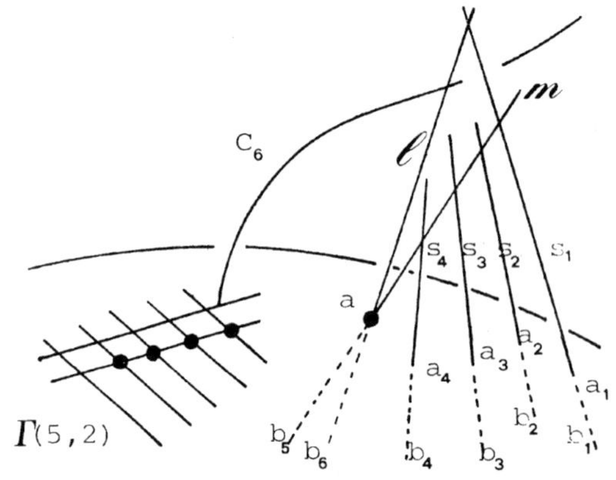

proof of type a

$Z=\pi^{-1}(\Gamma(5,2) \cup C_6 \cup (\cup_{i=1,\dots,4} s_i) \cup \cup \ell \cup m \cup \chi(a))$, with s_1, s_2, s_3, s_4, ℓ, m lines as in picture, $a = \ell \cap m \in Q$. Then, r(Z) is a generalization of N, and the i-claim is the forthcoming lemma 4.6.1.

4.6.1 Lemma. In the previous notations, there exists a subscheme B of $\pi^{-1}(Q)$ such that B is specialization of $Z\cap\pi^{-1}(Q)$, and moreover $H^0(I_B\otimes F_{9,9})=0$.

Proof. The scheme $Z\cap\pi^{-1}(Q)$ is the union of $\pi^{-1}(\Gamma(5,2)_{red})$, with the inverse image through π of a zero-dimensional scheme A'; hence it is enough to prove that a specialization A of A' satisfies $H^0(I_A\otimes\Lambda(4,7))=0$.

Fix three smooth cubics of type (1,2) on Q, D_1, D_2, D_3 such that $D_1\cap D_2\cap D_3$ is empty.

Set $s_i\cap Q=\{a_i,b_i\}$, $m\cap Q=\{a,b_5\}$, $\ell\cap Q=\{a,b_6\}$, and denote by $\eta(a)$ the first infinitesimal neighbourhood of a on Q. It is possible to specialize the configuration of lines $s_1,\dots,s_4$, ℓ,m so that $a\in D_1\cap D_2$, $a\notin D_3$, $b_1\in D_2$, $b_1\notin D_3$, $a_i\in D_3$ for $i=1,\dots,4$, $b_j\in D_3$ for $j=2,\dots,6$; moreover, the union of 11 points among $C_6\cap Q$ specializes to the union Y of 3 points on D_3, 5 points on D_2 and not on D_3, 3 points on D_1 and not on $D_2\cup D_3$. Now set: A_0 = the empty set, $A_1=(Y\cap D_1)\cup\{a\}$, $A_2=(Y\cap(D_1\cup D_2))\cup\eta(a)\cup\{b_1\}$, $A_3=Y\cup\eta(a)\cup\{b_1,\dots,b_6,a_1,\dots,a_4\}$.

We apply 3 times la méthode d'Horace on Q, exploiting D_h for h=1,2,3:

$$0\to I_{res_{D_h}A_h}\otimes\Lambda(1+(h-1),1+2(h-1))\to I_{A_h}\otimes\Lambda(1+h,1+2h)\to$$

$$\to I_{A_h\cap D_h}\otimes\Lambda(1+h,1+2h)_{|D_h}\to 0.$$

Since $\Lambda(1+h,1+2h)_{|D_h}\cong O_{\mathbb{P}^1}(4h-1)^{\oplus 3}$, and $res_{D_h}A_h=A_{h-1}$ (notice that $\eta(a)\cap D_2$ is a double point, and the residue is the simple point $a\in D_1$), choosing $A=A_3$ we get $H^0(I_A\otimes\Lambda(4,7))=0$.

R(10) **t(8)=19, t(10)-t(8)=7, n(10)=7, δ(10)=1, ε(10)=0.**

R(8) $\Rightarrow$ R(10)

proof of type a

$Z=\pi^{-1}\big(\Gamma(6,1)\cup C_{19}\big)\cup r(7,1,0)$.

Then, $r(Z)=\pi^{-1}(C_{19})$ is a generalization of Y_8, while the i-claim follows by [I], lemma 3.3.1 (for f=1, h=3, m=0, u=19, n=5, δ=ε=0) and lemma 3.3.8 (for a=4, b=7, f=18, e=0, g=2, h=1).

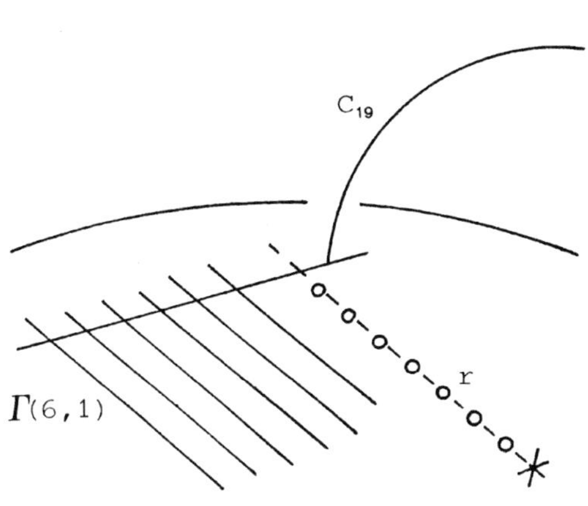

R(12) **$t(10)=26$, $t(12)-t(10)=9$, $n(12)=10$, $\delta(12)=1$, $\varepsilon(12)=0$.**

$R(10) \Rightarrow R(12)$

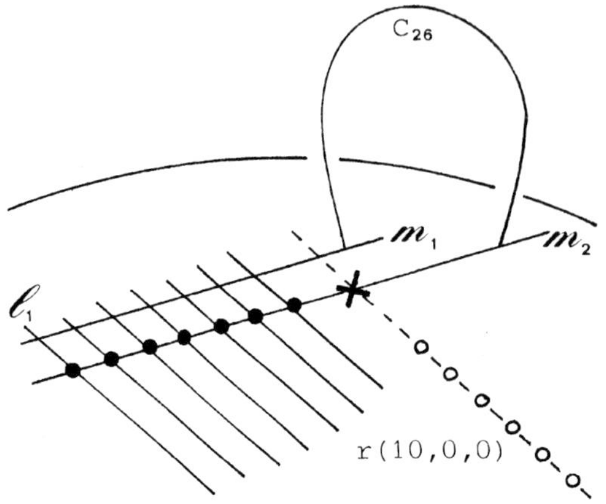

proof of type a

Z is the union of $\pi^{-1}\big(\Gamma(7,2) \cup \chi(\ell_1 \cap m_2) \cup C_{26}\big)$, of $r(10,0,0)$, and of a nilpotent supported on a d-point of $r \cap m_2$ (here ℓ_1, m_2 are lines as in picture). Then, $r(Z)$ is a generalization of Y_{10}, while the i-claim follows by [I], lemma 3.3.1 (for $f=1$, $h=5$, $m=0$, $u=50$, $n=10$, $\delta=\varepsilon=0$).

R(14) **$t(12)=35$, $t(14)-t(12)=11$, $n(14)=3$, $\delta(14)=3$, $\varepsilon(14)=0$.**

$R(10) \Rightarrow R(M,12)$, where $M= \pi^{-1}\big(C_{32} \cup 2 \text{ skew lines} \cup G(2,8)\big) \cup r(3,2,0)$

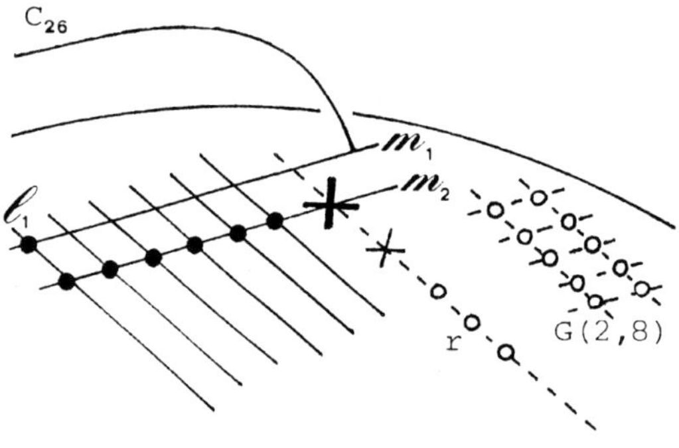

proof of type a

Z is the union of $\pi^{-1}\big(\Gamma(6,2) \cup \cup\chi(\ell_1 \cap m_2) \cup \chi(\ell_1 \cap m_1) \cup C_{26} \cup \cup G(2,8)\big) \cup r(3,1,0)$ and of a nilpotent supported on a d-point of $r \cap m_2$.

Then, $r(Z)$ is a generalization of Y_{10}, while the i-claim follows by the forthcoming lemma 4.6.2.

4.6.2 Lemma. If A denotes the union in $\mathbb{P}(\Lambda) = \pi^{-1}(Q)$ of 51 general t-points, of $\pi^{-1}(G(2,8))$ and of $r(3,1,0)$, with r line of type (1,0) on Q, then $H^0(I_A \otimes F_{7,11})=0$.

Proof. Let A° denotes the union of 44 general t-points, of $\pi^{-1}(G(2,8))$ and of 10 t-points and 1 d-point on r; A° is a specialization of A. We apply la méthode d'Horace for $U=\pi^{-1}(Q)$, $L=F_{7,11}$, $D=\pi^{-1}(r)$:

$$0 \to I_{res_{\pi^{-1}(r)}A^\circ} \otimes F_{6,11} \to I_{A^\circ} \otimes F_{7,11} \to I_{A^\circ \cap \pi^{-1}(r), \pi^{-1}(r)} \otimes F_{7,11}|_{\pi^{-1}(r)} \to 0.$$

Since $H^0(I_{A^\circ \cap \pi^{-1}(r), \pi^{-1}(r)} \otimes F_{7,11}|_{\pi^{-1}(r)})=0$ ([I] 3.2.1) and $H^0(I_{res_{\pi^{-1}(r)}A^\circ} \otimes F_{6,11})=0$ ([I] 3.3.7 with $a=1$, $h=5$, $p=2$, $q=8$, $v=44$) we have the thesis.

$R(M,12) \Rightarrow R(14)$

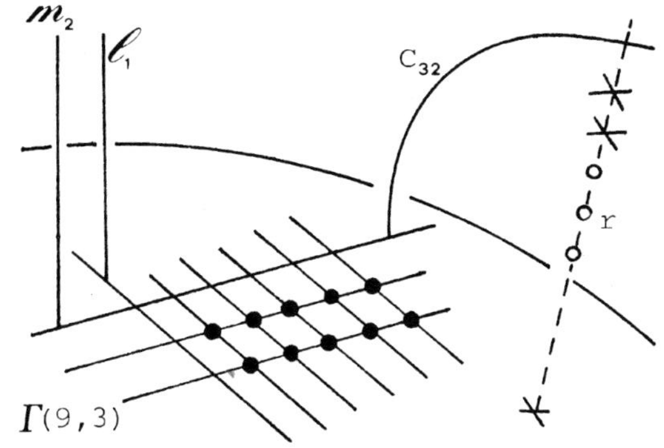

proof of type a
$Z = \pi^{-1}(\Gamma(9,3) \cup C_{32} \cup 2$ skew lines $\ell_1, m_2) \cup r(3,3,0)$, with ℓ_1, m_2 lines as in picture; a d-point of r lies on Q. Then, r(Z) is M, while the i-claim follows by [I], lemma 3.3.1 (for f=1, h=4, m=0, u=40, $n=\delta=\varepsilon=0$) and lemma 3.3.8 (for a=5, b=9, f=25, e=g=0, h=1).

R(16) **t(14)=46, t(16)-t(14)=12, n(16)=1, δ(16)=2, ε(16)=0.**
$R(12) \Rightarrow R(H,14)$, where $H = \pi^{-1}(C_{44} \cup$ a line $\ell) \cup 1$ d-point $\cup\, n(9,0,0) \cup$ $\cup\, m(9,0,0)$; we denote by a_i the t-points on the line n, by b_i the t-points on the line m; ℓ meets C_{44} in one point, and there esists a quadric Q containing n, m, the d-point and the lines a_ib_i for i=1...,9.

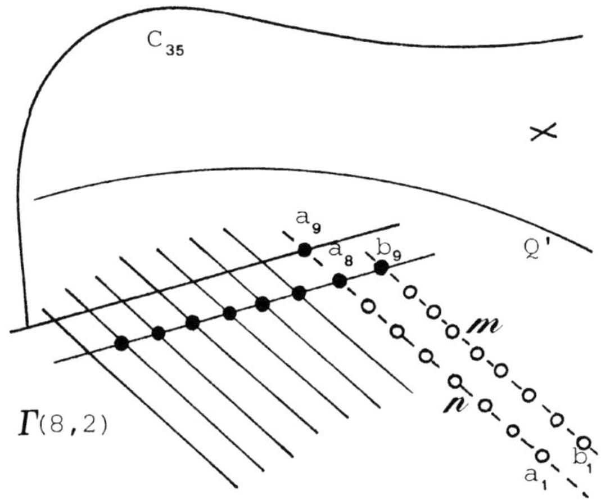

proof of type a
It is possible to choose a quadric Q' s.t. Q'contains n, m, the lines a_8b_9, and Q' does not contain the d-point and the lines a_ib_i for i=1...,7. In this proof we work over Q'.
Z is the union of $\pi^{-1}(\Gamma(8,2) \cup C_{35} \cup \{a_i\}_{i=1,..,7} \cup \{b_j\}_{j=1,...,8})$, of 1 d-point and of nilpotents supported on 3 t-points lying on $(m \cup n) \cap \Gamma(8,2)$ (see picture).

Then, r(Z) is a generalization of Y_{12}, while the i-claim is the following lemma 4.6.3.

4.6.3 Lemma. Let A be the union of the t-points $a_1,\dots,a_7,b_1,\dots,b_8$ with 69 general t-points in $\pi^{-1}(Q')$; then $H^0(I_A \otimes F_{7,13})=0$.

Proof. Let B_0 be the union of 12 general t-points; then $H^0(I_{B_0}\otimes F_{3,5})=0$ ([I] 3.3.1 for f=1, h=2). Choose a smooth curve Γ_i of type (1,2) with Γ_i through a_i,b_i,a_{i+4},b_{i+4}, for i=1,2,3, and Γ_4 through a_4,b_4,b_8. We apply la méthode d'Horace four times: $0 \to I_{res_{\pi^{-1}(\Gamma_i)}B_i}\otimes F_{3+(i-1),5+2(i-1)} \to I_{B_i}\otimes F_{3+i,5+2i} \to$ $\to I_{B_i\cap\pi^{-1}(\Gamma_i),\pi^{-1}(\Gamma_i)}\otimes F_{3+i,5+2i}\,|_{\pi^{-1}(\Gamma_i)} \to 0$, i=1,...,4, where B_i is defined as the union of B_{i-1} with the t-points a_i,b_i,a_{i+4},b_{i+4} and (4+4i) t-points on $\Gamma_i-(\Gamma_{i+1}\cup\ldots\cup\Gamma_4)$, and B_4 as the union of B_3 with $\{a_4,b_4,b_8\}$ and 21 t-points on Γ_4. Then $res_{\pi^{-1}(\Gamma_i)}\,B_i=B_{i-1}$, and $H^0(I_{B_i\cap\pi^{-1}(\Gamma_i),\pi^{-1}(\Gamma_i)}\otimes F_{3+i,5+2i}\,|_{\pi^{-1}(\Gamma_i)})=0$ for i=1,...,4 ([I] 3.2.2); since B_4 is a specialization of A, we have the thesis.

$R(H,14) \Rightarrow R(16)$

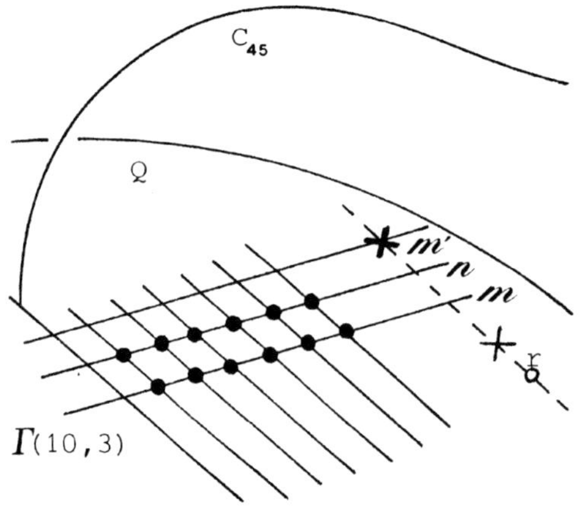

proof of type a

We work on the quadric Q. Z is the union of $\pi^{-1}(\Gamma(10,3) \cup C_{45}) \cup$ $\cup r(1,1,0)$ and of a nilpotent supported on a d-point of $r\cap m'$. Then, r(Z) is a generalization of H, while the i-claim follows by [I], lemma 3.3.8 (for a=6, b=11, g=1, f=29, e=0, h=1), and [I], 3.3.1 (for f=1, h=5, m=0, u=60, $n=\delta=\varepsilon=0$).

R(18) **t(16)=58, t(18)-t(16)=13, n(18)=0, δ(18)=7, ε(18)=0.**

$R(H,14) \Rightarrow R(S,16)$, where H was defined in R(16), and S is the disjoint union of $\pi^{-1}(C_{57})$, of 2 d-points and of $\pi^{-1}(\ell)$ where ℓ is a line.

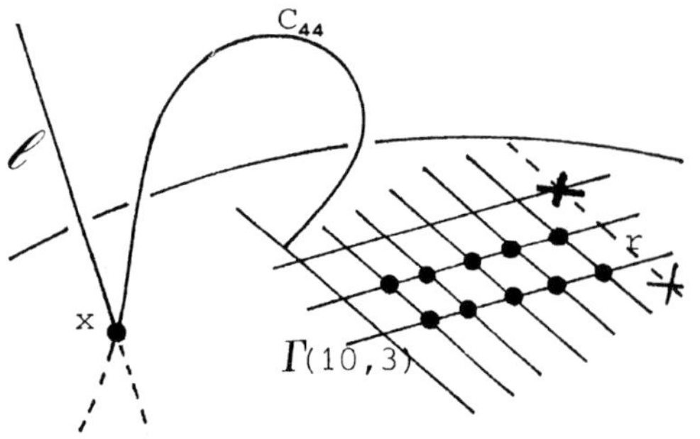

proof of type a

Z is the union of $\pi^{-1}(\Gamma(10,3) \cup C_{44} \cup$ $\cup$ a line $\ell \cup \chi(x))$ (x denoting the point $\ell\cap C_{44}$, see picture), of r(0,1,0) and of a nilpotent supported on a d-point lying on $r\cap\Gamma(10,3)$.

Then, r(Z)=H, while the i-claim follows by [I], lemma 3.3.8 (for a=6, b=11, g=0, f=30, e=0, h=1) and [I], 3.3.1 (for f=1, h=5, m=1, u=57, n=δ=ε=0).

R(S,16) $\Rightarrow$ R(18)

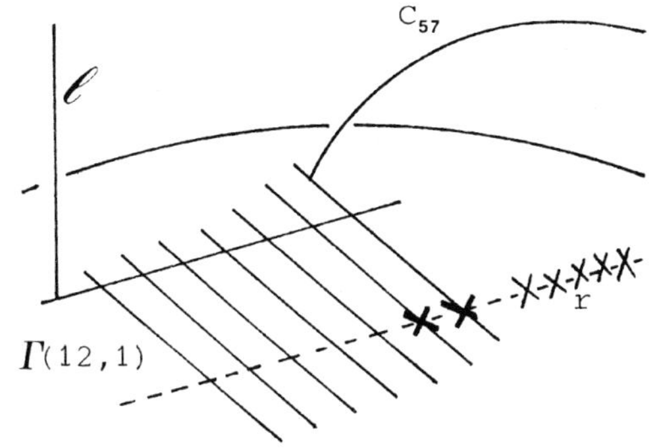

proof of type a

Z is the union of $\pi^{-1}(\Gamma(12,1) \cup C_{57} \cup \cup$ a line $\ell)$, of r(0,5,0), and of 2 nilpotents supported on two d-points lying on $\Gamma(12,1)$.

Then, r(Z) is a generalization of S, while the i-claim follows by the following lemma 4.6.4.

4.6.4 Lemma. Let A be the generic union of 114 t-points in $\pi^{-1}(Q)$ with r(0,5,0) where r is a line of type (0,1) on Q; then $H^0(I_A \otimes F_{7,18})=0$.

Proof. Let T_i be the union of $3i^2+7i+4$ general t-points and i d-points on r. We show that $H^0(I_{T_i} \otimes F_{2+i,3+3i})=0$; i=5 gives the thesis.

For i=0,T_0 is the union of 4 general t-points; then $H^0(I_{T_0} \otimes F_{2,3})=0$ ([I] 3.3.1 for f=1, h=1). Let Γ be a smooth curve of type (1,3), and define B_i as the union of $3i^2+i$ general t-points, and i-1 d-points on r, with 6i+4 t-points on Γ and 1 d-point on $\Gamma \cap r$; B_i is a specialization of T_i. Now apply la méthode d'Horace and use induction:

$$0 \rightarrow I_{res_{\pi^{-1}(\Gamma)} B_i} \otimes F_{1+i,3i} \rightarrow I_{B_i} \otimes F_{2+i,3+3i} \rightarrow I_{B_i \cap \pi^{-1}(\Gamma), \pi^{-1}(\Gamma)} \otimes F_{2+i,3+3i} |_{\pi^{-1}(\Gamma)} \rightarrow 0.$$

Since $res_{\pi^{-1}(\Gamma)} B_i = T_{i-1}$, and by induction assumption $H^0(I_{T_{i-1}} \otimes F_{1+i,3i})=0$, and since $H^0(I_{B_i \cap \pi^{-1}(\Gamma), \pi^{-1}(\Gamma)} \otimes F_{2+i,3+3i} |_{\pi^{-1}(\Gamma)})=0$ ([I] 3.2.2), we conclude that $H^0(I_{B_i} \otimes F_{2+i,3+3i})=0$.

SECTION 5: A FEW REMARKS.

The lemmas in section 2 concern the following problem: under which conditions $f_* I_{f^{-1}(Z)} \cong I_Z$ (f being a blowing up)? This is not always true: for example, let f: $X \rightarrow \mathbb{A}^2$ be the blowing up of the affine plane in the origin, and $I_Z=(x^2,y^2)$; then, $f_* I_{f^{-1}(Z)} \neq I_Z$ (this counterexample was given to me by R.Hartshorne, whom I thank). Hence, lemma 1.9.3.1 in my paper [I] is false under such general assumptions. This lemma 1.9.3.1 was used where elementary transformations were involved, and precisely in [I,2], 5.2, in [I] 1.9.3, 3.3.4, and finally inside the proofs of the following statements in [I]:

H(4): The inverse image in $\mathbb{P}(\Omega_3)$ of the generic union of 6 lines of $\mathbb{P}^3$ is L_4-settled.

H(5): The inverse image in $\mathbb{P}(\Omega_3)$ of the generic union of 8 lines of $\mathbb{P}^3$, union with 4 s-points lying on a line, is L_5-settled.

Luckily, using the tecniques of the present paper, the results in [I] and in [I,2] remain valid, and I take here the opportunity of explaining briefly how.

In [I] 1.9.3 and [I] 3.3.4 lemma 1.9.3.1 was repeatedly used in order to have isomorphisms of the following type: $h_*I_{h^{-1}(Z)} \cong I_Z$, and $\lambda_*I_{\lambda^{-1}(U),D} = I_{U,\mathbb{P}(F)}$, with h the blowing up of $\mathbb{P}(E)$ along $\mathbb{P}(F)$, and $\lambda := h|_D: D \to \mathbb{P}(F)$ (here E is a vector bundle on a projective smooth variety S, F is a vector bundle on a smooth divisor of S with rank(F) < rank(E), and F quotient sheaf of E, D is the exceptional divisor, Z is a closed subscheme of $\mathbb{P}(E)$, and U is a closed subscheme of $\mathbb{P}(F)$); due to the particular form of the schemes Z concerned, it is possible to use lemma 2.6 and lemma 2.4 of the present paper, instead of lemma 1.9.3.1. Something analogous, but more complicated, since the occurring nilpotents are more complicated, could be done in the remaining cases, but I think preferable to give here new and shorter proofs.

The proof of H(4) follows immediately by the forthcoming proposition 5.1, while the proof of H(5) can be rephrased avoiding elementary transformations as follows: let M be the subscheme of $\mathbb{P}^3$ generic union of 4 lines, of a singular reduced conic C with the singular point b on a plane H, of a singular reduced conic C' in H with singular point a, and of $\chi(a)$ and $\chi(b)$; let r be a generic line in H, fix a splitting $\Omega_{3|H} = \Omega_H \oplus L$, where $L \cong O_H(-1)$, and choose 4 s-points lying on r, 3 of which in $\mathbb{P}(L) \cap \pi^{-1}(r)$, and one generic in $\mathbb{P}(\Omega_H) \cap \pi^{-1}(r)$; then, exploiting $\pi^{-1}(H)$, we see that the union N of $\pi^{-1}(M)$ with the 4 s-points satisfies $H^0(L_5 \otimes I_N) = 0$ (the i-claim works as in [I], the r-claim follows again by the forthcoming proposition 5.1).

5.1 Proposition. Let T be the generic union in $\mathbb{P}^3$ of 4 lines, a singular reduced conic, and the first infinitesimal neighbourhood of the singular point of the conic. Then, $H^0(I_{\pi^{-1}(T)} \otimes L_4) = 0$.

Proof. Let M be the subscheme of $\mathbb{P}^3$ generic union of 2 lines of type (1,0) on a smooth quadric Q, of 4 lines r_1, r_2, s_1, s_2 such that r_1 and r_2 are skew, s_1 and s_2 are skew, r_i and s_j meet at a point P_{ij} and the four points of intersection are on Q, and of $\chi(P_{ij})$ for i,j=1,2. We exploit Q, in order to prove $H^0(\Omega_3(5)\otimes I_M)=0$; since M is a specialization of T, conclusion will follow by lemma 2.1. The r-claim $H^0(\Omega_3(5)\otimes I_Q\otimes I_{res_Q M})=0$ is easily proved exploiting a smooth quadric Q' containing the lines r_1, r_2, s_1, s_2, while for the i-claim is enough to prove that, denoting with Z the generic union of the first infinitesimal neighbourhoods of four points in Q, one has $H^0(\Omega_{3|Q}(3,5)\otimes I_Z)=0$. In the following I_Z denotes the ideal sheaf of Z in Q.

The cohomology of the exact sequence $0 \to O_Q(-1,-2)^{\oplus 2} \to \Omega_{3|Q} \to$ $\to O_Q(-2,0) \to 0$ (see proof of 1.1), twisted by $I_Z(3,5)$, gives the thesis, once shown that $H^0(I_Z(2,3))=0$ and $H^0(I_Z(1,5))=0$. We prove both at the same time. Set L:= $O_Q(2,3)$, resp. L:= $O_Q(1,5)$. We exploit C, where C is a smooth curve of type (1,2) on Q, resp. the union of 4 lines m_i of type (0,1), through the four points: the cohomology of the exact sequence $0 \to L\otimes I_C\otimes I_{res_C Z} \to L\otimes I_Z \to$ $\to L_{|C}\otimes I_{Z\cap C,C} \to 0$ gives the thesis, because $L_{|C} \cong O_{\mathbb{P}^1}(7)$, resp. $L_{|C} \cong \oplus_i O_{m_i}(1)$, and $Z\cap C$ consists of 4 double points; moreover, $res_C Z$ is the union of 4 general points, and in both cases $L\otimes I_C \cong O_Q(1,1)$.□

Finally, in [I,2] 5.2 it is stated that the union A in X of 4 general s-points with C, where C denotes the generic union of six lines $r_1,\dots,r_6$ with a proper 6-secant ℓ (proper means that ℓ meets each r_i in one point), satisfies $H^0(L_4\otimes I_A)=0$ (this was used to show that the map σ_4 is surjective for such a configuration of lines). A section of L_4 vanishing on A is obliged to vanish also on ℓ, since $\Omega_3(5)|_\ell \cong O_\ell(4)^{\oplus 2}\oplus O_\ell(3)$, so here we prove (using the notations of the previous sections) that $H^0(L_4\otimes I_J)=0$, where J is the union of C with ℓ and 1 t-point and 1 s-point on a line r meeting C; but the union of C with ℓ is a specialization of the generic rational curve of degree 7, hence we are in fact also giving another proof of R(4).

$R(1) \Rightarrow R(D,3)$, where D is the union of three lines and 4 t-points $P_1,\ldots,P_4$, with P_1,P_2,P_3 on a plane H containing a 3-secant ℓ for the three lines.

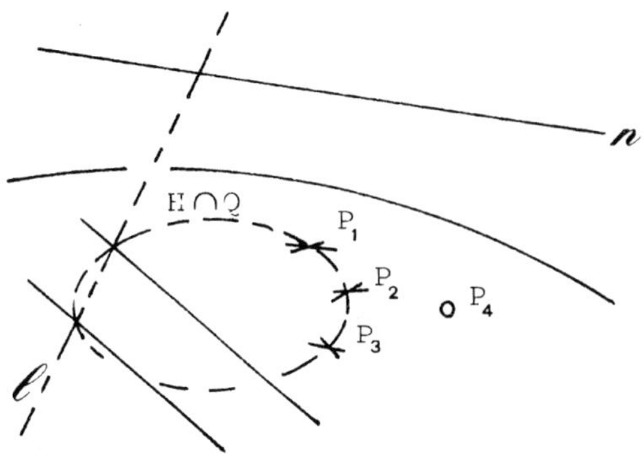

proof of type b

We choose Q not containing ℓ.

1st step: K' is the disjoint union of $p^{-1}(\Gamma(2,0))\cap W'$, of $p^{-1}(n)$, of 3 d-points in W', lying respectively on P_1, P_2, P_3, and of 1 t-point lying on $P_4 \in Q$; n is a line outside Q; hence $T^g=T=Y_1$. The scheme P' is the union of $\Gamma(2,0)$ and of P_1, P_2, P_3, P_4, $n\cap Q$.

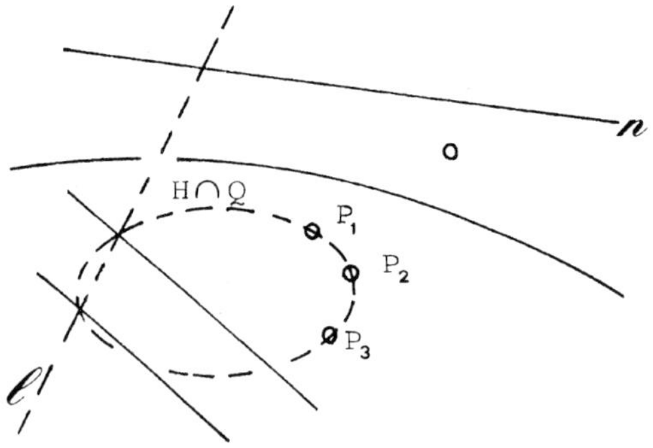

2nd step: K is the union of $\pi^{-1}(\Gamma(2,0) \cup n)$, of 3 t-points lying on P_1,P_2,P_3, and of 1 t-point outside Q.

The scheme P is the union of $\Gamma(2,0)$ and of P_1, P_2, P_3, $n\cap Q$.

$R(D,3) \Rightarrow R(J,4)$: here we use la méthode d'Horace for: U=X, $L=L_4$, $D=\pi^{-1}H$, where H denotes a plane of $\mathbb{P}^3$.

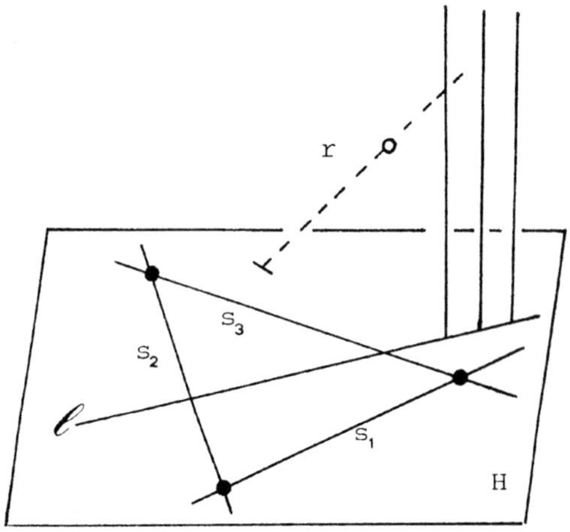

We fix a splitting $\Omega_3(5)|_H=\Omega_H(5)\oplus E$, with $E \cong \mathcal{O}_H(4)$, we chose Z specialization of J s.t. Z is union of the inverse images of: 4 lines ℓ, s_1,s_2,s_3 on H, $\chi(s_1\cap s_2)$, $\chi(s_2\cap s_3)$, $\chi(s_1\cap s_3)$, 3 lines outside H meeting ℓ, together with a t-point lying on r but outside H and an s-point in $\mathbb{P}(E)$; hence the r-claim is R(D,3), while the i-claim $H^0(L_4|_{\pi^{-1}(H)}\otimes \mathcal{I}_{Z\cap\pi^{-1}(H),\pi^{-1}(H)})=0$ is the same as in R(2).

REFERENCES.

[A-H] J.Alexander - A.Hirschowitz: "La méthode d'Horace éclatée: application à l'interpolation en degré quatre", Invent. Math. 107 (1992), 585-602.

[B-E] E.Ballico - Ph.Ellia: "The maximal rank conjecture for non special curves in $\mathbb{P}^3$", Invent. Math. 79 (1985), 541-555.

[B-E,2] E.Ballico - Ph.Ellia: "Beyond the maximal rank conjecture for curves in $\mathbb{P}^3$", in "Space curves", Proceedings Rocca di Papa 1985, Lect. Notes in Math. 1266 (1987), 1-23, Springer Verlag.

[C] G.Castelnuovo: "Sui multipli di una serie lineare di gruppi di punti appartenente ad una curva algebrica", Rend. Circ. Mat. Palermo 7 (1893), 89-110.

[D] J.D'Almeida: "Courbes de l'espace projectif: séries linéaires incomplètes et multisécantes", J. Reine Angew. Math. 370 (1986), 30-51.

[E-P] D.Eisenbud - S.Popescu: "Coble-Gale Duality, Adjoint Series, and Syzygies of Points", preprint.

[EGA I] A.Grothendieck - J.Dieudonné: "Eléménts de Géométrie Algébrique, I", Grudl. der Math. 166, Berlin-Heidelberg-New York (1971).

[EGA III-2] A.Grothendieck (rédigés avec la collaboration de J.Dieudonné): "Eléménts de Géométrie Algébrique, III, 2e partie", Pub. Math. 17, Institut des Hautes Etudes Scientifiques, Bures s/Yvette.

[G-L-P] L.Gruson - R.Lazarsfeld - Ch.Peskine: "On a theorem of Castelnuovo and the equations defining space curves", Invent. Math. 72 (1983), 491-506.

[H-H] R.Hartshorne - A.Hirschowitz: "Droites en position générale dans l'espace projectif", in "Algebraic Geometry", Proceedings La Rabida 1981, Lect. Notes in Math. 961 (1982), 169-188, Springer Verlag.

[Ha] R.Hartshorne: "Algebraic Geometry", Springer Verlag, Berlin-Heidelberg-New York 1977.

[H] A.Hirschowitz: "Sur la postulation générique des courbes rationnelles", Acta Math. 146 (1981), 209-230.

[H,2] A.Hirschowitz: "La Méthode d'Horace pour l'interpolation à plusieurs variables", Manuscripta Math. 50 (1985), 337-388.

[H,3] A.Hirschowitz: letter to R.Hartshorne (1983).

[H-S] A.Hirschowitz - C.Simpson: "La résolution minimale de l'idéal d'un arrangement général d'un grand nombre de points dans $\mathbb{P}^n$", Invent. Math. 126 (1996), 467-503.

[I] M.Idà: "On the homogeneous ideal of the generic union of lines in $\mathbb{P}^3$", J.Reine Angew. Math. (Crelles Journal) 403 (1990), 67-153.

[I,2] M.Idà: "Generating six skew lines in $\mathbb{P}^3$", in "Algebraic Curves and Projective Geometry", Proceedings Trento 1988, Lect.Notes in Math 1389 (1989), 112-127, Springer Verlag.

[I,3] M.Idà: "Maximal rank and minimal generation", Arch. Math. 52 (1989), 186-190.

[M] M.Maruyama: "Elementary transformations in the theory of algebraic vector bundles", in "Algebraic Geometry", Proceedings La Rabida 1981, Lect. Notes in Math. 961 (1982), 241-266, Springer Verlag.

[P] D.Perrin: "Courbes passant par m points généraux de $\mathbb{P}^3$", Mémoire de la Société Mathématique de France, n.28/29 (1987).

Equisingularity, Multiplicity, and Dependence

Steven L Kleiman

Department of Mathematics, Room 2-278 MIT,
77 Mass Ave, Cambridge, MA 02139-4307, USA
E-mail: Kleiman@math.mit.edu

ABSTRACT. This is a report on some recent work by Gaffney, Massey, and the author, characterizing the conditions A_f and W_f for a family of ICIS germs equipped with a function. First we introduce the work informally. Then we review the formal definitions of A_f and W_f, and state the theorems that characterize them by the constancy of Milnor numbers. Next we review the definition of the Buchsbaum–Rim multiplicity, and reformulate the theorems by the constancy of certain Buchsbaum–Rim multiplicities. Finally, we review the theory of integral dependence of elements on submodules of free modules, and apply it to prove the reformulated theorems.

1. Introduction

The conditions A_f and W_f are "relative" forms of the Whitney conditions A and B (or W). The latter are important as they are manageable algebraic-geometric conditions on the tangents to a variety X, yet they imply the local topological triviality of X along a given smooth subvariety Y such that $X - Y$ is smooth. When X carries a map f that has constant rank off Y, then A_f and W_f are defined as the corresponding conditions on the tangents to the level sets (or fibers) of f. Thom introduced A_f as the primary condition that would ensure the topological triviality of the pair X, f along Y.

By definition, A_f and W_f signify this. For simplicity, assume that f is a nonconstant function vanishing on Y, and embed X in $\mathbf{C}^n$ so that Y is a linear subspace through the origin 0. Then A_f holds at 0 if Y lies in every hyperplane obtained as a limit of hyperplanes H, each tangent to a level hypersurface of f at a point x of $X - Y$, as x aproaches 0. A more stringent condition, W_f signifies that the angle between H and Y approaches 0 as fast as x approaches 0.

Research supported in part by NSF grant 9400918-DMS.

It is a great pleasure for the author to thank Terence Gaffney for enthusiastically introducing him to the present theory, for patiently answering his countless questions, and for generously encouraging their fruitful collaboration. It is also a pleasure to thank David Massey for several helpful discussions. In addition, the author is sincerely grateful to Herwig Hauser and Bernard Teissier for reading an earlier draft and making valuable comments and suggestions.

The Thom–Mather second isotopy lemma readily yields the following isotopy theorem: if W_f holds at 0, then, after X is replaced by a neighborhood of 0, the pair X, f is topologically right trivial. More precisely, let X_y denote the fiber of a transverse projection to Y, and set $f_y := f|X_y$. Then there is a homeomorphism h from the product $X_0 \times Y$ onto X such that fh is equal to $f_0 \times 1_Y$. Thus, if W_f holds at 0, then, for $y \in Y$ near 0, the pairs X_y, f_y are topologically the same, or "topologically equisingular."

In the present article, we discuss and develop some recent work done by Gaffney, Massey, and the author in [**5**] and [**6**] on the algebraic-geometric significance of A_f and W_f. Here X is the total space of a complex analytic family of germs of isolated complete-intersection singularities (ICIS germs) X_y, parameterized by a smooth variety Y.

For convenience, we identify Y with the subvariety of X traced by the central points $0 \in X_y$ as y varies. We assume f vanishes on Y. We choose an embedding of X in $\mathbf{C}^a \times \mathbf{C}^b$ such that Y represents the germ of $0 \times \mathbf{C}^b$. We choose an extension of f over a neighborhood of X, and we denote the extension by f too. Needless to say, our results are independent of these choices.

What does A_f mean by itself? Let Z_y be the level hypersurface through 0 of f_y. It turns out that A_f is closely related to the vanishing cycles associated to the individual X_y and Z_y. In the case where X represents the germ of the whole affine space $\mathbf{C}^a \times \mathbf{C}^b$ and Y is the critical set $\Sigma(f)$, Lê and Saito [**18**] proved that if the number of vanishing cycles, the Milnor number $\mu(Z_y)$ at 0, is constant in y, then A_f holds.

We prove a definitive generalization of this celebrated theorem of Lê and Saito to ICIS germs. (However, Green and Massey, [**7**] and [**21**], showed that other information about the vanishing cycles implies A_f for families with generalized isolated singularities.) Namely, we characterize A_f by the constancy in y of the Milnor numbers $\mu(X_y)$ and $\mu(Z_y)$. Notice that these Milnor numbers refer to the ambient topology of X_y and Z_y in $\mathbf{C}^a$, but the isotopy theorem does not.

We characterize W_f similarly, by the constancy of the sequences of Milnor numbers $\mu_i(X_y)$ and $\mu_i(Z_y)$ of the sections of X_y and Z_y by general linear spaces of codimension i for $i = 0, \ldots, a - k$. The appearance of these linear sections is a reflection of the Lipschitz-like nature of the vector fields that we integrate to prove the isotopy theorem. (The corresponding theorem for W was proved for hypersurface germs in part by Teissier and in part by Briançon and Speder, and for general ICIS germs by Gaffney [**4**, Thm. 1].)

We also characterize A_f and W_f by the constancy of certain Buchsbaum–Rim multiplicities. Say X is defined in $\mathbf{C}^a \times \mathbf{C}^b$ by the vanishing of $f_1, \ldots, f_k$, and form the Jacobian matrix of $f_1, \ldots, f_k, f$ with respect to the first a variables. For each y, the a columns of this matrix generate a module over the local ring $\mathcal{O}_{X_y,0}$. This is a submodule $\mathcal{M}_y$ of finite colength of the free module of rank $k+1$. We characterize A_f by the constancy in y of the Buchsbaum–Rim multiplicity $e(\mathcal{M}_y)$. Let $\mathbf{m}_y$ be the maximal ideal of X_y; its appearance is a more direct reflection of the Lipschitz-like nature of the vector fields. We characterize W_f by the constancy of the Buchsbaum–Rim multiplicity of the product, $e(\mathbf{m}_y\mathcal{M}_y)$.

Thus our theorems provide necessary and sufficient conditions for A_f (resp., for W_f) to hold. These conditions require the constancy in y of certain numerical invariants of the individual pairs X_y, f_y. Therefore, when these invariants are constant, then, in whatever way the individual pairs are glued together to form a pair X, f, necessarily A_f (resp., W_f) holds. Conversely, if one pair X, f exists for which A_f (resp., W_f) holds, then these invariants of X_y, f_y have the same values for all $y \in Y$ near 0. Thus these numerical invariants may be considered as indicators of "A_f-equisingularity" (resp., "W_f-equisingularity").

The invariants depend only on the germs of X_y and f_y at 0. Yet, of course, there is a significant difference between these germs and the larger representatives themselves. Assume that 0 is the only critical point of the germ of f_y. Still, the representative f_y may have an additional critical point no matter how close y is to 0. However, f_y has no additional critical point if our numerical invariants are constant. Furthermore, if there is no such critical point, then A_f holds. Conversely, if A_f holds, then, for all y close to 0, each additional critical point of f_y is also a critical point of f.

Our treatments of A_f and W_f run in parallel. Not only are the statements similar, but, to a fair extent, the proofs are similar. First, we prove that the two numerical characterizations of A_f (resp., of W_f) are equivalent by proving that the constancy of the Milnor numbers is equivalent to the constancy of the Buchsbaum–Rim multiplicities. This proof involves the theorem of Lê and Greuel, which re-expresses the Milnor numbers algebraically, and some theorems of Buchsbaum and Rim, which re-express the multiplicity $e(\mathcal{M}_y)$ as a length (resp., and a more recent theorem, which re-expresses $e(\mathbf{m}_y\mathcal{M}_y)$ as a linear combination of the polar multiplicities of X_y).

We prove the characterization of A_f (resp., of W_f) by Buchsbaum–Rim multiplicities using the theory of integral dependence of elements on submodules of free modules. Let $\mathcal{M}$ be the module, over the local ring of X at 0, generated by the a columns of the Jacobian matrix above. Let g_j be the column vector of partial derivatives of $f_1, \ldots, f_k, f$ with respect to the jth coordinate variable on $\mathbf{C}^b$. Finally, let $\mathbf{m}_Y$ be the ideal of Y in X.

We characterize A_f (resp., W_f) by the integral dependence of $g_1, \ldots, g_b$ on $\mathcal{M}$ (resp., on $\mathbf{m}_Y\mathcal{M}$). We prove this characterization of A_f using a remarkable result, Lemma (5.7) in [**6**], concerning the geometry of the relative conormal variety. Technically, this result is the new ingredient, which permits perfecting the treatment of A_f in [**5**]. We prove the characterization of W_f via a direct computation with analytic inequalities.

The restrictions $g_j|X_y$ are always integrally dependent on $\mathcal{M}_y$ (resp., on $\mathbf{m}_y\mathcal{M}_y$) for all y in a dense Zariski open subset of Y, as A_f (resp., W_f) holds for all y in such a subset by a celebrated theorem proved by Hironaka (resp., by Henry, Merle, and Sabbah). It follows that, if the Buchsbaum-Rim multiplicity

$e(\mathcal{M}_y)$ (resp., $e(\mathbf{m}_y\mathcal{M}_y)$) is constant, then the g_j are dependent on $\mathcal{M}$ (resp., on $\mathbf{m}_Y\mathcal{M}$) by virtue of the "principle of specialization of integral dependence."

This principle is the main algebraic result, Theorem (1.8), in [**5**]. For ideals, it was discovered, named, and proved by Teissier [**24**, 3.2, p. 330] and [**25**, App. I]. To prove it for modules, we follow Teissier's approach, but first we must generalize certain basic results in the theory of multiplicity for ideals, including upper semicontinuity, Rees's characterization of integral dependence, and Böger's generalization of it (Teissier rediscovered the latter in the case at hand). Upper semicontinuity of the Buchsbaum–Rim multiplicity is proved in Proposition (1.1) of [**5**]. The generalizations of Rees's theorem and of Böger's theorem are proved in Theorems (6.7a)(iii) and (10.9) in [**14**]. Moreover, Böger's theorem is given a new proof in [**16**]; see its subsections (1.4) and (1.7). This proof is simpler, shorter, and more direct. It was inspired by Lemma (5.7) in [**6**].

The present article is meant to give a feeling for the nature and the spirit of the work in [**5**] and its extension in [**6**]. The latter article also surveys a lot of other work in equisingularity theory, and discusses its historical development. This is a masterful report, and may be highly recommended.

The present discussion is, of course, limited, and so achieves greater focus. Some results in [**5**] are not discussed here, and some are discussed in restricted generality. Some proofs are abridged, and some are omitted. Also, the general theory is recalled in the special case at hand. However, some noteworthy results were not explicitly discussed before, and they are stated and proved here.

In Section 2, we review the formal definitions of A_f and W_f. We discuss isotopy, and state the two theorems, which characterize A_f and W_f by the constancy of Milnor numbers. In Section 3, we review the formal definition of the Buchsbaum–Rim multiplicity. Then we reformulate the theorems in terms of the constancy of Buchsbaum–Rim multiplicities, and we prove that the two formulations are equivalent. In Section 4, we review the theory of integral dependence of elements on submodules of free modules, and we characterize A_f and W_f in terms of integral dependence. Finally, we apply this theory to prove the reformulated theorems.

2. Equisingularity

Let X be a complex-analytic germ at 0 in $\mathbf{C}^a \times \mathbf{C}^b$. Say that, on a (Euclidean) neighborhood of 0, we have

$$X : f_1, \ldots, f_k = 0,$$

where each f_i is an analytic function $f_i(x, y)$ of the two sets of variables,

$$x = (x_1, \ldots, x_a) \text{ and } y = (y_1, \ldots, y_b).$$

Assume that X is a reduced complete intersection of codimension k with $k < a$.

For fixed y, let $X_y \subset \mathbf{C}^a$ denote the locus of x such that $(x, y) \in X$; so X_y is the locus of zeros of the $f_i(x, y)$ as x varies. Assume that, if X_y is nonempty, then $0 \in X_y$. Let Y be the locus of y with $(0, y) \in X$, and identify Y with $0 \times Y$. View Y as the parameter space and X as the total space of the family of X_y. Assume that the X_y represent germs at 0 of isolated complete-intersection singularities (ICIS germs) of codimension k.

Let f be a nonconstant analytic function on X. Set

$$f_y := f|X_y \text{ and } Z := X \cap f^{-1}0 \text{ and } Z_y := X_y \cap f^{-1}0.$$

Assume that f vanishes on Y, so $Y \subset Z$.

If $x \in X$ is a simple point of the level hypersurface $f^{-1}fx$, then x must be a simple point of X. Let $\Sigma(f)$ denote the critical set, the union of the singular sets of the various level hypersurfaces. Then, in other words, $\Sigma(f)$ contains the singular set of X.

In turn, $\Sigma(f)$ is contained in the union of the critical sets of the various restrictions f_y. Denote this union by $\Sigma_Y(f)$. Note that $\Sigma_Y(f)$ consists of all the singular points of all the level hypersurfaces of all the restrictions f_y for all the y in Y. Replacing X by a smaller representative of the same germ, we may also assume that every component of $\Sigma_Y(f)$ contains 0. Similarly, if we assume that Z_0 represents a germ with an isolated singularity at 0, then we may also assume that the projection $\Sigma_Y(f) \to Y$ is finite. Note that we work only with the reduced structures on the sets $\Sigma(f)$ and $\Sigma_Y(f)$.

The *Thom condition* A_f can be formulated succinctly using the *relative conormal variety* $C(X, f)$ and the *absolute conormal variety* $C(Y)$. These varieties are defined as follows. Both are closures in $X \times \mathbf{P}^{a+b-1}$, or more correctly, in the restriction to X of the projectivized cotangent bundle of $\mathbf{C}^a \times \mathbf{C}^b$. The former closure is that of the set of pairs (x, H) such that x is a point in $X - \Sigma(f)$ and H is a hyperplane tangent at x to the level surface $f^{-1}fx$. The latter closure is that of the set of pairs (x, H) such that $x \in Y \subset H$. In these terms, A_f is said to be satisfied by the pair $(X - \Sigma(f), Y)$ at 0 if the fiber of $C(X, f)$ over $0 \in X$ lies in $C(Y)$.

Suppose the germs of $\Sigma(f)$ and Y at 0 are equal. If A_f holds at 0, then A_f holds at every y in a neighborhood U of 0 in Y. This statement is not obvious from the definition, but follows from Proposition (4.2)(2) below.

Hence, $C(Y)$ contains the preimage of U in $C(X, f)$, which will be denoted by $C(X, f)|U$. Since the intersection $C(X, f) \cap C(Y)$ always projects into Y, we conclude that A_f holds at 0 if and only if, after we replace X by a smaller representative of the same germ, we obtain the set-theoretic equation,

$$C(X, f) \cap C(Y) = C(X, f)|Y.$$

In other words, A_f *holds at* 0 *if and only if, along the fiber of* $C(X, f)$ *over* $0 \in X$, *the ideal of the intersection* $C(X, f) \cap C(Y)$ *has the same radical as the ideal of the preimage* $C(X, f)|Y$; compare with the *Remarque* on p. 550 in [**19**].

The condition A_f can also be expressed analytically in terms of the "angular distance" $\operatorname{dist}(Y, T_x f^{-1} fx)$ from Y to the tangent space at x to the level hypersurface. Namely, A_f holds at 0 if and only if this distance approaches 0 as x approaches 0 along any analytic path $\phi: (\mathbf{C}, 0) \to (X, 0)$ such that $\phi(u)$ lies in $X - \Sigma(f)$ for $u \neq 0$. Now, this distance approaches 0 if and only if the inequality,

$$\operatorname{dist}(Y, T_x f^{-1} fx) \leq c \cdot \operatorname{dist}(x, Y)^e,$$

holds where the constant c and the exponent e depend on the path ϕ.

The *Whitney condition relative to* f, denote W_f, is defined by requiring the preceding analytic inequality to hold for some constant c independent of ϕ and for $e = 1$. This condition generalizes Teissier's condition of 'c-equisingularity' (see [**19**, top, p. 550]). It reduces to Whitney's Condition B, in the equivalent form of Verdier's condition W [**27**, Sect. 1], when f is constant (and so vanishes).

There is also a characterization of W_f in terms of the conormal varieties $C(X, f)$ and $C(Y)$, which strengthens the corresponding characterization of A_f in an interesting way. Indeed, Lê and Teissier proved (a more general version of) the following result in their Proposition 1.3.8 on p. 550 of [**19**] (see Proposition (6.1) in [**5**] for another treatment of the general case): W_f *holds at* 0 *if and only if, along the fiber of* $C(X, f)$ *over* $0 \in X$, *the ideal of the intersection* $C(X, f) \cap C(Y)$ *is integral over the ideal of the preimage* $C(X, f)|Y$.

The condition W_f implies that the pair X, f is topologically right trivial over Y. Indeed, Thom, Mather, and Teissier, Verdier, Gaffney and others introduced and developed the requisite methods to prove this triviality by integrating vector fields. Namely, it is possible to take the constant tangent vector field to Y and carefully lift it to X so that the lift is Lipschitz-like (Fr. *rugueux*), hence integrable, and is tangent to the level hypersurfaces of f, so that the integral gives an appropriate continuous flow on X.

Thus we obtain the following isotopy theorem: *If the critical set* $\Sigma(f)$ *is equal to* Y *and if the pair* $(X - Y, Y)$ *satisfies* W_f *at* 0, *then, after* X *is replaced by a neighborhood of* 0, *there is a homeomorphism* $h: X_0 \times Y \to X$ *such that* $fh = f_0 \times 1_Y$ *and* h *is* C^∞ *off* Y.

Let X^i be the section of X by a general linear space of codimension i containing Y, and set $f^i := f|X^i$; so $X^0 = X$ and $f^0 = f$. If W_f holds, then W_{f^i} holds for $i = 0, \ldots, a - k - 1$ because the requisite analytic inequality follows from that for W_f. Hence the pair X^i, f^i is also topologically trivial by the isotopy theorem.

The next theorem characterizes W_f in terms of the Milnor numbers $\mu_i(X_y)$ and $\mu_i(Z_y)$ of the sections of X_y and Z_y by a general linear space through 0 of codimension i for $i = 0, \ldots, a - k$. By convention, $\mu_{a-k}(X_y)$ and $\mu_{a-k-1}(Z_y)$ are the ordinary multiplicities at 0 diminished by 1, and $\mu_{a-k}(Z_y) = 1$. However, for $\mu_i(Z_y)$ to be defined, Z_y too must have an isolated singularity at 0. For y near 0, it does if the germs of $\Sigma(f)$ and Y at 0 are equal, and if W_f, or simply A_f, holds; indeed, then the germs of $\Sigma(f)$ and $\Sigma_Y(f)$ at 0 are equal by Lemma (4.3) below.

Theorem (2.1) *The following three conditions are equivalent:*

(i) *the germs of* $\Sigma(f)$ *and* Y *at* 0 *are equal, and the pair* $(X - \Sigma(f), Y)$ *satisfies* W_f *at* 0;

(ii) *for the* y *in a neighborhood of* 0 *in* Y, *the level hypersurface* Z_y *has an isolated singularity at* 0, *and the sequences of Milnor numbers,* $\{\mu_i(X_y)\}$ *and* $\{\mu_i(Z_y)\}$, *are constant in* y;

(ii′) *for the* y *in a neighborhood of* 0 *in* Y, *the level hypersurface* Z_y *has an isolated singularity at* 0, *and the sequence of sums of Milnor numbers,* $\{\mu_i(X_y) + \mu_i(Z_y)\}$, *is constant in* y.

This theorem is part of Theorem (6.4) in [**5**]. The latter also asserts that (i) and (ii) are equivalent to the following condition: *the germs of* $\Sigma_Y(f)$ *and* Y *at* 0 *are equal, and both pairs* $(X - Y, Y)$ *and* $(Z - Y, Y)$ *satisfy the absolute Whitney condition* W *at* 0.

This additional condition is implied by (i). Indeed, the two germs are equal by Lemma (4.3) below because W_f implies A_f. Furthermore, (i) implies the requisite analytic inequalities. Indeed, $T_x f^{-1} f x \subset T_x X$ and, if $x \in Z$, then $T_x f^{-1} f x = T_x Z$.

The additional condition implies (ii); indeed, this implication is virtually the assertion of Théorème (10.1) on p. 223 of [**22**]. Trivially, (ii) implies (ii′).

Finally, (ii′) implies (i), but our proof is more involved, and runs as follows: in Theorem (3.1) below, we replace (ii′) with an equivalent condition involving a Buchsbaum–multiplicity, and then at the end of Section 4, we prove that latter implies (i). On the other hand, there is a direct proof that the additional condition implies (i); indeed, Briançon, Maisonobe and Merle gave such a proof in [**1**, Thm. 4.3.2, p. 543] in a more general setting using a different approach.

The next theorem characterizes A_f in terms of Milnor numbers.

Theorem (2.2) *The following four conditions are equivalent:*

(i) *the germs of* $\Sigma(f)$ *and* Y *at* 0 *are equal, and the pair* $(X - \Sigma(f), Y)$ *satisfies* A_f *at* 0;

(ii) *the germs of* $\Sigma_Y(f)$ *and* Y *at* 0 *are equal;*

(iii) *for the* y *in a neighborhood of* 0 *in* Y, *the level hypersurface* Z_y *has an isolated singularity at* 0, *and the Milnor numbers,* $\mu(X_y)$ *and* $\mu(Z_y)$, *are constant in* y;

(iii′) *for the* y *in a neighborhood of* 0 *in* Y, *the level hypersurface* Z_y *has an isolated singularity at* 0, *and the sum of Milnor numbers,* $\mu(X_y) + \mu(Z_y)$, *is constant in* y.

Notice that (iii) implies (iii′) trivially. The converse holds because $\mu(X_y)$ and $\mu(Z_y)$ are each upper semicontinuous by [**20**, bot. p. 126]. We prove the rest of the theorem in two steps too: first, we replace (iii′) with an equivalent condition involving a Buchsbaum–Rim multiplicity, obtaining Theorem (3.2); then at the end of Section 4, we prove Theorem (3.2).

In the case where where X represents the germ of $\mathbf{C}^a \times \mathbf{C}^b$, Lê and Saito [**18**] proved that (iii) implies (i). They used Morse theory, but Teissier reproved their theorem almost immediately using more algebraic-geometric methods. Teissier's work served as a model for most of the work in this report.

Combining the preceding two theorems, we obtain the following corollary, which asserts the equivalence of W_f, W_{f^i}, and A_{f^i}.

Corollary (2.3) *If $\Sigma(f) = Y$, then the following conditions are equivalent:*

(i) *the pair $(X - Y, Y)$ satisfies W_f at 0;*
(ii) *the pair $(X^i - Y, Y)$ satisfies W_{f^i} at 0 for every i;*
(iii) *the pair $(X^i - Y, Y)$ satisfies A_{f^i} at 0 for every i.*

3. Multiplicity

In this section, we reformulate Theorems (2.1) and (2.2). Instead of Milnor numbers, we use certain Buchsbaum–Rim multiplicities. Thus we obtain Theorems (3.1) and (3.2). In the next section, we discuss the proofs of these reformulated theorems.

The Buchsbaum–Rim multiplicity was introduced by Buchsbaum and Rim [**2**] in 1963, and the theory has been developed more recently by Kirby and Rees [**13**], by Henry and Merle [**10**], and by Thorup and the author [**14**] and [**15**]. For our purposes here, it suffices to work over the local ring $\mathcal{O}$ of a complex-analytic germ, say one of dimension d. Let $\mathcal{E}$ be a free $\mathcal{O}$-module, and $\mathcal{M}$ a submodule of finite colength; that is, the vector-space dimension $\dim_{\mathbf{C}}(\mathcal{E}/\mathcal{M})$ is finite.

The Buchsbaum–Rim multiplicity generalizes the ordinary multiplicity. In the case where $\mathcal{E}$ is the ring $\mathcal{O}$ and where $\mathcal{M}$ is an ideal $\mathcal{I}$, Samuel defined the multiplicity $e(\mathcal{I})$ in 1951 to be the rectified leading coefficient e of the Hilbert–Samuel polynomial,

$$\dim_{\mathbf{C}}(\mathcal{E}/\mathcal{I}^n) = e\,n^d/d! + \cdots \quad \text{for } n \gg 0.$$

Buchsbaum and Rim considered the case in which $\mathcal{E}$ is free of arbitrary (finite) rank r. They generalized Samuel's definition essentially as follows. Form the symmetric algebra $\mathcal{O}[\mathcal{E}]$; it is just the polynomial algebra in r variables with coefficients in $\mathcal{O}$. Form the *Rees* algebra $\mathcal{O}[\mathcal{M}]$; it is just the subalgebra generated by $\mathcal{M}$ placed in degree 1. Both these algebras are graded; denote their nth graded pieces by $\mathcal{O}[\mathcal{E}]_n$ and $\mathcal{O}[\mathcal{M}]_n$. For example, if $\mathcal{E} = \mathcal{O}$ and $\mathcal{M} = \mathcal{I}$, then $\mathcal{O}[\mathcal{E}]_n = \mathcal{O}$ and $\mathcal{O}[\mathcal{M}]_n = \mathcal{I}^n$.

Buchsbaum and Rim formed the quotient of these two graded pieces, and proved that its dimension is eventually given by a polynomial in n of degree $d + r - 1$; thus,

$$\dim_{\mathbf{C}}(\mathcal{O}[\mathcal{E}]_n/\mathcal{O}[\mathcal{M}]_n) = e\,n^{d+r-1}/(d+r-1)! + \cdots \quad \text{for } n \gg 0.$$

Then they defined the *multiplicity* $e(\mathcal{M})$ to be e.

To use the Buchsbaum–Rim multiplicity, return to the setup described at the beginning of Section 2. Fix an extension of f over a neighborhood of X in $\mathbf{C}^a \times \mathbf{C}^b$ on which $f_1, \ldots, f_k$ are defined, and abusing notation, denote the extension too by f. Form the Jacobian matrix with respect to the first set of variables:

$$\begin{bmatrix} \partial f_1/\partial x_1 & \ldots & \partial f_1/\partial x_a \\ \vdots & \ddots & \vdots \\ \partial f_k/\partial x_1 & \ldots & \partial f_k/\partial x_a \\ \partial f/\partial x_1 & \ldots & \partial f/\partial x_a \end{bmatrix}.$$

Its columns generate a module over the affine ring $\mathcal{O}_X$,

$$JM_x(f_1, \ldots, f_k; f) \subset \mathcal{E} := \mathcal{O}_X^{k+1},$$

called the *Jacobian module* of $f_1, \ldots, f_k; f$ with respect to the x-variables.

Given $y \in Y$, form the image $\mathcal{M}_y$ of $JM_x(f_1, \ldots, f_k; f)$ in the free module $\mathcal{E}_y$ of rank $k+1$ over the local ring $\mathcal{O}_{X_y,0}$; so

$$\mathcal{M}_y := JM_x(f_1, \ldots, f_k; f)|X_y \subset \mathcal{E}_y.$$

In addition, denote the maximal ideal of $\mathcal{O}_{X_y,0}$ by $\mathbf{m}_y$.

Suppose for a moment that each Z_y has an isolated singularity at 0. Since

$$\Sigma_Y(f) = \operatorname{Supp}\big(\mathcal{E}/JM_x(f_1, \ldots, f_k; f)\big),$$

the module $\mathcal{M}_y$ has finite colength. Hence the Buchsbaum–Rim multiplicities $e(\mathcal{M}_y)$ and $e(\mathbf{m}_y\mathcal{M}_y)$ are defined.

We can now reformulate Theorems (2.1) and (2.2); we obtain Theorems (3.1) and (3.2). After stating them, we discuss the proof that the two formulations are equivalent.

Theorem (3.1) *The following two conditions are equivalent:*

(i) *the germs of $\Sigma(f)$ and Y at 0 are equal, and the pair $(X - \Sigma(f), Y)$ satisfies W_f at 0;*

(ii) *for the y in a neighborhood of 0 in Y, the level hypersurface Z_y has an isolated singularity at 0, and the multiplicity $e(\mathbf{m}_y\mathcal{M}_y)$ is constant in y.*

Theorem (3.2) *The following three conditions are equivalent:*

(i) *the germs of $\Sigma(f)$ and Y at 0 are equal, and the pair $(X - \Sigma(f), Y)$ satisfies A_f at 0;*

(ii) *the germs of $\Sigma_Y(f)$ and Y at 0 are equal;*

(iii) *for the y in a neighborhood of 0 in Y, the level hypersurface Z_y has an isolated singularity at 0, and the multiplicity $e(\mathcal{M}_y)$ is constant in y.*

The new theorems are equivalent to the old because of the following lemma. Indeed, the summands in (3.3.2) are upper semicontinuous; so they are constant if and only if $e(\mathbf{m}_y\mathcal{M}_y)$ is. For future use, note that therefore the lemma yields this: *if $e(\mathbf{m}_y\mathcal{M}_y)$ is constant, then so is $e(\mathcal{M}_y)$.*

Lemma (3.3) *For each $y \in Y$, we have these two equations:*

$$e(\mathcal{M}_y) = \mu(X_y) + \mu(Z_y); \tag{3.3.1}$$

$$e(\mathbf{m}_y \mathcal{M}_y) = \sum_{i=0}^{a-k} \binom{a-1}{i} \big(\mu_i(X_y) + \mu_i(Z_y)\big). \tag{3.3.2}$$

Moreover, each sum $\mu_i(X_y) + \mu_i(Z_y)$ is upper semicontinuous in y.

To prove Equation (3.3.1), note that X_y is a complete intersection of dimension $a-k$ and that $\mathcal{M}_y$ is a submodule of $\mathcal{O}_{X_y,0}^{k+1}$ generated by a elements, namely, the columns of the Jacobian matrix. Denote the ideal of maximal minors of this matrix by $\mathcal{J}$. Then some theorems of Buchsbaum and Rim [**2**, 2.4, 4.3, 4.5] yield

$$e(\mathcal{M}_y) = \dim_{\mathbf{C}}\big(\mathcal{O}_{X_y,0}/\mathcal{J}\mathcal{O}_{X_y,0}\big). \tag{3.3.3}$$

The right side is equal to the sum $\mu(X_y) + \mu(Z_y)$ by the theorem of Lê [**17**, Thm. 3.7.1, p. 130] and Greuel [**8**, Kor. 5.5, p. 263]. Thus Equation (3.3.1) holds.

To prove Equation (3.3.2), for each i, let P_i be a general linear space of codimension i in $\mathbf{C}^a$, and let Π^i be the i-dimensional "relative polar subscheme" of f_y with P_i as pole. By definition, Π^i is the closure in X_y of the locus of simple points x of the level hypersurface surface $X_y \cap f^{-1}fx$ such that there exists a tangent hyperplane at x that contains P_i. Algebraically, Π^i is cut out of X_y by the maximal minors of the Jacobian matrix of the map $\mathbf{C}^a \to \mathbf{C}^{k+1} \times \mathbf{C}^i$ with components $f_1, \ldots, f_k, f$ and p_i, where $p_i \colon \mathbf{C}^a \to \mathbf{C}^i$ is the linear map with kernel P_i. Hence Π^i is Cohen–Macaulay as X_y is.

From the polar multiplicity formula [**14**, Thm. (9.8)(i)] (compare with [**10**, 4.2.7] and [**4**, §3]), it follows that

$$e(\mathbf{m}_y \mathcal{M}_y) = \sum_{i=0}^{a-k} \binom{a-1}{i} m(\Pi^i),$$

where $m(\Pi^i)$ is the ordinary multiplicity at 0 of Π^i. In particular, $m(\Pi^{a-k})$ is simply the multiplicity of X_y at 0; so it is equal to $\mu_{a-k}(X_y) + 1$. For any i, since Π^i is Cohen–Macaulay,

$$m(\Pi^i) = \dim_{\mathbf{C}}(\mathcal{O}_{\Pi^i,0}/\mathcal{I}_i), \tag{3.3.4}$$

where $\mathcal{I}_i$ is the ideal of any linear space L_i of codimension i in $\mathbf{C}^a$ that is transverse to Π^i. It follows that $m(\Pi^i)$ is upper semicontinuous.

Remarkably, although Π^i is defined using P_i, nevertheless P_i is transverse to Π^i simply because P_i is general. This important result was proved by Teissier in [**26**, (4.1.8), p. 569] as a consequence of his general idealistic Bertini theorem. The result was also proved, at about the same time, by Henry and Merle [**9**, Cor. 2, p. 195]. The result is reproved in Lemma (6.2) of [**5**] in a new way, using the theory of the W_f condition in the spirit of the current work. By this transversality result, we may take L_i to be P_i. Then, for $i < a-k$, the right side of Equation (3.3.4) is equal to $\mu_i(X_y) + \mu_i(Z_y)$ by the theorem of Lê and Greuel. The asserted formula follows immediately, and the proof is complete.

4. Dependence

In this section, we discuss the proofs of Theorems (3.1) and (3.2), and thereby complete the proofs of Theorems (2.1) and (2.2). Our main technical tool is the theory of integral dependence of elements on modules, which generalizes the older theory for ideals. We begin by reviewing this theory.

Let $\mathcal{O}$ be the local ring of a complex-analytic germ $(X, 0)$. Let $\mathcal{E}$ be a free $\mathcal{O}$-module, $\mathcal{M}$ a submodule, and $g \in \mathcal{E}$ an element. By definition, g is *integrally dependent* on $\mathcal{M}$ if, when g is viewed as an element of degree 1 in the symmetric algebra $\mathcal{O}[\mathcal{E}]$, then g is integrally dependent on the Rees algebra $\mathcal{O}[\mathcal{M}]$ in the usual sense; namely, g satisfies an equation of integral dependence,

$$g^n + r_1 g^{n-1} + \cdots + r_n = 0,$$

where $n \geq 1$ and $r_i \in \mathcal{O}[\mathcal{M}]$, both depending on g. Of course, each r_i may be replaced by its homogeneous piece of degree i if desired.

There are two useful criteria for integral dependence. The first is a form of the valuative criterion, but it is also known in the trade as the "curve criterion."

(Curve criterion) *For $g \in \mathcal{E}$ to be integrally dependent on $\mathcal{M}$ it is necessary that, for every map germ $\phi\colon (\mathbf{C}, 0) \to (X, 0)$, the pullback $\phi^* g$ lie in the pullback $\phi^*\mathcal{M}$, viewed in the free $\mathcal{O}_{\mathbf{C},0}$-module $\phi^*\mathcal{E}$, or put more concisely,*

$$\phi^* g \in \phi^*\mathcal{M} \subset \phi^*\mathcal{E};$$

conversely, it is sufficient that this condition obtain for every nonconstant ϕ whose image meets any given dense Zariski open subset of X.

The second criterion is an analytic inequality, which re-expresses integral dependence in terms of speeds of vanishing.

(Analytic criterion) *For $g \in \mathcal{E}$ to be integrally dependent on $\mathcal{M}$ it is necessary that, for any finite set of generators g_i of $\mathcal{M}$, there exist a Euclidean neighborhood U of 0 in X and a constant c such that $|g(x)| \leq c \max |g_i(x)|$ for any $x \in U$; conversely, it is sufficient that this condition obtain for some finite set of generators g_i of $\mathcal{M}$.*

Indeed, the condition in the "curve criterion" is equivalent to the condition in the "analytic criterion" by Proposition 1.11 on p. 306 of [**3**]. In fact, on p. 303 in [**3**], Gaffney takes the former as the defining condition of integral dependence. On p. 305, he proves that this definition is equivalent to Rees's definition [**23**, p. 435]. Finally, Theorem 1·5 in [**23**, p. 437] yields the curve criterion.

In fact, neither [**3**] nor [**23**] treat the present version of the curve criterion, with its weaker sufficiency condition. However, it is not difficult to extend that work: if $\phi^* g \notin \phi^*\mathcal{M}$, then ϕ can be tweaked, preserving this relation, so that its image does meet the given open set (see the proof of Prop. 1.7 on p. 304 in [**3**]).

From now on, work in the setup described at the beginning of Section 2. The next lemma is the main algebraic result, Theorem (1.8), in [**5**]. Let $\mathcal{E}$, $\mathcal{M}$, and g be as above, but assume that they arise from a free module, a submodule, and an element defined over the affine ring of X. In addition, for each $y \in Y$, form the restriction $\mathcal{E}_y$ of the free module, the image $\mathcal{M}_y \subset \mathcal{E}_y$ of the submodule, and the image $g_y \in \mathcal{E}_y$ of the element. Finally, assume that $\mathcal{M}_y$ has finite colength.

Lemma (4.1) (Principle of specialization of integral dependence) *Assume that the Buchsbaum–Rim multiplicity $e(\mathcal{M}_y)$ is constant in y. Then g is integrally dependent on $\mathcal{M}$ if g_y is integrally dependent on $\mathcal{M}_y$ for all y in a dense Zariski open subset of Y.*

Of course, if g is integrally dependent on $\mathcal{M}$, then g_y is integrally dependent on $\mathcal{M}_y$ for all y in a neighborhood of 0, but the latter condition is strictly weaker than the former. Indeed, a simple example is given in Example (1.3) of [**5**]. Thus the conclusion of the lemma is stronger than it might seem at first.

To prove Theorems (3.1) and (3.2), take $\mathcal{M}$ to be the Jacobian module,

$$\mathcal{M} := JM_x(f_1, \ldots, f_k; f) \subset \mathcal{E} := \mathcal{O}_X^{k+1},$$

the column space of the Jacobian matrix; see Section 3. For convenience, let $\mathcal{M}$ also denote the induced module over the local ring $\mathcal{O}_{X,0}$. For $1 \le j \le b$, let g_j be the column vector,

$$g_j := \begin{bmatrix} \partial f_1/\partial y_j \\ \vdots \\ \partial f_k/\partial y_j \\ \partial f/\partial y_j \end{bmatrix}.$$

Finally, denote the ideal of Y in X by $\mathbf{m}_Y$; so $\mathbf{m}_Y := (x_1, \ldots, x_a)\mathcal{O}_X$.

The next result characterizes A_f (resp., W_f) by the integral dependence of the g_j on the submodule $\mathcal{M}$ (resp., $\mathbf{m}_Y\mathcal{M}$) of $\mathcal{E}$.

Proposition (4.2) *Assume that $\Sigma(f) = Y$.*

(1) *Then $(X - Y, Y)$ satisfies W_f at 0 if and only if the columns $g_1, \ldots, g_b$ are all integrally dependent on $\mathbf{m}_Y\mathcal{M}$.*

(2) *Then $(X - Y, Y)$ satisfies A_f at 0 if and only if the columns $g_1, \ldots, g_b$ are all integrally dependent on $\mathcal{M}$.*

Indeed, (1) is part of Proposition (6.1) in [**5**]. The characterization is proved by developing the analytic inequalities involved in the definition of W_f until the condition in the analytic criterion is met.

Part (2) is not explicitly stated in either [**5**] or [**6**], but may be derived as follows. First note the formula (Gaffney, priv. comm., 1990),

$$C(X, f) := \mathrm{Projan}(\mathcal{O}_X[\mathcal{M}, g_1, \cdots, g_b]).$$

Indeed, both sides are closed subvarieties of $X \times \mathbf{P}^{a+b-1}$. Both are equal, over $X - \Sigma(f)$, to the set of pairs (x, H) such that H is a hyperplane tangent at x

to the level surface $f^{-1}fx$. The left side is, by definition, the closure of this set. The right side is also the closure of this open subset of itself, because its algebra is, by construction, a subalgebra of the symmetric algebra $\mathcal{O}[\mathcal{E}]$.

Let V be an arbitrary component of the preimage of Y in $C(X, f)$. Since $\Sigma(f) = Y$, the dimension of V is $a+b-1$ by Gaffney and Massey's Lemma (5.7) in [**6**]; see also Theorem 4.2 in [**21**] and the corollary in [**16**, (1.2)].

Assume that $g_1, \cdots, g_b$ are integrally dependent on $\mathcal{M}$. Then, after X is replaced by a neighborhood of 0, the inclusion of $\mathcal{O}_X[\mathcal{M}]$ into $\mathcal{O}_X[\mathcal{M}, g_1, \cdots, g_b]$ induces a finite map,

$$\gamma: C(X, f) \to X \times \mathbf{P}^{a-1},$$

since $\mathcal{M}$ is generated by a elements. Hence $\dim \gamma(V) = a + b - 1$. Since $\gamma(V)$ is contained in $Y \times \mathbf{P}^{a-1}$, therefore these two sets are equal. Hence V maps onto Y. Since A_f holds at every y in a dense Zariski open set U of Y by Hironaka's Theorem 2 on p. 247 in [**12**], the preimage $V|U$ lies in $C(Y)$. Since $V|U$ is nonempty and so dense in V, therefore $V \subset C(Y)$. Thus A_f holds at 0.

The converse is a special case of (1) of the following lemma, and so the proof of the proposition is complete.

Both parts of the lemma are used below to complete the proofs of Theorems (3.1) and (3.2). Moreover, both are interesting in their own right.

Lemma (4.3) (1) *If* $(X - \Sigma(f), Y)$ *satisfies* A_f *at* 0, *then each* g_j *is integrally dependent on* $\mathcal{M}$.

(2) *If each* g_j *is integrally dependent on* $\mathcal{M}$, *then the germs of* $\Sigma(f)$ *and* $\Sigma_Y(f)$ *at* 0 *are equal.*

To prove (2), suppose that the germ of $\Sigma_Y(f)$ is strictly larger than that of $\Sigma(f)$. Then there is a path $\phi: (\mathbf{C}, 0) \to (X, 0)$ whose image lies in the former set, but outside the latter. Now,

$$\Sigma_Y(f) = \mathrm{Supp}(\mathcal{E}/\mathcal{M}) \text{ and } \Sigma(f) = \mathrm{Supp}\big(\mathcal{E}/(\mathcal{M}, g_1, \ldots, g_b)\big).$$

Hence, for some j, the pullback $\phi^* g_j$ does not lie in the pullback $\phi^*\mathcal{M}$. So this g_j is not integrally dependent on $\mathcal{M}$ by the curve criterion. Thus (2) holds.

To prove (1), take a $\phi: (\mathbf{C}, 0) \to (X, 0)$ whose image lies outside $\Sigma_Y(f)$, so outside $\Sigma(f)$. The gradients of $f_1, \ldots, f_k, f$ define hyperplanes tangent to the level hypersurfaces of f. Each hyperplane must approach along ϕ a hyperplane that contains Y since A_f holds. Consider the last b components of each gradient. Each such component must, therefore, vanish at 0 along ϕ to order higher than the order of one, or more, of the first a components. Denote the maximal ideal of $(\mathbf{C}, 0)$ by $\mathbf{m}$. Then, therefore,

$$\phi^* g_1, \ldots, \phi^* g_b \in \mathbf{m}\mathcal{M} \subset \mathcal{M}.$$

So, by the curve criterion, g_j is integrally dependent on $\mathcal{M}$. Thus (1) holds.

To prove Theorem (3.2), let $\mathcal{J}$ be the ideal of maximal minors of the Jacobian matrix in Section 3. Then

$$\Sigma_Y(f) = \operatorname{Supp}(\mathcal{E}/\mathcal{M}) = \operatorname{Supp}(\mathcal{O}_X/\mathcal{J}).$$

Now, if either (ii) or (iii) holds, then each Z_y has an isolated singularity; hence, we may assume that $\Sigma_Y(f)$ is finite over Y. Therefore $\mathcal{O}_X/\mathcal{J}$ is flat over Y. Hence the following number is constant in y:

$$e'(y) := \dim_{\mathbf{C}}(\mathcal{O}_X/\mathcal{J})(y).$$

For each y, this number $e'(y)$ is a sum of positive numbers, one for each point z in $\Sigma_Y(f)$ lying over y, and the number corresponding to $z = 0$ is equal to $e(\mathcal{M}_y)$ by Equation (3.3.3). Hence, we have

$$e(\mathcal{M}_0) = e'(0) = e'(y) \geq e(\mathcal{M}_y),$$

with equality at the end for every y if and only if $\Sigma_Y(f) = Y$. Therefore, (ii) and (iii) are equivalent.

Assume (i). Then (ii) follows directly from Lemma (4.3).

Conversely, assume (ii). Then, since $\Sigma_Y(f)$ and $\Sigma(f)$ and Y are nested, all three represent the same germ. Furthermore, (iii) holds; so $e(\mathcal{M}_y)$ is constant. Now, to prove that A_f holds, we use Proposition (4.2)(2). We have to show that the g_j are integrally dependent on $\mathcal{M}$. However, they are so by the principle of specialization of integral dependence. Indeed, $(X - \Sigma(f), Y)$ satisfies A_f at every y in a dense Zariski open subset U of Y by Hironaka's Theorem 2 on p. 247 in [**12**]. Hence, for $y \in U$, the restrictions $g_j|X_y$ are integrally dependent on $\mathcal{M}_y$ by Proposition (4.2)(2) again. Thus Theorem (3.2) is proved.

Finally, consider Theorem (3.1). We have left to prove that (ii) implies (i). So assume (ii); that is, $e(\mathbf{m}_y\mathcal{M}_y)$ is constant. Then $e(\mathcal{M}_y)$ is constant too; we noted this consequence before stating Lemma (3.3). Hence, by Theorem (3.2), the germs of $\Sigma(f)$ and Y at 0 are equal. To prove that W_f holds, we use Proposition (4.2)(1) and the principle of specialization of integral dependence much as we just did for A_f; we need only replace Hironaka's theorem by Henry, Merle, and Sabbah's Théorème 5.1 on p. 255 of [**11**]. (They attribute this result to Navarro, who didn't publish it.) Thus Theorem (3.1) is proved.

References

[1] J. Briançon, P. Maisonobe and M. Merle, Localisation de systèmes différentiels, stratifications de Whitney et condition de Thom, *Invent. Math.* **117** (1994), 531–550.

[2] D.A. Buchsbaum and D.S. Rim, A generalized Koszul complex. II. Depth and multiplicity, *Trans. Amer. Math. Soc.* **111** (1963), 197–224.

[3] T. Gaffney, Integral closure of modules and Whitney equisingularity, *Invent. Math.* **107** (1992), 301–322.

[4] T. Gaffney, Multiplicities and equisingularity of ICIS germs, *Invent. Math.* **123** (1996), 209–220.

[5] T. Gaffney and S.L. Kleiman, Specialization of integral dependence for modules, *alg-geom*/9610003.

[6] T. GAFFNEY AND D. MASSEY, Trends in equisingularity theory, *in* "the proceedings of the Liverpool conference in honor of CTC Wall, singularities volume."

[7] M.D. GREEN AND D.B. MASSEY, Vanishing cycles and Thom's a_f conditions, *Preprint* 1996.

[8] G.M. GREUEL, Der Gauss–Manin Zusammenhang isolierter Singularitäten von vollständigen Durchschnitten, *Dissertation*, Göttingen (1973), *Math. Ann.* **214** (1975), 235–266.

[9] J.P.G. HENRY AND M. MERLE, Limites d'espaces tangents et transversalité de variétés polaires, *in* "Proc. La Rábida, 1981." J. M. Aroca, R. Buchweitz, M. Giusti and M. Merle (eds.) *Springer Lecture Notes in Math.* **961** (1982), 189–199.

[10] J.P.G. HENRY AND M. MERLE, Conormal Space and Jacobian module. A short dictionary, *in* "Proceedings of the Lille Congress of Singularities," J.-P. Brasselet (ed.), London Math. Soc. Lecture Notes **201** (1994), 147–174.

[11] J.P.G. HENRY, M. Merle, and C. Sabbah, *Sur la condition de Thom stricte pour un morphisme analytique complexe, Ann. Scient. Éc. Norm. Sup.* **17** (1984), 227–268.

[12] H. HIRONAKA, Stratification and flatness, *in* "Real and complex singularities, Nordic Summer School, Oslo, 1976," Sijthoff and Noordhoff, 1977, 199–265.

[13] D. KIRBY AND D. REES, *Multiplicities in graded rings I: The general theory,* in "Commutative algebra: syzygies, multiplicities, and birational algebra" W. J. Heinzer, C. L. Huneke, J. D. Sally (eds.), *Contemp. Math.* **159** (1994), 209–267.

[14] S. KLEIMAN AND A. THORUP, A geometric theory of the Buchsbaum–Rim multiplicity, *J. Algebra* **167** (1994), 168–231.

[15] S. KLEIMAN AND A. THORUP, Mixed Buschsbaum–Rim Multiplicities, *Amer. J. Math.* **118** (1996), 529–569.

[16] S. KLEIMAN AND A. THORUP, Conormal geometry of maximal minors, alg-geom/970818.

[17] D.T. LÊ, Calculation of Milnor number of isolated singularity of complete intersection, *Funct. Anal. Appl.* **8** (1974), 127–131.

[18] D.T. LÊ AND K. SAITO, La constance du nombre de Milnor donne des bonnes stratifications, *C. R. Acad. Sci. Paris* **277** (1973), 793–795.

[19] D.T. LÊ AND B. TEISSIER, Limites de'espaces tangent en géométrie analytique, *Comment. Math. Helvetici* **63** (1988), 540–578.

[20] E.J.N. LOOIJENGA, Isolated singular points on complete intersections, London Mathematical Society lecture note series **77**, Cambridge University Press, 1984.

[21] D.B. MASSEY, Critical points of functions on singular spaces, *Preprint* 1998.

[22] V. NAVARRO, Conditions de Whitney et sections planes, *Invent. Math.* **61** (1980), 199–226.

[23] D. REES, Reduction of modules, *Math. Proc. Camb. Phil. Soc.* **101** (1987), 431–449.

[24] B. TEISSIER, Cycles évanescents, sections planes et conditions de Whitney, in "Singularités à Cargèse," Astérisque **7–8** (1973), 285–362.

[25] B. TEISSIER, *Résolution simultanée et cycles évanescents,* iin "Sém. sur les singularités des surfaces." Proc. 1976–77. M. Demazure, H. Pinkham and B. Teissier (eds.) *Springer Lecture Notes in Math.* **777** (1980), 82–146.

[26] B. TEISSIER, Multiplicités polaires, sections planes, et conditions de Whitney, in "Proc. La Rábida, 1981." J. M. Aroca, R. Buchweitz, M. Giusti and M. Merle (eds.) *Springer Lecture Notes in Math.* **961** (1982), 314–491.

[27] J.-L. VERDIER, Stratifications de Whitney et théorème de Bertini–Sard, *Invent. Math.* **36** (1976), 295–312.

Picard Bundles and Syzygies of Canonical Curves

Giuseppe Pareschi

Dipartimento di Matematica, Universitá di Roma "La Sapienza"
P.le A.Moro 5, I-00185, Roma, Italy. pareschi@mat.uniroma1.it

To Professor Mario Fiorentini

Introduction. The purpose of this note is to show a relation between the syzygies of a curve completely embedded by its canonical bundle and the Ext-cohomology of Picard bundles on its Jacobian. The spirit of the result is to try to interpret geometrically Green's conditions N_p on the Jacobian rather than on the curve. In order to describe the result, we need some preliminary material and notation:

(a) First order deformations of vector bundles on abelian varieties. Given a vector bundle F on an abelian variety X there are two obvious ways to deform it: the first one (as in any variety) is by tensoring it with a line bundle in $\mathrm{Pic}^0 X$ and the second one is by translating it. So we have the two families of vector bundles $\{F \otimes \alpha\}_{\alpha \in \mathrm{Pic}^0 X}$ and $\{T_a^* F\}_{a \in X}$. Correspondingly, we have the Kodaira-Spencer maps

$$\Phi : H^1(\mathcal{O}_X) \to \mathrm{Ext}^1(F, F) \quad \text{and} \quad \Psi : T_{X,0} \to \mathrm{Ext}^1(F, F)$$

(where $T_{X,0}$ is the tangent space of X at the identity point 0).

(b) Picard bundles. Let C be a curve and L a line bundle of degree $d \geq 2g(C) - 1$ on C. Let $\mathcal{P}$ be a Poincaré line bundle on $C \times \mathrm{Pic}^0 C$ and let p and q the projections. For us a *Picard bundle* will be a vector bundle on $\mathrm{Pic}^0 C$ of the form

$$\mathcal{E}_L = q_*(p^*(L) \otimes \mathcal{P}).$$

By Riemann-Roch and Grauert's theorem, if $\alpha \in \mathrm{Pic}^0 C$ parametrizes a line bundle of degree zero (which, by abuse of language, we will call α too), the fibre $\mathcal{E}_L(\alpha)$ is $H^0(L \otimes \alpha)$. We refer to [Sc],[K1],[Mu],[EL] and references therein for basic results about these bundles. In particular Kempf [K1] and Mukai [Mu] have shown that the curve C is non-hyperelliptic if and only if the map

$$\Phi \oplus \Psi : H^1(\mathcal{O}_{\mathrm{Pic}^0 C}) \oplus T_{\mathrm{Pic}^0 C, 0} \to \mathrm{Ext}^1(\mathcal{E}_L, \mathcal{E}_L)$$

is an isomorphism, i.e. at the first order level, the space of all deformations of $\mathcal{E}_L$ is the product of the two families above. The reader is referred to [K1] and [Mu] for more on the geometric meaning of this result (compare also the end of Section 4 below).

(c) Higher Ext*'s.* Returning to (a), it turns out that $\mathrm{Ext}^\bullet(F,F)$ is naturally a graded module over the cohomology algebra $H^\bullet(\mathcal{O}_X)$ and the two maps Φ and Ψ are the degree-one pieces of maps of graded modules

$$\Phi^\bullet : H^\bullet(\mathcal{O}_X) \to \mathrm{Ext}^\bullet(F,F) \quad \text{and} \quad \Psi^\bullet : T_{X,0} \otimes H^{\bullet-1}(\mathcal{O}_X) \to \mathrm{Ext}^\bullet(F,F)$$

This is shown in Section 1 below.

(d) Syzygies of canonical curves. Let $R = \bigoplus_{i\geq 0} H^0(K^{\otimes i})$ be the canonical ring of C. It is naturally a graded module over the polynomial ring $S = \bigoplus_{i\geq 0} \mathrm{Sym}^i(H^0(K))$. Let us consider a minimal resolution of R as graded S-module

$$L^\bullet \qquad 0 \to E_{g-2} \to E_{g-1} \to \cdots \to E_1 \to E_0 \to R \to 0$$

where

$$E^j = \bigoplus S(-k)^{\oplus b_{jk}}.$$

The Betti numbers b_{jk} are intrinsic invariants of C so they must be related to the intrinsic geometry of C. A precise relation of this type has been conjectured by Green ([G]). To state it let us recall some terminology: one says that C satisfies property N_0 if $E_0 = S$ (i.e. $b_{0k} = 0$ for $k \neq 0$ and $b_{00} = 1$). If K is very ample, this means that the canonical curve is projectively normal. Next, one says that C satisfies property N_1 if it satisfies N_0 and $b_{1k} = 0$ for $k \neq 2$. This means that the homogeneous ideal of the canonical curve is generated by quadrics. Inductively, C is said to satisfy N_p if it satisfies N_{p-1} and moreover $b_{pk} = 0$ for $k \neq p+1$. In a word, C satisfies N_p if the resolution $L^\bullet$ is linear up to the p-th step. In this language Noether's theorem states that C satisfies N_0 if and only if it is not hyperelliptic, while Petri's theorem states that if C satisfies N_0, then it satisfies N_1 if and only if it is not trigonal or isomorphic to a plane quintic. Green conjectures is that C satisfies N_p if and only if the Clifford index of C is strictly greater than p, a statement recovering the cases $p = 0, 1$. We refer to [G],[L1],[S],[V1],[V2] for the definition of Clifford index, discussions and results.

Our purpose here is not to add any evidence to the truth or not of Green's conjecture, but just to make a remark about the nature of conditions N_p. The result is

Theorem A. *Let $\mathcal{E}$ be any Picard bundle on $\mathrm{Pic}^0 C$. Then C satisfies property N_p if and only if the map of graded $H^\bullet(\mathcal{O}_{\mathrm{Pic}^0 C})$-modules*

$$\Phi^\bullet \oplus \Psi^\bullet : H^\bullet(\mathcal{O}_{\mathrm{Pic}^0 C}) \oplus (T_{\mathrm{Pic}^0 C,0} \otimes H^{\bullet-1}(\mathcal{O}_{\mathrm{Pic}^0 C})) \to \mathrm{Ext}^\bullet(\mathcal{E},\mathcal{E})$$

is surjective in any degree k such that $1 \leq k \leq p+1$.

It is worth to remark (see [K1] and [Mu], compare also Theorem 2.1 below) that in any case $k \cong H^0(\mathcal{O}_{\mathrm{Pic}^0 C}) \cong \mathrm{Hom}(\mathcal{E},\mathcal{E})$ (i.e. $\mathcal{E}$ is *simple*) and the map $\Phi^\bullet \oplus \Psi^\bullet$ is always injective in degree 1.

Actually, an interesting variant of the statement will follow from the proof of Theorem A. More precisely it will follow that condition N_p holds if and only if $\mathrm{Ext}^k(\mathcal{E},\mathcal{E})$ have the expected (i.e. minimal) dimension for any k such that $1 \leq k \leq p+1$. We refer to the beginning of Section 4 for details. Since $\mathrm{Ext}^\bullet(\mathcal{E},\mathcal{E})$ is self-dual, this implies also that $\mathrm{Ext}^k(\mathcal{E},\mathcal{E})$ has the expected dimension also for any k such that $g-p-1 \leq k \leq g-1$. In particular, for *general* curves of genus g, Green's conjecture turns out to be equivalent to the statement that all the graded components of $\mathrm{Ext}^\bullet(\mathcal{E},\mathcal{E})$ have the expected dimension. To put the result into perspective, let us recall that in the papers [K1] and [K2] are proved two results, somehow complementary, on the first-order deformations respectively of of symmetric products of curves and of Picard bundles. The first one deals with the derivative $du : T_{C^{(d)}} \to u^*T_{J(C)}$ of the Abel-Jacobi map $u : C^{(d)} \to J(C)$ and states, in particular, that $H^1(du)$ is injective if and only if C satisfies N_0. This have been generalized to higher syzygies by Lazarsfeld in [L2]. The second one is the aforementioned result about first order deformations of Picard bundles, proved independentey also by Mukai in [Mu] (see point (b) above). Therefore Theorem A can be seen as the analogue for Picard bundles of Lazarsfeld's generalization. (It should also be mentioned that also a part of the statement of Theorem A in degree two can be found in Kempf's paper ([K1], Theorem 6.7).

We will work over an algebraically closed field k.

The debt of the present paper to Kempf's article [1] does not need to be acknowledged. This note is dedicated to Professor Mario Fiorentini with deep friendship and gratitude.

1. The maps $\Phi^\bullet$ and $\Psi^\bullet$. Let F be a locally free sheaf an abelian variety X.

(a) There is the canonical isomorphism

$$H^j(\mathcal{H}om(F,F)) \cong \mathrm{Ext}^j(F,F). \tag{1.1}$$

Thus $\mathrm{Ext}^j(F,F)$ is naturally a graded $H^\bullet(\mathcal{O}_X)$-module via cup product. Explicitely, the map

$$\Phi^\bullet : H^\bullet(\mathcal{O}_X) \to \mathrm{Ext}^\bullet(F,F) \tag{1.2}$$

is defined considering the omothety map $\mathcal{O}_X \to \mathcal{H}om(F,F)$ and taking cohomology. Of course $H^\bullet(\mathcal{O}_X)$ is the exterior algebra $\Lambda^\bullet H^1(\mathcal{O}_X)$. As we said, the degree-one map Φ^1 is the Kodaira-Spencer map of the family of $\{F \otimes \beta\}_{\beta \in \mathrm{Pic}^0 X}$.

(b) We have a another map of graded-$H^\bullet(\mathcal{O}_X)$-modules

$$\Psi^\bullet : T_{X,0} \otimes H^{\bullet-1}(\mathcal{O}_X) \to \mathrm{Ext}^\bullet(F,F) \tag{1.3}$$

($T_{X,0}$ is the tangent space of X at 0) defined as follows. One has the bundle of principal parts of F: $P^1(F) = p_{1*}((p_2^*F)_{|\Delta^{(2)}})$ (p_i are the projections on $X \times X$ and $\Delta^{(2)}$ is the first infinitesimal neighborhood of the diagonal in $X \times X$). It sits in the canonical extension

$$0 \to E \otimes \Omega^1_X \to P^1(E) \to E \to 0 \tag{1.4}$$

Applying $\mathcal{H}om(F,\cdot)$ and using that Ω^1_X is trivial and isomorphic to $\Omega^1_{X,0} \otimes \mathcal{O}_X$, one gets

$$0 \to \Omega^1_{X,0} \otimes \mathcal{H}om(F,F) \to \mathcal{H}om(F,P^1(F)) \to \mathcal{H}om(F,F) \to 0 \tag{1.5}$$

The coboundaries of the short exact sequence (1.5) give a map $H^{\bullet-1}(\mathcal{H}om(F,F)) \to \Omega^1_{X,0} \otimes H^{\bullet}(\mathcal{H}om(F,F))$. Composing with (1.2) and contracting one gets $\Psi^{\bullet}$. As shown in Sect.8 of [K1], the degree-one map Ψ^1 is the Kodaira-Spencer map of the family $\{T_x^* F\}_{x\in X}$, where T_x denotes the traslation $y \mapsto y + x$.

2. Computing $\mathrm{Ext}^{\bullet}(\mathcal{E}_{\mathbf{L}}, \mathcal{E}_{\mathbf{L}})$. Let K be the canonical bundle of C and let us consider the bundle M_K, defined as the kernel of the evaluation map $H^0(K) \otimes \mathcal{O}_X \to K$:

$$0 \to M_K \to H^0(K) \otimes \mathcal{O}_C \to K \to 0 \tag{2.1}$$

Here is the basic computation of the paper:

Theorem 2.1. *For any $i \geq 0$ there is a canonical exact sequence*

$$0 \to H^1(\overset{p-1}{\Lambda} M_K^{\vee}) \to \mathrm{Ext}^p(\mathcal{E},\mathcal{E}) \to H^0(\overset{p}{\Lambda} M_K^{\vee}) \to 0 \tag{2.2}$$

Proof. The method of proof is the one of Kempf's work [K1]. Here we refine Kempf's computations using the general construction of the article [P], where these ideas are applied to a totally different context.

STEP 1. We will work on $C \times C \times \mathrm{Pic}^0 C$. Let us denote p_i the three projections and p_{ij} the three intermediate projections. Moreover let $\mathcal{P}$ be a Poincaré line bundle on $C \times \mathrm{Pic}^0 C$ and let Δ be the diagonal in $C \times C$. Finally, let L be any line bundle of degree $\geq 2g-1$ on C. For any integer k let us consider on $C \times C \times \mathrm{Pic}^0 C$ the line bundles

$$M^k_{L^+, K\otimes L^{\vee -}} = p_{13}^*(p_1^*(L) \otimes \mathcal{P}) \otimes p_{23}^*(p_2^*(K \otimes L) \otimes \mathcal{P}^{\vee})) \otimes p_{12}^*(\mathcal{O}_{C\times C}(-k\Delta)) \tag{2.3}$$

where $\mathcal{P}$ is a Poincaré bundle on $C \times X$. Since $\deg L \geq 2g-1$ we have that

$$H^1(M^0_{L^+, K\otimes L^{\vee -}}{}_{|C\times C\times\{\alpha\}}) = H^0(L\otimes\alpha) \otimes H^1(K \otimes L^{\vee} \otimes \alpha^{\vee})$$

which is, by Serre duality, isomorphic to $\mathrm{Hom}(H^0(L\otimes\alpha), H^0(L\otimes\alpha))$. On the other hand, again since $\deg L \geq 2g-1$, we have that $H^i(M^0_{L^+, K\otimes L^{\vee -}}{}_{|C\times C\times\{\alpha\}}) = 0$ for any $\alpha \in \mathrm{Pic}^0 C$ and $i = 0, 2$. Therefore, by relative duality and Grauert's theorem, we have that $R^i p_{3*} M^0_{L^+, K\otimes L^{\vee -}} \cong \mathcal{H}om(\mathcal{E},\mathcal{E})$ if $i = 1$ and zero otherwise. Thus the Leray spectral sequence of p_3 degenerates giving isomorphisms

$$H^j(\mathcal{H}om(\mathcal{E},\mathcal{E})) \cong H^{j+1}(M^0_{L^+, K\otimes L^{\vee -}}) \tag{2.4}$$

Hence we are reduced to compute the cohomology of $M^0_{L^+, K\otimes L^{\vee -}}$. To do that we will apply the projection p_{12} onto $C \times C$ and study the Leray spectral sequence.

STEP 2. For future reference, let us first record the following key results about duality between abelian varieties

Theorem 2.2. ([M], p.127) *Let A be an abelian variety of dimension q, $\mathrm{Pic}^0 A$ the dual variety and $\mathcal{Q}$ a Poincaré line bundle on $A \times \mathrm{Pic}^0 A$. Then*

$$R^k p_{A*}\mathcal{Q} = \begin{cases} H^q(\mathcal{O}_{\mathrm{Pic}^0 A}) \otimes \mathcal{O}_0 & \text{for } k = q \\ 0 \quad \text{otherwise} \end{cases}$$

(where $\mathcal{O}_0$ is the one-dimensional skyscraper sheaf on the point 0 of A).

Corollary 2.3. ([K1] Cor.2.2) *Let T be a variety and $\pi : T \to A$ a morphism. Then we have functorial isomorphisms*

$$R^i p_{T*}((\pi \times id_{\mathrm{Pic}^0 A})^*(\mathcal{Q})) \cong \mathcal{T}or_{q-i}^{\mathcal{O}_A}(\mathcal{O}_T, H^q(\mathcal{O}_{\mathrm{Pic}^0 A}) \otimes \mathcal{O}_0)$$

(where $\mathcal{O}_T$ is seen as $\mathcal{O}_A$-module via π.)

Corollary 2.3 can be applied to our line bundle $M^k_{L^+, K \otimes L^{\vee -}}$ on $C \times C \times \mathrm{Pic}^0 C$ as follows: let us take $A = \mathrm{Alb}C$, $T = C \times C$ and $\pi = d \circ (a \times a)$, where $d : A \times A \to A$ is the difference map $(x, y) \to x - y$ and $a : C \to A$ is a fixed Albanese (Abel-Jacobi) map. Since $\mathrm{Pic}^0 C = \mathrm{Pic}^0 A$ and the Poincaré bundle $\mathcal{P}$ on $C \times \mathrm{Pic}^0 C$ is the pullback via a of a Poincaré bundle $\mathcal{Q}$ on $A \times \mathrm{Pic}^0 A$, i.e. $\mathcal{P} = (a, id_{\mathrm{Pic}^0 A})^*(\mathcal{Q})$, it is easily seen that, by the Theorem of the cube ([M] p.91),

$$p_{13}^*\mathcal{P} \otimes p_{23}^*\mathcal{P}^\vee \cong (\pi \times id_{A^\vee})^*\mathcal{Q} \tag{2.5}$$

Therefore

$$M^k_{L^+, K \otimes L^{\vee -}} = p_1^*(L) \otimes p_2^*(K \otimes L^\vee) \otimes p_{12}{}^*(\mathcal{O}_{C \times C}(-k\Delta)) \otimes (\pi \times id_{\mathrm{Pic}^0 A})^*(\mathcal{Q}) \tag{2.6}$$

Applying Cor.2.2 and projection formula we get

$$R^i p_{12*} M^k_{L^+, K \otimes L^{\vee -}} \cong \mathcal{T}or_{q-i}^{\mathcal{O}_A}(\mathcal{O}_{C \times C}, \mathcal{O}_0) \otimes p_1^*(L) \otimes p_2^*(K \otimes L^\vee) \otimes p_{12}^*(\mathcal{O}_{C \times C}(-k\Delta)) \tag{2.7}$$

where $\mathcal{O}_{C \times C}$ is seen as an $\mathcal{O}_A$-module via π and we have made, in order to make the notation less heavy, the identification $H^q(\mathcal{O}_{\mathrm{Pic}^0 C}) \cong k$. So (2.7) yields that the sheaves $R^h p_{12*} M^k_{L^+, K \otimes L^{\vee -}}$ are supported on the diagonal $\Delta = \pi^{-1}(0)$ and in fact, denoting $\Delta : C \to C \times C$ the diagonal embedding, we have that

$$R^i p_{12*} M^k_{L^+, K \otimes L^{\vee -}} \cong \mathcal{T}or_{q-i}^{\mathcal{O}_A}(\mathcal{O}_{C \times C}, \mathcal{O}_0) \otimes \Delta_*(K^{\otimes k+1}) \tag{2.8}$$

In particular, their cohomology vanishes for $j \geq 2$. Therefore we have achieved a first result: the spectral sequence $H^j(R^h p_{12*} M^k_{L^+, K \otimes L^{\vee -}}) \Rightarrow H^{j+h}(M^k_{L^+, K \otimes L^{\vee -}})$ degenerates as follows

$$\begin{aligned} \cdots \xrightarrow{\delta^k_{j-1}} H^1(R^{j-1} p_{12*} M^k_{L^+, K \otimes L^{\vee -}}) \to H^j(M^k_{L^+, K \otimes L^{\vee -}}) \to H^0(R^j p_{12*} M^k_{L^+, K \otimes L^{\vee -}}) \xrightarrow{\delta^k_j} \\ \xrightarrow{\delta_j} H^1(R^j p_{12*} M^k_{L^+, K \otimes L^{\vee -}}) \to H^{j+1}(M^k_{L^+, K \otimes L^{\vee -}}) \to H^0(R^{j+1} p_{12*} M^k_{L^+, K \otimes L^{\vee -}}) \xrightarrow{\delta_{j+1}} \end{aligned} \tag{2.9}$$

Then Theorem 2.1 will follow from the following

Claim. *(i)* $R^j p_{12*} M^0_{L+,K\otimes L^\vee -} = \Delta_*(\Lambda^{j-1} M_K^\vee)$; *(ii) the connecting maps* δ_j^k *in (2.9) are zero.*

STEP 3. Here we will prove (i) of the Claim. Let $\tilde{\Delta} = \Delta \times \mathrm{Pic}^0 C$. Since $p_{13}^*\mathcal{P} \otimes p_{23}^*\mathcal{P}^\vee_{|\tilde{\Delta}}$ is trivial and $(p_{13}^* L \otimes p_{23}^*(K \otimes L^\vee))(-k\Delta))_\Delta = K^{\otimes k+1}$ we have the basic sequences

$$0 \to M^{k+1}_{L+,K\otimes L^\vee -} \xrightarrow{\cdot\tilde{\Delta}} M^k_{L+,K\otimes L^\vee -} \xrightarrow{\rho^k} p_{12}^*(\Delta_*(K^{\otimes k+1})) \to 0. \tag{2.10}$$

Applying p_{12*} one gets a long cohomology sequence

$$\cdots \xrightarrow{\theta^i_k} R^{i-1} p_{12*} M^k_{L+,K\otimes L^\vee -} \to \Lambda^i H^0(K)^\vee \otimes \Delta_*(K^{\otimes k+1}) \to R^i p_{12*} M^{k+1}_{L+,K\otimes L^\vee -} \xrightarrow{\theta^{i+1}_k} \cdots \tag{2.11$_{i,k}$}$$

(recall that – by duality of abelian varieties – we have that $H^1(\mathcal{O}_{\mathrm{Pic}^0 X}) \cong T_{\mathrm{Pic}^0(\mathrm{Pic}^0 X),0} \cong T_{\mathrm{Alb}X,0} \cong H^0(K)^\vee$). Applying the (functorial) isomorphism (2.7) to sequences (2.11) we get the long homology sequence of $\mathcal{T}or$'s

$$\begin{aligned}\cdots \to \mathcal{T}or^{\mathcal{O}_A}_{g-i}(\mathcal{O}_{C\times C}, \mathcal{O}_0) \otimes \Delta_*(K^{\otimes k+1}) \to \mathcal{T}or^{\mathcal{O}_A}_{g-i}(\mathcal{O}_\Delta, \mathcal{O}_0) \otimes \Delta_*(K^{\otimes k+1}) \to \\ \to \mathcal{T}or^{\mathcal{O}_A}_{g-i-1}(\mathcal{O}_{C\times C}, \mathcal{O}_0) \otimes \Delta_*(K^{\otimes k+2}) \to \cdots\end{aligned} \tag{2.12}$$

associated to the sequence $0 \to \mathcal{O}_{C\times C}(-(k+1)\Delta) \xrightarrow{\cdot\Delta} \mathcal{O}_{C\times C}(-k\Delta) \to K^{\otimes k} \to 0$, tensored with K. Now, since all sheaves are supported on Δ and since the maps $\mathcal{T}or^{\mathcal{O}_A}_h(\cdot\Delta, \mathcal{O}_0)$ are zero when restricted to Δ, the long sequences (2.11) are in fact chopped into short sequences

$$0 \to R^{i-1} p_{12*} M^k_{L+,K\otimes L^\vee -} \to \overset{g-i+1}{\Lambda} H^0(K) \otimes \Delta_*(K^{\otimes k+1}) \to R^i p_{12*} M^{k+1}_{L+,K\otimes L^\vee -} \to 0. \tag{2.13}$$

Therefore for $i = g$ we get

$$R^i p_{12*} M^k_{L+,K\otimes L^\vee -} \cong K^{\otimes k+1}$$

proving (i) of the Claim for $i = g$ (since the determinant $\Lambda^{g-1} M_K^\vee$ is equal to K). Next, considering $i = g-1$, we have the exact sequence

$$0 \to R^{g-1} p_{12*} M^k_{L+,K\otimes L^\vee -} \to H^0(K) \otimes \Delta_*(K^{\otimes k+1}) \to \Delta_*(K^{\otimes k+2}) \to 0 \tag{2.14}$$

and, via the isomorphism (2.7), the third arrow in (2.14) is the evaluation map $H^0(K) \to K$ tensored with $K^{\otimes k}$. Thus

$$R^{g-1} p_{12*} M^k_{L+,K\otimes L^\vee -} \cong \Delta_*(M_K \otimes K^{\otimes k+1}) \cong \Delta_*(\overset{g-2}{\Lambda} M_K^\vee \otimes K^{\otimes k}) \tag{2.15}$$

where the last isomorphism follows from duality in the exterior algebra and (2.14) is identified to the exact sequence

$$0 \to M_K \otimes K^{\otimes k+1} \to H^0(K) \otimes K^{\otimes k+1} \to K^{\otimes k+2} \to 0$$

i.e. our basic sequence (2.1) twisted by $K^{\otimes k+1}$. Arguing inductively one gets in a similar fashion (compare [K1] Section 6) that

$$R^i p_{12*} M^k_{L^+, K\otimes L^{\vee -}} \cong \Delta_*(\overset{g-i}{\Lambda} M_K \otimes K^{\otimes k+1}) \cong \Delta_*(\Lambda^{i-1} M_K^\vee \otimes K^{\otimes k}) \tag{2.16}$$

and the exact sequences (2.13) are identified with the exact sequences:

$$0 \to \overset{g-i}{\Lambda} M_K \otimes K^{\otimes k+1} \to \overset{g-i}{\Lambda} H^0(K) \otimes K^{\otimes k+2} \to \overset{g-i+1}{\Lambda} K^{\otimes k+2} \to 0 \tag{2.17}$$

obtained taking exterior products in (2.1) and tensoring with $K^{\otimes k}$. Thus we have proved (i) of the Claim.

STEP 4. Here we will prove (ii) of the Claim, which will conclude the proof of Theorem 2.1. The proof of (ii) of the Claim will be by descending induction on k. If k is high enough , so that $H^1(\Lambda^{g-i} M_K \otimes K^{\otimes k+1}) = 0$ for any j, (actually it can be shown that $k = 1$ suffices) the claim is obvious. To prove the induction step, note that, since the long cohomology sequences $(2.11)_{i,k}$ are chopped into the short exact sequences (2.13) (i.e., written in a different way, (2.17)), the corresponding hypercohomologies degenerate fitting in a commutative exact diagram

$$\begin{array}{ccccccc}
 & & & & & & \downarrow 0 \\
\overset{\delta^k_{i-1}}{\to} & H^1(R^{i-1}_{p_{12}*} M^k_{L^+, K\otimes L^{\vee -}}) & \to & H^i(M^k_{L^+, K\otimes L^{\vee -}}) & \to & H^0((R^i_{p_{12}*} M^k_{L^+, K\otimes L^{\vee -}})) & \overset{\delta^k_i}{\to} \\
 & \downarrow & & \downarrow & & \downarrow & \\
0 \to & H^1(R^{i-1}_{p_{12}*}(M^k_{L^+, K\otimes L^{\vee -}}{}_{|\tilde{\Delta}})) & \to & H^i(M^k_{L^+, K\otimes L^{\vee -}}{}_{|\tilde{\Delta}}) & \twoheadrightarrow & H^0(R^i_{p_{12}*}(M^k_{L^+, K\otimes L^{\vee -}}{}_{|\tilde{\Delta}})) & \to 0 \\
 & \downarrow & & \downarrow & & \downarrow & \\
0 \overset{\delta^{k+1}_i}{\to} & H^1(R^i_{p_{12}*} M^{k+1}_{L^+, K\otimes L^{\vee -}}) & \to & H^{i+1}(M^{k+1}_{L^+, K\otimes L^{\vee -}}) & \to & H^0(R^{i+1}_{p_{12}*} M^{k+1}_{L^+, K\otimes L^{\vee -}}) & \overset{\delta^{k+1}_{i+1}}{\to} 0 \\
 & \downarrow 0 & & & & &
\end{array}$$

where:
(a) the middle horizontal short exact sequence is given by Künneth formula (recall that $M^k_{L^+, K\otimes L^{\vee\vee}} = p_{12}^*(\Delta_*(K^{\otimes k+1}))$); *(b)* the third horizontal sequence is exact by induction; *(c)* the middle vertical exact sequence is the long cohomology sequence of (2.10), *(d)* the left and right vertical exact sequence are the cohomology sequences of (2.13) (or, equivalently, of (2.17)).

Then, by the snake lemma $\delta^k_i = 0$. Therefore $\delta^k_i = 0$ for any i and k. □

3. Syzygies of canonical curves. Here we will recall some more basic facts – largely due to Green – about syzygies of canonical curves (again we refer to [G] and [L1] for details). We keep the notation of (d) in the Introduction:

(a) it is easy to see that in any case $b_{pk} = 0$ for $k \neq p+1, p+2$, i.e. at each step condition N_p can fail af most by one. This follows e.g. from Castelnuovo-Mumford regularity.

(b) (see e.g. [L1] p.511). We have that $b_{p,p+2} = 0$ (i.e. N_p holds) if and only if the following complex $\mathcal{K}_p^{\bullet}$ is exact in the middle

$$\mathcal{K}_p^{\bullet}: \quad \stackrel{p+1}{\Lambda} H^0(K) \otimes H^0(K) \to \stackrel{p}{\Lambda} H^0(K) \otimes H^0(K^{\otimes 2}) \to \stackrel{p-1}{\Lambda} H^0(K) \otimes H^0(K^{\otimes 3})$$

(c) Taking wedge products of sequence (2.1) we get

$$0 \to \stackrel{p+1}{\Lambda} M_K \otimes K \to \stackrel{p+1}{\Lambda} H^0(K) \otimes K \to \stackrel{p}{\Lambda} M_K \otimes K^{\otimes 2} \to 0. \tag{3.1}$$

The fact that the complex $\mathcal{K}_p^{\bullet}$ is exact means that the map

$$\stackrel{p+1}{\Lambda} H^0(K) \otimes H^0(K) \to H^0(\stackrel{p}{\Lambda} M_K \otimes K^{\otimes 2}) \tag{3.2}$$

is surjective, and this is in turn equivalent to $h^1(\Lambda^{p+1} M_K \otimes K) = \binom{g}{p+1}$ (since, as it is easy to see, (a) yields that $H^1(\Lambda^p M_K \otimes K^{\otimes 2}) = 0$ for any j). Actually, it will more convenient for us the dual version. Applying $\mathcal{H}om(\cdot, K)$ to (3.1) one gets

$$0 \to \stackrel{p}{\Lambda} M_K^{\vee} \otimes K^{\vee} \to \stackrel{p+1}{\Lambda} H^0(K)^{\vee} \otimes \mathcal{O}_C \to \stackrel{p+1}{\Lambda} M_K^{\vee} \to 0 \tag{3.3}$$

Then N_p holds if and only if the map $H^1(\Lambda^p M_K^{\vee} \otimes K^{\vee}) \to \Lambda^{p+1} H^0(K)^{\vee} \otimes H^1(\mathcal{O}_C)$ is injective. This is equivalent to the fact that the injective map $\Lambda^{p+1} H^0(K)^{\vee} \to H^0(\Lambda^{p+1} M_K^{\vee})$ is an isomorphism i.e. $h^0(\Lambda^{p+1} M_K^{\vee}) = \binom{g}{p+1}$. So N_p means that $H^0(\Lambda^{p+1} M_K^{\vee})$ has the minimal possible dimension.

4. Proof of Theorem A. Proposition 2.1 shows already that $\mathrm{Hom}(\mathcal{E},\mathcal{E}) = k$ and that $\mathrm{Ext}^p(\mathcal{E},\mathcal{E})$ has the expected dimension if and only if $H^1(\Lambda^{p-1} M_K^{\vee})$ and $H^0(\Lambda^p M_K^{\vee})$ do, that is, by the above remarks, if and only if C satisfies N_{p-1} and N_p. Moreover, since $N_p \Rightarrow N_{p-1}$, we can summarize the above remarks as follows: in any case

$$\dim(\mathrm{Ext}^p(\mathcal{E},\mathcal{E}) \geq \binom{g}{p-1} - \chi(\stackrel{p-1}{\Lambda} M_K^{\vee}) + \binom{g}{p}$$

and we have equality if and only if C satisfies condition N_p.

To recover the statement of Theorem A, a few remarks are in order. First of all, by Theorem 2.1 we have the exact sequence

$$0 \to H^1(\stackrel{\bullet-1}{\Lambda} M_K^{\vee}) \to \mathrm{Ext}^{\bullet}(\mathcal{E},\mathcal{E}) \to H^0(\stackrel{\bullet}{\Lambda} M_K^{\vee}) \to 0 \tag{4.1}$$

Let us recall that there is a canonical identification

$$H^1(\mathcal{O}_C) \cong T_{0,\mathrm{Pic}^0 C} \tag{4.2}$$

and a canonical identification of algebras

$$\stackrel{\bullet}{\Lambda} H^0(K)^{\vee} \cong H^{\bullet}(\mathcal{O}_{\mathrm{Pic}^0 C}) \tag{4.3}$$

(given by duality: $\mathrm{Pic}^0(\mathrm{Pic}^0C) = \mathrm{Alb}C$). Then considering the H^0 of the third arrow in sequences (3.3) and using (4.3) one gets a map

$$H^\bullet(\mathcal{O}_{\mathrm{Pic}^0C}) \to H^0(\overset{\bullet}{\Lambda} M_K^\vee), \tag{4.4}$$

while considering the H^1 one gets a map

$$T_{\mathrm{Pic}^0C,0} \otimes H^{\bullet-1}(\mathcal{O}_{\mathrm{Pic}^0C}) \to H^1(\overset{\bullet-1}{\Lambda} M_K^\vee) \tag{4.5}$$

(it can be seen that they are in fact maps of graded $H^\bullet(\mathcal{O}_{\mathrm{Pic}^0C})$-modules). To conclude the proof of Theorem A we need only to show the folloing claim: via sequence (4.1), the map (4.4) lifts to $\Phi^\bullet$ and the map (4.5) is $\Psi^\bullet$. In fact, assuming the claim true, since (4.5) is always surjective, N_p holds if and only if $\Phi^\bullet$ is surjective (i.e. an isomorphism) in degree $\leq p+1$ i.e. if and only if $\Phi^\bullet \oplus \Psi^\bullet$ is surjective in degree $\leq p+1$. (Note that in this case $\Phi^\bullet$ gives a splitting of (4.1) up to degree $p+1$.) To prove that (4.4) lifts to the map $\Phi^\bullet$ is easy: this amounts to say that the third arrow of (4.1) is a map of graded $H^\bullet(\mathcal{O}_{\mathrm{Pic}^0C})$-modules, where $\mathrm{Ext}^\bullet(\mathcal{E},\mathcal{E})$ has the module structure given by the map $\Phi^\bullet$ and $H^0(\Lambda^\bullet M_K^\vee)$ is equipped by the module structure given by (4.4) itself, and this follows immediately from the way Theorem 2.1 was proved.

It remains to verify that (4.5) is – via the injection of (4.1) – really the map $\Psi^\bullet$. First of all let us note that it is enough to prove that the two maps coincide in degree 1: indeed, by construction, both (4.5) and $\Psi^\bullet$ are obtained from the respective maps in degree 1 by composition with cup product:

$$\begin{array}{ccc} T_{0,\mathrm{Pic}^0C} \otimes H^{j-1}(\mathcal{O}_{\mathrm{Pic}^0C}) & \to & \mathrm{Ext}^1(\mathcal{E},\mathcal{E}) \otimes H^{j-1}(\mathcal{O}_{\mathrm{Pic}^0C}) \\ & \searrow & \downarrow \\ & & \mathrm{Ext}^j(\mathcal{E},\mathcal{E}) \end{array}$$

The fact that $\Psi^\bullet$ and (4.5) are the same map in degree 1 is proved in [K1] (Prop.8.3). A perhaps more natural proof, although less homogeneous with the methods and notation of this paper, follows from Mukai's work [Mu]. Let us quickly outline it: the point is to show that via the canonical isomorphism $H^1(\mathcal{O}_C) \cong T_{0,\mathrm{Pic}^0C}$, the map (4.5) in degree one is the Kodaira-Spencer map of the family $\{\mathcal{E}_{L\otimes\alpha}\}_{\alpha\in\mathrm{Pic}^0C}$. This is what we want, since by construction $\mathcal{E}_{L\otimes\alpha} = T^*_{\{\alpha\}}\mathcal{E}_L$ and $\Psi^\bullet$ is the Kodaira-Spencer map of the family of translations $\{T^*_{\{\alpha\}}\mathcal{E}\}_{\alpha\in\mathrm{Pic}^0C}$. To prove what claimed, one uses the Mukai-Fourier transform: via an Abel-Jacobi embedding a of C in $\mathrm{Alb}C$ one can see L as a sheaf on $\mathrm{Alb}C$. Then the "Fourier functor" induces a natural map $\mathrm{Ext}^p_{\mathcal{O}_{\mathrm{Alb}C}}(L,L) \to \mathrm{Ext}^p_{\mathcal{O}_{\mathrm{Pic}^0C}}(\mathcal{E}_L,\mathcal{E}_L)$ which turns out to be an isomorphism for any p ([Mu] Cor.2.5). Plugging such an isomorphism for $p=1$ into the beginning of the spectral sequence of local-to-global Ext one has

$$0 \to H^1(\mathcal{O}_C) \to \mathrm{Ext}^1(\mathcal{E}_L,\mathcal{E}_L) \to H^0(\mathcal{N}_{C|\mathrm{Alb}C}) \to 0 \tag{4.6}$$

where $\mathcal{N}$ means normal bundle (compare [Mu], Lemma 4.9). (By the way, this is another way of finding the degenerate spectral sequence (2.9)). The deformation theoretic meaning of (4.6) is well known: the injection is the Kodaira-Spencer map of our claim, and corresponds to deforming the line bundle L on the curve C, while $H^0(\mathcal{N}_{C|\mathrm{Alb}C})$ is the tangent space to the Hilbert scheme of C in Alb$C$ and corresponds to moving C inside Alb$C$. Going trough the ways (4.5) for $j = 1$ and (4.6) are obtained, it is easy to show that they are the *same* exact sequence (note that $M_K^\vee = \mathcal{N}_{C|\mathrm{Alb}C}$) and this proves what claimed. □

References.

[EL] Ein, L., Lazarsfeld, R.: *Stability and restrictions of Picard bundles, with an application to the normal bundles of elliptic curves*, in *Complex Projective Geometry* (Ellingsrud, G., et al. eds), Cambridge Univ. Press (1992) 149-157

[G] Green, M. *Koszul cohomology and the geometry of projective varieties*, J. Diff. Geom. 19 (1984) 125-171

[K1] Kempf, G.: *Towards the inversion of abelian integrals,I* Ann. of Math. 110 (1979) 243-273

[K2] Kempf, G.: *Deformations of symmetric products*, in *Riemann surfaces and related topics* (Kra,I., Maskit, B. eds) Princeton Univ. Press (1981) 343-369

[L1] Lazarsfeld, R.: *A sampling of vector bundles techinques in the study of linear series* in *Riemann surfaces* (Cornalba, M. et al. eds) World Scientific (1989)

[L2] Lazarsfeld, R.: *Cohomology on symmetric products, syzygies of canonical curves and a theorem of Kempf*, in *Einstein metrics and Yang-Mills connections* (Mabuchi, T. and Mukai, S. eds) Marcel Dekker (1993) 89-90

[M] Mumford, D.: *Abelian varieties*, Second edition, Oxford Univ. Press (1974)

[Mu] Mukai, S.: *Duality between $D(X)$ and $D(\hat{X})$ with its application to Picard sheaves* , Nagoya math.J. 81 (1981) 153-175

[P] Pareschi, G. *Gaussian maps, hyperplane sections and generic vanishing of the H^1*, preprint

[S] Schreyer, F.: *A standard basis approach to syzygies of canonical curves*, J. reine angew. Math. 421 (1991) 83-123

[Sc] Schwarzenberger, R.: *Jacobian and symmetric products*, Ill. J. Math. 7 (1963), 257-268

[V1] Voisin, C.: *Courbes tétragonales et cohomology de Koszul*, J. reine angew. Math. 387 (1988) 111-121

[V2] Voisin, C.: *Deformation des syzygies et theorie de Brill-Noether*, Proc. London Math. Soc. 67 (1993) 493-515

Variations on Green's Theorem Concerning the Hilbert Functions

DORIN POPESCU
Univ. Bucharest, Inst. Of Math., P.O. Box 1-764, Bucharest, Romania

DEDICATED TO M. FIORENTINI

ABSTRACT. In connection with Green Theorem, we give a numerical property of stable and p-Borel ideals and we show extremal properties of lex among these ideals. Also we study some applications of some therems of type Green (see [11],[9]) to several conjectures from Higher Castelnuovo Theory and Cayley-Bacharach Theory.

INTRODUCTION

Let K be an infinite field, A a homogeneous K-algebra, h a generic form of A of degree $s \geq 1$ and $H(A, -)$ the Hilbert function associated to A.

Theorem 0.1. *(Green [10]) If $s = 1$ then*

$$H(A/(h), d) \leq H(A, d)_{<d>}$$

for all positive integer d.

The numerical function $\mathbb{N} \to \mathbb{N}$, $a \to a_{<d>}$ is given as follows: if

$$a = \binom{k(d)}{d} + \binom{k(d-1)}{d-1} + \cdots + \binom{k(j)}{j} \tag{1}$$

with $k(d) > k(d-1) > \cdots > k(j) \geq j \geq 1$ (such expansion exists always and it is unique and called the d-th Macaulay expansion (see [12], or [4, 4.2.6])) then $a_{<d>}$ denotes

$$\binom{k(d)-1}{d} + \binom{k(d-1)-1}{d-1} + \cdots + \binom{k(j)-1}{j}$$

In [11, 3.7] Green's Theorem was extended for arbitrary s, when $\operatorname{char} K = 0$. Recently Gasharov [9] succeded to show that the bound given in [11] works also when $p = \operatorname{char} K > 0$.

The author thanks the Alexander von Humboldt Foundation , University of Bucharest and Institute of Mathematics for support.

Theorem 0.2. *([11],[9])*

$$H(A/(h),d) \leq \sum_{i=0}^{s-1} H(A,d)_{<<d,i>>}.$$

The numerical function $\mathbb{N} \to \mathbb{N}, a \to a_{<<d,i>>}$ is given as follows: if (1) is the d-th Macaulay expansion of a then we denote

$$a_{<d,i>} = \binom{k(d)-i-1}{d-i} + \cdots + \binom{k(t)-i-1}{t-i},$$

where $t = j$ if $j > i$ and $t = i+1$ if $j \leq i$ and $a_{<<d,i>>} = a_{<d,i>} + \binom{k(i)-i}{0}$, where as usual $\binom{k(i)-i}{0} = 1$ if $k(i) \geq i$ and 0, otherwise.

Suppose $A = R/I$, $R = K[X_1, \ldots, X_n]$ for a homogeneous ideal $I \subset R$. The proof of the above theorem starts with the following

Proposition 0.3. *([11, 3.1]) If I is a lex ideal (see 1.1 below) then*

$$H(A/(h),d) = \sum_{i=0}^{s-1} H(A,d) << d,i >>.$$

Then extending in [11, 3.4] (the key Lemma of [11]) some ideas of Bayer [2] and Bigatti [3] concerning the so called the extremal properties of lex ideals it was possible in [11, 3.6] to give a proof of the above theorem when I is a strongly stable ideal (see 1.1 below). Since the generic initial ideals are strongly stable when $p = 0$ (see [5, 15.23]), the proof ends applying a Macaulay Theorem (see [4, 4.2.4], or [5, 15.26]).

When $p > 0$ then the generic initial ideals are only p-Borel ideals (see 1.1 below and [5, 15.23]). These ideals have a very reach structure but they were not too much studied. The resolution of a p-Borel principal Cohen-Macaulay ideal is given in [1], but the description of the homological structure of general p-Borel ideals is still a dream. However we hoped that the extremal properties of lex ideals preserve also somehow in the frame of p-Borel ideals and we will be able to prove Theorem 0.2 in the same way when $p > 0$. Our Corollary 1.7 shows that indeed lex ideals have extremal properties in the frame of p-Borel ideals but only in the form from [2], [3] and not in the strong form from [11, 3.4] (see here Remark 1.8). In fact these extremal properties are just a variation of Green Theorem (here they follow from Green Theorem and on the other hand they imply this theorem as Bigatti showed in [3]). Our Corollary 1.6 shows that the extremal properties of lex among stable ideals hold even in the strong form of [11, 3.4], given there for strongly stable ideals.

The p-Borel ideals (as well the stable ideals) I have also a nice numerical property

$$\dim_K I_d \leq (\dim_K (I_d \cap K[X_1, \ldots, X_{n-1}]))^{<n-2>}$$

if $n \geq 3$ (see 1.7, 1.6). The Macaulay operator $\mathbb{N} \to \mathbb{N}$, $a \to a^{<d>}$ is given as follows: if (1) is the d-th Macaulay expansion of a then $a^{<d>}$ denotes

$$\binom{k(d)+1}{d+1} + \ldots + \binom{k(j)+1}{j+1}.$$

This section ends with a small numerical application of the above result (see 1.9).

In [11, 4.2,4.3,4.4] are studied some applications of Theorem 0.2 to some conjectures from Higher Castelnuovo Theory and Cayley-Bacharach Theory (see [6], [13], [7]) in the generic case. Section 2 (partially joint work with J.Herzog) extends these applications in a new unified form and our Theorem 2.2 (announced already in [11, Remark 4.3]) gives a partial answer to Conjecture (V_m) of [6] in the generic case.

1. p-BOREL IDEALS

Let K be an infinite field, $p = \operatorname{char} K$ and $R = K[X_1, \ldots, X_n]$, $n \geq 2$ the polynomial ring with the standard grading $R = \sum_{d \geq 0} R_d$, where R_d is the linear space spanned by all monomials of degree d. Throughout the paper we consider the deglex order ¡ on the set of monomials of R. Given a monomial X^a, $a = (a_1, \ldots, a_n)$ we denote by $m(X^a)$ the greatest i, $1 \leq i \leq n$ such that $a_i > 0$.

Definition 1.1. *A set $J \subset R_d$ of monomials is called*

1) lex or lexsegment if for any $u \in J$ and each monomial $v \in R_d$ with $v > u$ it follows $v \in J$;

2) stable, if $X_i(u/X_{m(u)}) \in J$ for any $u \in J$ and $1 \leq i \leq m(u)$;

3) strongly stable, if $X_i(u/X_j) \in J$ for any $u \in J$, $1 \leq i < j \leq n$ such that $X_j | u$;

4) p-Borel, if $X_i^r(u/X_j^r) \in J$ for any $u \in J$, $1 \leq i < j \leq n$ and $r_i \leq q_i$, $r = \sum_i r_i p^i$, $q = \sum_i q_i p^i$ being the p-adic expansions of r, resp. q the integer defined by the property $X_j^q | u$, $X_j^{q+1} \not| u$.

Let $< J >$ be the K-space of R_d spanned by J. We will say that $< J >$ has one of the above properties if J does. A monomial ideal I is said to be lex, or stable (resp. strongly stable, or p-Borel) if all its homogeneous components I_d have this property.

Lemma 1.2. *(after Macaulay [12]) Let $u = X_1^{i_1} \cdots X_{n-1}^{i_{n-1}}$ be a monomial from $\{X_1, \ldots, X_{n-1}\}^d$ and a the number of all monomials $w \in \{X_1, \ldots, X_n\}^d$ with $w > u$, $w \neq u$. Then*

$$a = \binom{n+d-i_1-2}{n-1} + \binom{n+d-i_1-i_2-3}{n-2} + \ldots + \binom{i_{n-1}+1}{2}$$

is the $(n-1)$-th Macaulay expansion of a.

Proof. We have

$$\{w \in \{X_1, \ldots, X_n\}^d | w > u, w \neq u\} = X_1^{i_1+1}\{X_1, \ldots X_n\}^{d-i_1-1} \cup$$
$$X_1^{i_1} X_2^{i_2+1}\{X_2, \ldots, X_n\}^{d-i_1-i_2-1} \cup \ldots \cup$$

$$X_1^{i_1}\cdots X_{n-3}^{i_{n-3}}X_{n-2}^{i_{n-2}+1}\{X_{n-2},X_{n-1},X_n\}^{d-i_1-\ldots-i_{n-2}-1},$$

where negative powers of a set are empty by convention. It follows

$$a=\binom{n+d-i_1-2}{d-i_1-1}+\ldots+\binom{d-i_1-\ldots-i_{n-2}+1}{d-i_1-\ldots-i_{n-2}-1},$$

which is enough. □

Let s, $1\le d$ be an integer, $E=\cup_{i=0}^{s-1}\{X_1,\ldots,X_{n-1}\}^{d-i}X_n^i$ and $r<|E|$ a positive integer. Let v be the $(r+1)$-th monomial of E in the lex order.

Lemma 1.3. *$P^{(r,s)}=\{w\in\{X_1,\ldots,X_n\}^d|w>v,w\neq v\}$, $P^{(r,s)}\cup\{v\}$ are lexsegments.*

Lemma 1.4. *Suppose $n>2$, $s=1$. Then $|P^{(r,1)}|=r^{<n-2>}$.*

Proof. By Lemma 1.2 we have

$$|P^{(r,1)}|=\binom{n+d-i_1-2}{n-1}+\binom{n+d-i_1-i_2-3}{n-2}+\ldots+\binom{i_{n-1}+1}{2}$$

and

$$r=|P^{(r,1)}\cap\{X_1,\ldots,X_{n-1}\}^d|=\binom{n+d-i_1-3}{n-2}+\ldots+\binom{i_{n-1}}{1}.$$

Clearly these expansions are Macaulay and we are done. □

Theorem 1.5. *Let $I\subset R$ be a homogeneous ideal. The following statements are equivalent:*

i) $H(R/(I,X_n^s),d)\le\sum_{i=0}^{s-1}H(R/I,d)_{<<d,i>>}$,

ii) For every lex ideal L such that $\dim_K L_d\le\dim_K I_d$ it holds

$$\dim_K(L_d\cap<E>)\le\dim_K(I_d\cap<E>),$$

iii) $\dim_K I_d\le|P^{(r,s)}|$, where $r=\dim_K(I_d\cap<E>)$.

Proof. i) ⇒ ii). We have

$$H(R/(I,X_n^s),d)\le\sum_{i=0}^{s-1}H(R/I,d)_{<<d,i>>}\le\sum_{i=0}^{s-1}H(R/L,d)_{<<d,i>>}=H(R/(L,X_n^s),d),$$

where the second inequality follows by [11, 4.1] because $H(R/I,d)\le H(R/L,d)$ by hypothesis and the last equality is given by Proposition 0.3.

ii) ⇒ iii). Suppose that $|P^{(r,s)}|<\dim_K I_d$. As $J:=P^{(r,s)}\cup\{v\}$ is a lexsegment by Lemma 1.3 and $|J|\le\dim_K I_d$ we obtain by ii)

$$r+1=|J\cap E|\le\dim_K(I_d\cap<E>)=r.$$

Contradiction!

iii) ⇒ i). We have

$$H(R/(I,X_n^s),d)=H(R/(P^{(r,s)},X_n^s),d)=\sum_{i=0}^{s-1}H(R/P^{(r,s)},d)_{<<d,i>>}$$

$$\leq \sum_{i=0}^{s-1} H(R/I,d)_{<<d,i>>},$$

where the first equality holds because $r = |P^{(r,s)} \cap E|$, the second one because of Proposition 0.3 and the last inequality holds because of [11, 4.1] since $H(R/P^{(r,s)},d) \leq H(R/I,d)$ by iii). □

Corollary 1.6. *Suppose $n > 2$. If X_n^s is generic for I (in particular if I is stable by [11, 1.4]), then the equivalent statements of Theorem 1.5 hold.*

For the proof note that i) from the above theorem holds by Theorem 0.2.

Corollary 1.7. *If I is stable, or p-Borel then the following statements hold:*
i) For every lex ideal L such that $\dim_K L_d \leq \dim_K I_d$ it holds

$$\dim_K(L_d \cap K[X_1,\dots,X_{n-1}]) \leq \dim_K(I_d \cap K[X_1,\dots,X_{n-1}]),$$

ii)

$$\dim_K I_d \leq (\dim_K(I_d \cap K[X_1,\dots,X_{n-1}]))^{<n-2>}.$$

Proof. If X_n is generic for I then the corollary follows by 1.6 for $s=1$ and 1.4. If I is stable then this is true as we saw in 1.6. If I is p-Borel then I is Borel-fixed by [5, 15.23] and it is enough to see that X_n is generic for I. Choose an arbitrary generic linear form f for I. We may choose f such that $f(0,\dots,0,X_n) \neq 0$. Then the K-automorphism ψ of R given by $X_i \to X_i$, $i < n$, $X_n \to f$ is in the Borel subgroup and so $\psi(I) = I$. Hence X_n is generic for I. □

Remark 1.8. *i) X_n^2 is not generic for p-Borel ideals in general. Take $I = (X_1^p,\dots, X_n^p)$, if $p = \operatorname{char} K > 0$, $s = 2$ and $L = X_1^{p-1}(X_1,\dots,X_n)$. We have $\dim_K(L_p \cap < E >) = n > n-1 = \dim_K(I_p \cap < E >)$ but $\dim_K L_p = \dim_K I_p = n$. Thus the statement ii) from Theorem 1.5 fails and so X_n^2 is not generic for I by Corollary 1.6.*

ii) The above Corollary holds even for monomial ideals generated by a set $J \subset R_d$ such that $X_i^r(u/X_n^r) \in J$ for any $u \in J$, $1 \leq i < n$ with $r_i \leq q_i$, $r = \sum_i r_i p^i$, $q = \sum_i q_i p^i$ being the p-adic expansions of r, resp. q - the positive integer defined by the property $X_j^q | u$, $X_j^{q+1} \nmid u$ (the same proof works). Clearly this class of ideals includes both the p-Borel ideals and the stable ideals.

Lemma 1.9. *Let a,b,t be three positive integers. Then $(ab)^{<t>} \geq a^{<t>} b^{<t>}$.*

Proof. Let $n = t+2$, $R = K[X_1,\dots,X_n]$, where K is an infinite field of characteristic $p > 0$. Choose $q,m >> 1$ and consider $P^{(a,1)} \subset \{X_1,\dots,X_n\}^q$, $P^{(b,1)} \subset \{X_1,\dots,X_n\}^m$. Let e be a positive integer such that $mp^e >> q$. Then $I = P^{(a,1)}(P^{(b,1)})^{p^e}$ satisfies $|I| = |P^{(a,1)}||(P^{(b,1)})| = a^{<t>}b^{<t>}$ by Lemma 1.4 and $|I \cap \{X_1,\dots,X_{n-1}\}^{q+mp^e}| = ab$. Since I is p-Borel by construction it follows

$$a^{<t>}b^{<t>} = |I| \leq (ab)^{<t>}$$

by Corollary 1.7. □

Remark 1.10. *The above Lemma is the multiplicative form of* [8, Lemma 4.4,2)] *and probably can be proved directly like there. We included it here just as a small application of our Corollary 1.7.*

2. An application of Theorem 0.2

Let n, s, $s \geq 2$ be positive integers, K an infinite field and $\mathcal{F}$ the set of all zero dimensional factor algebras B of $K[X_1, \ldots, X_n]$ such that

(*) If $H(B,s) = \binom{k(s)}{s} + \ldots + \binom{k(j)}{j}$, $k(s) > \ldots > k(j) \geq j \geq 1$ is the s-th Macaulay expansion of $H(B,s)$ then $H(B,d) \leq \binom{k(s)}{d} + \ldots + \binom{k(j)}{d-s+j}$ for all integers $d \geq s$.

Note that $\mathcal{F} \neq \emptyset$ because zero dimensional complete intersections of quadrics belong to $\mathcal{F}$. On the other hand it is easy to see that zero dimensional complete intersections of cubics are not in $\mathcal{F}$.

Proposition 2.1. *If $B \in \mathcal{F}$ and h is a generic form of degree s of B then $B/(h) \in \mathcal{F}$ too.*

Proof. Denote $A = B/(h)$ and suppose $d > s$. By Theorem 0.2, [11, Corollary 3.8] and [11, Lemma 4.1] it follows

$$H(A,d) \leq \sum_{i=0}^{s-1} H(B,d)_{<<d,i>>} \leq \sum_{i=0}^{s-1}[\binom{k(s)}{d} + \ldots + \binom{k(j)}{d-s+j}]_{<<d,i>>} \leq$$

$$\leq \sum_{i=0}^{s-1}[\binom{k(s)}{d} + \ldots + \binom{k(j)}{d-s+j}]_{<d,i>} + s - j.$$

because $\binom{2s-d-j}{1} \leq \binom{s-j}{1} = s - j$. We claim that

$$\alpha := s-j+\sum_{i=1}^{s-1}[\binom{k(s)}{d}+\ldots+\binom{k(j)}{d-s+j}]_{<d,i>} \leq \beta+\binom{k(s)-1}{d-1}+\ldots+\binom{k(j+1)-1}{d-s+j},$$

where $\beta := \binom{k(j)-2}{d-s+j-1} + \ldots + \binom{k(j)-j}{d-s+1}$. Indeed, $\alpha = \beta + \sum_{i=j+1}^{s} \alpha_i + s - j$, where

$$\alpha_i := \binom{k(i)-2}{d-s+i-1}+\ldots+\binom{k(i)-i}{d-s+1}+\binom{k(i-1)-i}{d-s}+\ldots+\binom{k(j)-i}{d-s-i+j+1}$$

if $d > i + s - j - 1$ and

$$\alpha_i := \binom{k(i)-2}{d-s+i-1}+\ldots+\binom{k(i)-i}{d-s+1}+\binom{k(i-1)-i}{d-s}+\ldots+\binom{k(s+i-d)-i}{1}$$

if $d \leq i+s-j-1$, $j < i \leq s$. But $\alpha_i < \binom{k(i)-1}{d-s+i-1}$ in both cases, which gives our claim. Then

$$H(a,d) \leq [\binom{k(s)-1}{d} + \binom{k(s)-1}{d-1}] + \ldots + [\binom{k(j+1)-1}{d-s+j+1} + \binom{k(j+1)-1}{d-s+j}] +$$
$$\binom{k(j)-1}{d-s+j} + \beta = \binom{k(s)}{d} + \ldots + \binom{k(j+1)}{d-s+j+1} + \binom{k(j)-1}{d-s+j} + \ldots + \binom{k(j)-j}{d-s+1},$$

which is enough because

$$H(A,s) = H(B,s) - 1 = \binom{k(s)}{s} + \ldots + \binom{k(j+1)}{j+1} + \binom{k(j)-1}{j} + \ldots + \binom{k(j)-j}{1}.$$

□

Theorem 2.2. *Let B be a zero dimensional complete intersection of quadrics, A a factor ring of B defined by generic s-forms of B, $s \geq 2$, and let $H(A,s) = \binom{k(s)}{s} + \ldots + \binom{k(j)}{j}$, $k(s) > \ldots > k(j) \geq j \geq 1$ be the s-th Macaulay expansion of $H(A,s)$. Then*

i) $H(A,d) = H(B,d) = \binom{n}{d}$ for $d < s$,

ii) $H(A,d) \leq \binom{k(s)}{d} + \ldots + \binom{k(j)}{d-s+j}$ for $d \geq s$,

ii) $\dim_K A \leq 2^{k(s)} + \ldots + 2^{k(j)} + \sum_{i=1}^{s-1}[\binom{n}{i} - H(A,s)_{(s,s-i)}]$, where $H(A,t)_{(s,t)}$ denotes the sum $\binom{k(s)}{s-t} + \ldots + \binom{k(t)}{0}$, $1 \leq t < s$.

Proof. i) is trivial and ii) follows by induction using Proposition 2.1.

ii) By summation on d in i) and ii) we have

$$\dim_K A \leq \sum_{i=0}^{s-1}\binom{n}{i} + 2^{k(s)} + \ldots + 2^{k(j)} - \sum_{i=0}^{s-1}\binom{k(s)}{i} - \sum_{i=0}^{s-2}\binom{k(s-1)}{i} - \ldots - \binom{k(j)}{0} =$$
$$2^{k(s)} + \ldots + 2^{k(j)} + \sum_{i=1}^{s-1}[\binom{n}{i} - H(A,s)_{(s,s-i)}].$$

□

Remark 2.3. *i) If $A = B/(h)$ is a factor of a zero dimensional complete intersection B by only one generic s form h of B then $H(A,s) = \binom{n}{s} - 1 = \binom{n-1}{s} + \ldots + \binom{n-s}{1}$ is the s-th Macaulay expansion of $H(A,s)$. Thus $H(A,s)_{(s,s-i)} = \binom{n-1}{i} + \ldots + \binom{n-s+i-1}{0} = \binom{n}{i}$ and so by Theorem 2.2 iii) $\dim_K A \leq 2^{n-1} + \ldots 2^{n-s} = 2^n - 2^{n-s}$. This was already done in* [11, Proposition 4.5] *in connection with* [6, Conjecture $\mathrm{III}_{k,r}$], [13, Conjecture 3.4], [7, Conjecture CB10].

ii) If $s = 2$ and $H(A,2) = \binom{k(2)}{2} + \binom{k(1)}{1}$ is the 2-th Macaulay expansion of $H(A,2)$ then

$$\dim_K A \leq 2^{k(2)} + 2^{k(1)} + \binom{n}{1} - H(A,2)_{(2,1)} = 2^{k(2)} + 2^{k(1)} + n - k(2) - 1.$$

This was already done in [11, Corollary 4.4] *in connection with* [6, Conjecture $\mathrm{II}_{m,r}$], [13, Conjecture 3.2].

References

[1] A.Aramova, J.Herzog. p-Borel principal ideals, Ill. J. Math. 41, n° 1(1997),103-121.

[2] D. Bayer. The division algorithm and the Hilbert scheme, Thesis, Harvard University, Cambridge , MA.

[3] A.M. Bigatti. Aspetti Combinatorici e Computazionali dell'Algebra Commutativa, Thesis, Genova, 1995.

[4] W.Bruns, J.Herzog. Cohen-Macaulay rings, Cambridge University Press, 1993.

[5] D.Eisenbud. Commutative Algebra with a View Toward Algebraic Geometry, Springer-Verlag,1995.

[6] D.Eisenbud, M. Green, J.Harris. Higher Castelnuovo Theory, Asterisque, 218(1993), 187-202.

[7] D.Eisenbud, M.Green, J.Harris. Cayley-Bacharach Theorems and Conjectures, Bull. AMS, 33(3),(1996),295-324.

[8] V. Gasharov. Extremal properties of Hilbert functions, Preprint 1996.

[9] V.Gasharov. Hilbert functions and homogeneous generic forms II, Preprint 1997.

[10] M.Green. Restriction of linear series to hyperplanes, and some results of Macaulay and Gotzmann, in "Algebraic curves and projective geometry", Springer Lect. Notes, 1389, Berlin,1989,76-86.

[11] J.Herzog, D.Popescu. Hilbert functions and generic forms, Compositio Math., 1997.

[12] F.S.Macaulay. Some properties of enumeration in the theory of modular systems, Proc. London Math.Soc. 26(1927),531-555.

[13] G.Valla. Problems and results on Hilbert functions of graded algebras, in "Summer School on Commutative Algebra", vol 1, Bellaterra, July 16-26, 1996, Centre de Reserca Matematica.

Dorin Popescu
University of Bucharest, Institute of Mathematics
P.O.Box 1-764, Bucharest 70700, Romania
E-mail: dorin @ stoilow.imar.ro

On the Dimension Filtration and Cohen–Macaulay Filtered Modules

PETER SCHENZEL

Martin Luther Universität Halle-Wittenberg
Halle (Saale), Germany

ABSTRACT. For a finitely generated A-module M we define the dimension filtration $\mathcal{M} = \{M_i\}_{0 \le i \le d}, d = \dim_A M$, where M_i denotes the largest submodule of M of dimension $\le i$. Several properties of this filtration are investigated. In particular, in case the local ring $(A, \mathfrak{m})$ possesses a dualizing complex, then this filtration occurs as the filtration of a spectral sequence related to duality. Furthermore, we call an A-module M a Cohen-Macaulay filtered module provided all of the quotient modules M_i/M_{i-1} are either zero or i-dimensional Cohen-Macaulay modules. We describe a few basic properties of these kind of generalized Cohen-Macaulay modules. In the case A posesses a dualizing complex it turns out - as one of the main results - that M is a Cohen-Macaulay filtered A-module if and only if for all $0 \le i < d$ the module of deficiency $K^i(M)$ is either zero or an i-dimensional Cohen-Macaulay module. Furthermore basic properties of Cohen-Macaulay filtered modules with respect to localizations, completion, passing to a non-zero divisor, flat extensions are investigated.

1. INTRODUCTION

Let $(A, \mathfrak{m})$ denote a local Noetherian ring. For a finitely generated A-module M with $d = \dim_A M$ and an integer $0 \le i \le d$ define M_i the largest submodule of M such that $\dim_A M_i \le i$. Because M is a Noetherian A-module the submodules M_i are well-defined. They form an increasing family of submodules. We call $\mathcal{M} = \{M_i\}_{0 \le i \le d}$ the dimension filtration of M. In the first Section of the paper we describe in more details the structure of the submodules M_i. It turns out, see 2.2, that they are described in terms of the reduced primary decomposition of 0 in M. For further investigations we introduce the notion of a distinguished system of parameters $\underline{x} = x_1, \ldots, x_d, d = \dim_A M$, see the Definition 2.5. It turns out that $M_i = 0 :_M (x_{i+1}, \ldots, x_d)$ for a distinguished system of parameters $\underline{x} = x_1, \ldots, x_d$, see 2.7. Moreover under the additional assumption that $\operatorname{Supp}_A M$ is a catenary subset of $\operatorname{Spec} A$ it follows that the dimension filtration localizes for all $\mathfrak{p} \in \operatorname{Supp}_A M$, see 2.5.

Suppose that the local ring $(A, \mathfrak{m})$ possesses a dualizing complex $D_A^{\cdot}$, see [H, Chapter V] for the definition and basic results. We normalize it in such a way that $D_A^{\cdot}$ is a bounded complex with finitely generated cohomology modules and

$$D_A^{-i} \simeq \oplus_{\mathfrak{p} \in \operatorname{Spec} A} E_A(A/\mathfrak{p})$$

for all $i \in \mathbb{N}$. Here $E_A(A/\mathfrak{p})$ denotes the injective hull of $A/\mathfrak{p}$. For a finitely generated d-dimensional A-module M the homology module

$$K^i(M) := H^{-i}(\operatorname{Hom}_A(M, D_A^{\cdot}))\; 0 \le i < d$$

is called the i-th module of deficiency. Moreover $K(M) = H^{-d}(\operatorname{Hom}_A(M, D_A^{\cdot}))$ is called the canonical module of M. This is the generalization of the canonical module of a ring $(A, \mathfrak{m})$ introduced by J. Herzog and E. Kunz in [HK], see also [S1, 3.1] for the generalization. The basic property of the dualizing complex says that the natural homomorphism of complexes

$$M \to \operatorname{Hom}_A(\operatorname{Hom}_A(M, D_A^{\cdot}), D_A^{\cdot})$$

induces an isomorphism in cohomology for a finitely generated A-module M. That is, the 0-th cohomology of the complex at the right hand side is isomorphic to M. Now there is a spectral sequence in order to compute the cohomology of this complex, see Section 3 for more details. In particular, it induces a filtration $\mathcal{F} = \{F^{-i}\}_{0 \le i \le d}$ on the A-module M.

Theorem 1.1. *Both of the filtrations $\mathcal{M}$ and $\mathcal{F}$ coincide, i.e. $M_i = F^{-i}$ for all $0 \le i \le d$.*

For the proof of Theorem 1.1 see Theorem 3.4. So the dimension filtration $\mathcal{M}$ occurs in a natural way as a by-product of the duality of the dualizing complex. Note that the initial terms of the spectral sequence are deficiency modules of the deficiency modules $K^i(K^j(M))$ of M. Note that the deficiency modules $K^i(M), i = 0, \ldots, d-1$, measure the non-Cohen-Macaulayness of M in the sense that M is a Cohen-Macaulay module if and only if $K^i(M) = 0$ for all $i = 0, \ldots, d-1$, as follows by the local duality theorem.

There are several approaches to study generalized Cohen-Macaulay modules from different point of views. We add here another one saying that a finitely generated A-module M is called a Cohen-Macaulay filtered module (CMF for short) whenever all the qoutients $\mathcal{M}_i = M_i/M_{i-1}$ of the dimension filtration are either zero or an i-dimensional Cohen-Macaulay module. Since a Cohen-Macaulay module M is unmixed it is a CMF module with $M = M_d, d = \dim_A M$, and $M_i = 0$ for all $0 \le i < d$.

Examples of non-Cohen-Macaulay CMF modules are approximately Cohen-Macaulay modules, see Section 3 for the definition. It extends the notion of an approximately Cohen-Macaulay ring introduced by S. Gôto in [G]. In Section 3 we describe some basic properties of CMF modules. Among them their permanence properties with respect to localizations, completion and passing to a non-zero divisor. In particular it follows that a finite direct sum of Cohen-Macaulay modules is a CMF module. For more examples we refer to Section 6 of the paper. There we prove also by an example, see 6.1, that in general the CMF property does not descend from the completion $\hat{M}$ to M, where M denotes a finitely generated A-module.

There is a cohomological characterization of CMF modules in terms of the modules of deficiency. This is another main observation of the present investigations.

Theorem 1.2. *Suppose that the local ring $(A, \mathfrak{m})$ possesses a dualizing complex $D_A^{\cdot}$. Let M be a finitely generated A-module. Then M is a CMF module if and only if for all $0 \le i < \dim_A M$ the module of defiency $K^i(M)$ is either zero or an i-dimensional Cohen-Macaulay module. Under this conditions the canonical module $K(M)$ is a Cohen-Macaulay module.*

This result will be shown in 5.5. Note the following: While the Cohen-Macaulayness is described by the vanishing of the modules of deficiency the property of being a CMF module is described in terms of the Cohen-Macaulay property of the modules of deficiency. Moreover it turns out that the canonical module $K(M)$ of a CMF module M is a Cohen-Macaulay module. Note that if $K(M)$ is a Cohen-Macaulay module, then in general M is not a Cohen-Macaulay module. So Theorem 1.2 provides another sufficient condition for $K(M)$ being a Cohen-Macaulay module.

In Section 6 we conclude with the behaviour of the CMF property under flat base extensions of the ground ring. As a basic reference of all of the unexplained terminology we use H. Matsumura's textbook [M]. For the results about Cohen-Macaulay rings and modules see also [BH]. Furthermore, a short introduction into the theory about dualizing complexes the interested reader might found also in [S2].

2. The Dimension Filtration

Let $(A, \mathfrak{m})$ denote a local Noetherian ring. Let M be a finitely generated A-module and $d = \dim_A M$. For an integer $0 \le i < d$ let M_i denote the largest submodule of M such that $\dim_A M_i \le i$. Because of the maximal condition of a Noetherian A-module the submodules M_i of M are well-defined. Moreover it follows that $M_{i-1} \subseteq M_i$ for all $1 \le i \le d$.

Definition 2.1. The inreasing filtration $\mathcal{M} = \{M_i\}_{0\le i\le d}$ of submodules of M is called the dimension filtration of M. Put $\mathcal{M}_i = M_i/M_{i-1}$ for all $1 \le i \le d$.

As a first part of our investigations we give a more detailed description of the modules M_i. Note that $M_0 = H^0_{\mathfrak{m}}(M)$, where $H^0_{\mathfrak{m}}(\cdot)$ denotes the section functor with support in $\{\mathfrak{m}\}$. In order to generalize this observation let $0 = \cap_{j=1}^n N_j$ denote a reduced primary decomposition of 0 in M. That is, $0 \ne \cap_{j=1, j\ne k}^n N_j$ for all $k = 1, \ldots, n$, and N_j is a $\mathfrak{p}_j$-coprimary submodule of M such that the prime ideals $\mathfrak{p}_j$ are pairwise different and $\operatorname{Ass}_A M = \{\mathfrak{p}_1, \ldots, \mathfrak{p}_n\}$. Hence $M_0 = \cap_{\dim A/\mathfrak{p}_j > 0} N_j$.

Both of these representations of M_0 will be generalized to $M_i, 0 \le i \le d$, in the following result. To this end let

$$\mathfrak{a}_i = \prod_{\mathfrak{p}\in \operatorname{Ass} M, \dim A/\mathfrak{p}\le i} \mathfrak{p}.$$

In the case that $\{\mathfrak{p} \in \operatorname{Ass} M \mid \dim A/\mathfrak{p} \le i\} = \varnothing$ put $\mathfrak{a}_i = A$.

Proposition 2.2. *Let M be a finitely generated A-module. Then*

$$M_i = H^0_{\mathfrak{a}_i}(M) = \cap_{\dim A/\mathfrak{p}_j > i} N_j$$

for all $0 \le i \le d$. Here $0 = \cap_{j=1}^n N_j$ denotes a reduced primary decomposition of 0 in M.

Proof. The equality of the last two modules in the statement follows by easy arguments about the primary decomposition of the zero submodule 0 of M. Now let us prove that $M_i = H^0_{\mathfrak{a}_i}(M)$ for all $0 \le i \le d$. Clearly we have $\operatorname{Supp} H^0_{\mathfrak{a}_i}(M) = \operatorname{Supp} M \cap V(\mathfrak{a}_i)$. Therefore it follows that $M_i \subseteq H^0_{\mathfrak{a}_i}(M)$ because any element of M_i is annihilated by an ideal of dimension $\le i$. By the maximality of M_i this proves the equality. □

The previous result provides information about the associated prime ideals of M_i and $\mathcal{M}_i$ respectively.

Corollary 2.3. *Let $\mathcal{M} = \{M_i\}_{0\le i\le d}$ denote the dimension filtration of M. Then*

a) $\operatorname{Ass}_A M_i = \{\mathfrak{p} \in \operatorname{Ass} M \mid \dim A/\mathfrak{p} \le i\}$,
b) $\operatorname{Ass}_A M/M_i = \{\mathfrak{p} \in \operatorname{Ass} M \mid \dim A/\mathfrak{p} > i\}$, *and*
c) $\operatorname{Ass}_A \mathcal{M}_i = \{\mathfrak{p} \in \operatorname{Ass} M \mid \dim A/\mathfrak{p} = i\}$

for all $0 \le i \le d$.

Proof. The two first equalities are obviously true by view of 2.2. Note that

$$\operatorname{Ass}_A H^0_{\mathfrak{a}_i}(M) = \{\mathfrak{p} \in \operatorname{Ass}_A M \mid \mathfrak{p} \in V(\mathfrak{a}_i)\}.$$

The third equality is a consequence of the embedding $\mathcal{M}_i \subseteq M/M_{i-1}$ and the short exact sequence

$$0 \to M_{i-1} \to M_i \to \mathcal{M}_i \to 0.$$

Here we use the containement relation

$$\operatorname{Ass}_A M_i \subseteq \operatorname{Ass}_A M_{i-1} \cup \operatorname{Ass}_A \mathcal{M}_i$$

for the associated prime ideals of the corresponding modules. □

In a certain sense the quotients $\mathcal{M}_i, 0 \le i \le d$, of the dimension filtration $\mathcal{M} = \{M_i\}_{0\le i\le d}$ of M are a measure for the unmixedness of M. Note that the A-module M is unmixed if

$$\dim A/\mathfrak{p} = \dim_A M \text{ for all } \mathfrak{p} \in \operatorname{Ass}_A M.$$

In this case $\mathcal{M}_i = 0$ for all $i < \dim_A M = d$ and $M_d = M$. So the filtration is discret in the case M is unmixed.

More general let $\mathcal{M} = \{M_i\}_{0\le i\le d}$ be the dimension filtration of M. Then $M_i = 0$ for all $i < \operatorname{depth}_A M$. This follows by 2.3 and the fact

$$\operatorname{depth}_A M \le \dim A/\mathfrak{p} \text{ for all } \mathfrak{p} \in \operatorname{Ass}_A M,$$

see [M, Theorem 17.2] for this inequality.

In the following we consider the question whether the dimension filtration behaves well under localizations.

Proposition 2.4. *Let $\mathcal{M} = \{M_i\}_{0\le i\le d}$ be the dimension filtration of a finitely generated A-module M. Suppose that* $\operatorname{Supp}_A M$ *is a catenary subset of* $\operatorname{Spec} A$. *Let* $\mathfrak{p} \in \operatorname{Supp} M$ *denote a prime ideal. Define*

$$M_i' = M_{i+\dim A/\mathfrak{p}} \otimes_A A_\mathfrak{p} \textit{ for all } 0 \le i \le \dim_{A_\mathfrak{p}} M_\mathfrak{p} = t.$$

Then $\mathcal{M}' = \{M_i'\}_{0\le i\le t}$ is the dimension filtration of the $A_\mathfrak{p}$-module $M_\mathfrak{p}$.

Proof. First we mention that there is the bound

$$\dim_A M_i' \le (i + \dim A/\mathfrak{p}) - \dim A/\mathfrak{p} = i$$

for all $i \in \mathbb{Z}$. Next we recall the following statement about associated prime ideals

$$\operatorname{Ass}_{A_\mathfrak{p}} M_\mathfrak{p} = \{\mathfrak{q}A_\mathfrak{p} \mid \mathfrak{q} \in \operatorname{Ass}_A M, \mathfrak{q} \subseteq \mathfrak{p}\},$$

see [M, Theorem 6.2]. Now let $0 = \cap_{j=1}^n N_j$ be a reduced primary decomposition of 0 in M, where N_j is $\mathfrak{q}_j$-coprimary. Suppose that $\mathfrak{q}_j \subseteq \mathfrak{p}$ for all $j = 1, \dots, m$ and $\mathfrak{q}_j \not\subseteq \mathfrak{p}$ for all $j = m+1, \dots, n$. Then $0 = \cap_{j=1}^m (N_j \otimes_A A_\mathfrak{p})$ is a reduced primary decomposition of 0 in $M_\mathfrak{p}$ as an $A_\mathfrak{p}$-module. Therefore, by view of 2.2, it yields that

$$(M_\mathfrak{p})_i = \cap_{\dim A_\mathfrak{p}/\mathfrak{q}_jA_\mathfrak{p} > i}(N_j \otimes_A A_\mathfrak{p}).$$

Moreover by the localization of $M_{i+\dim A/\mathfrak{p}}$ we get the following equality

$$M_i' = \cap_{\dim A/\mathfrak{q}_j > i+\dim A/\mathfrak{p}}(N_j \otimes_A A_\mathfrak{p}).$$

Because $\operatorname{Supp}_A M$ is supposed to be a catenary subset of $\operatorname{Spec} A$ we get that

$$\dim A/\mathfrak{q}_j = \dim A/\mathfrak{p} + \dim A_\mathfrak{p}/\mathfrak{q}_jA_\mathfrak{p}.$$

First this proves that $d = t + \dim A/\mathfrak{p}$. Because of the above statement about the associated prime ideals it shows finally that $M_i' = (M_\mathfrak{p})_i$ for all $0 \le i \le t$, as required. □

In the following we consider a variation of the notion of a system of parameters of an A-module M.

Definition 2.5. Let $\underline{x} = x_1, \dots, x_d, d = \dim_A M$, denote a system of parameters of M. Then $\underline{x} = x_1, \dots, x_d$ is called a distinguished system of parameters of M provided $(x_{i+1}, \dots, x_d)M_i = 0$ for all $i = 0, \dots, d-1$.

In the next result let us prove the existence of distinguished systems of parameters of an A-module M.

Lemma 2.6. *Any finitely generated A-module M admits a distinguished system of parameters.*

Proof. First we show the existence of a parameter x_d of M such that $x_d M_i = 0$ for all $i = 0, \ldots, d-1$. To this end note that $\dim_A M_i \leq i < d$ for all $i = 0, \ldots, d-1$. Put $\mathfrak{b} = \prod_{i=0}^{d-1} \operatorname{Ann}_A M_i$. Then $\mathfrak{b} \nsubseteq \mathfrak{p}$ for any associated prime ideal $\mathfrak{p} \in \operatorname{Ass}_A M$ with $\dim A/\mathfrak{p} = d$. Therefore there is an element $x_d \in \mathfrak{b}$ and $x_d \notin \mathfrak{p}$ for all $\mathfrak{p} \in \operatorname{Ass}_A M$ with $\dim A/\mathfrak{p} = d$. Whence x_d is a parameter with the desired property. Now pass to the factor module $M/x_d M$ and choose a parameter x_{d-1} of $M/x_d M$ such that $x_{d-1} M_i = 0$ for all $i = 0, \ldots, d-2$. Then an induction finishes the proof of the claim. □

It turns out that whenever $\underline{x} = x_1, \ldots, x_d$ is a distinguished system of parameters of M, the elements $x_1, \ldots, x_i$ generate an ideal of definition of $\mathcal{M}_i$. This follows since $M_i/\underline{x}M_i$ is an A-module of finite length. Therefore, whenever $\mathcal{M}_i \neq 0$, then $x_1, \ldots, x_i$ is a system of parameters of $\mathcal{M}_i$.

Lemma 2.7. *A system of parameters $\underline{x} = x_1, \ldots, x_d$ of M is a distinguished system of parameters if and only if $M_i = 0 :_M (x_{i+1}, \ldots, x_d)$ for $i = 0, \ldots, d-1$.*

Proof. Let $\underline{x} = x_1, \ldots, x_d$ denote a system of parameters of M such that

$$M_i = 0 :_M (x_{i+1}, \ldots, x_d) \text{ for all } i = 0, \ldots, d-1.$$

Then $(x_{i+1}, \ldots, x_d)M_i = 0$, i.e. $\underline{x}$ is a distinguished system of parameters.

Conversely let $\underline{x}$ be a distinguished system of parameters. Then

$$M_i \subseteq 0 :_M (x_{i+1}, \ldots, x_d) \text{ for all } i = 0, \ldots, d-1$$

as follows by the definition. Moreover there is the following expression for the associated prime ideals

$$\operatorname{Ass}_A(0 :_M (x_{i+1}, \ldots, x_d)) = \{\mathfrak{p} \in \operatorname{Ass}_A M \mid \mathfrak{p} \in V(x_{i+1}, \ldots, x_d)\}.$$

Let $\mathfrak{p}$ denote an associated prime ideal of $0 :_M (x_{i+1}, \ldots, x_d)$. Then we obtain $\mathfrak{p} \in \operatorname{Supp}_A M/(x_{i+1}, \ldots, x_d)M$ and therefore $\dim A/\mathfrak{p} \leq d - (d-i) = i$. That is, $\dim_A(0 :_M (x_{i+1}, \ldots, x_d)) \leq i$. Because of the maximality of M_i the equality $M_i = 0 :_M (x_{i+1}, \ldots, x_d)$ follows now. □

3. A Supplement to Duality

In this section let $(A, \mathfrak{m})$ denote a local ring possessing a dualizing complex $D_A^{\cdot}$. That is a bounded complex of injective A-modules D_A^i whose cohomology modules $H^i(D_A^{\cdot}), i \in \mathbb{Z}$, are finitely generated A-modules. We refer to [H, Chapter V, §2] or to [S2, 1.2] for basic results about dualizing complexes. Note that the natural homomorphism of complexes

$$M \to \operatorname{Hom}_A(\operatorname{Hom}_A(M, D_A^{\cdot}), D_A^{\cdot})$$

induces an isomorphism in cohomlogy for any finitely generated A-module M. Moreover there is an integer $l \in \mathbb{Z}$ such that

$$\operatorname{Hom}_A(k, D_A^{\cdot}) \simeq k[l],$$

where $k = A/\mathfrak{m}$ denotes the residue field of A. Without loss of generality assume that $l = 0$. Then the dualizing complex $D_A^{\cdot}$ has the property

$$D_A^{-i} \simeq \oplus_{\mathfrak{p} \in \operatorname{Spec} A, \dim A/\mathfrak{p} = i} E_A(A/\mathfrak{p}),$$

where $E_A(A/\mathfrak{p})$ denotes the injectice hull of $A/\mathfrak{p}$ as A-module. Therefore $D_A^i = 0$ for $i < -\dim A$ and $i > 0$. The following modules were introduced in [S1, 3.1], see also [S2, 1.2].

Definition 3.1. Let M denote a finitely generated A-module and $d = \dim_A M$. For an integer $i \in \mathbb{Z}$ define

$$K^i(M) := H^{-i}(\operatorname{Hom}_A(M, D_A^{\cdot})).$$

The module $K(M) := K^d(M)$ is called the canonical module of M. For $i \neq d$ the modules $K^i(M)$ are called the modules of deficiency of M. Note that $K^i(M) = 0$ for all $i < 0$ or $i > d$.

By the local duality theorem, see [H, Chapter V, §6] or [S2, Theorem 1.11], there are the following canonical isomorphisms

$$H_{\mathfrak{m}}^i(M) \simeq \operatorname{Hom}_A(K^i(M), E), i \in \mathbb{Z},$$

where $E = E_A(A/\mathfrak{m})$. Recall that all of the $K^i(M), i \in \mathbb{Z}$, are finitely generated A-modules. Moreover M is a Cohen-Macaulay module if and only if $K^i(M) = 0$ for all $i \neq d$. Whence the modules of deficiencies of M measure the deviation of M from being a Cohen-Macaulay module. The canonical module $K(M)$ of M is a Cohen-Macaulay module provided M is a Cohen-Macaulay module. The converse does not hold in general, see [S2, Lemma 1.9] for the precise statements.

For an arbitrary A-module X and an integer $i \in \mathbb{N}$ let

$$(\operatorname{Ass}_A X)_i = \{\mathfrak{p} \in \operatorname{Ass}_A M \mid \dim A/\mathfrak{p} = i\}.$$

For the proof of the next result see [S1, 3.1] and [S2, Lemma 1.9].

Proposition 3.2. *Let M denote a d-dimensional A-module. Then the following results are true:*

a) $\dim_A K^i(M) \leq i$ *for all* $0 \leq i < d$ *and* $\dim_A K(M) = d$.
b) $\operatorname{Ass}_A K(M) = (\operatorname{Ass}_A M)_d$.
c) $(\operatorname{Ass}_A K^i(M))_i = (\operatorname{Ass}_A M)_i$ *for all* $0 \leq i < d$.
d) *Let M be a Cohen-Macaulay module. Then $K(M)$ is also a Cohen-Macaulay module.*

As mentioned above for a finitely generated A-module M the induced homomorphisms of the cohomology of the natural map

$$M \to \operatorname{Hom}_A(\operatorname{Hom}_A(M, D_A^{\cdot}), D_A^{\cdot})$$

are isomorphisms. In order to compute the cohomology of the complex at the right hand side there is the following spectral sequence

$$E_1^{pq} = H^q(\operatorname{Hom}_A(\operatorname{Hom}_A(M, D_A^{\cdot}), D_A^p)),$$

see [E, Appendix 3, Part II] or [W, Section 5] for the details about spectral sequences used here in this section. Because D_A^p is an injective A-module the corresponding E_2-term has the following form

$$E_2^{pq} = H^p(\mathrm{Hom}_A(H^{-q}(\mathrm{Hom}_A(M, D_A^{\cdot})), D_A^{\cdot})).$$

With regard to our previous notation it follows that $E_2^{pq} = K^{-p}(K^q(M))$. Now we have to prove the following basic observation.

Lemma 3.3. *Let M denote a finitely generated A-module. Let $\mathfrak{p} \in \mathrm{Supp}_A M$ be a prime ideal with $t = \dim A/\mathfrak{p}$. Then there are the following isomorphisms*

$$K^i(K^j(M)) \otimes_A A_{\mathfrak{p}} \simeq K^{i-t}(K^{j-t}(M \otimes_A A_{\mathfrak{p}}))$$

for any pair $(i, j) \in \mathbb{Z}^2$.

Proof. First note that there is an isomorphism of dualizing complexes

$$D_A^{\cdot} \otimes_A A_{\mathfrak{p}} \simeq D_{A_{\mathfrak{p}}}^{\cdot}[t],$$

see e.g. [H, Chapter V, Proposition 7.1]. Now by the definition of the K^i's write

$$K^i(K^j(M)) \simeq H^{-i}(\mathrm{Hom}_A(H^{-j}(\mathrm{Hom}_A(M, D_A^{\cdot})), D_A^{\cdot})).$$

The localization functor $\cdot \otimes_A A_{\mathfrak{p}}$ is exact, i.e. it commutes with cohomology. Moreover let X denote a bounded complex of A-modules whose cohomology modules are finitely generated A-modules. Then there is the following isomorphism of complexes

$$\mathrm{Hom}_A(X, D_A^{\cdot}) \otimes_A A_{\mathfrak{p}} \simeq \mathrm{Hom}_{A_{\mathfrak{p}}}(X \otimes_A A_{\mathfrak{p}}, D_{A_{\mathfrak{p}}}^{\cdot})[t],$$

see [H, Chapter II]. Putting together all of these ingredients the statement of the proposition follows now. □

Let M denote a finitely generated A-module with $d = \dim_A M$. Let us return to the above spectral sequence. Consider the stage $p + q = 0$, the onliest place in which non-zero cohomology occurs. Then the limit terms $E_\infty^{p,-p}$, $-d \le p \le 0$, are the quotients of a filtration

$$F^0 \subseteq F^{-1} \subseteq \ldots \subseteq F^{-d+1} \subseteq F^{-d} = M$$

of M. That is we have $F^p/F^{p+1} \simeq E_\infty^{p,-p}$ for all $-d \le p \le 0$. The natural question about the filtration $\mathcal{F} = \{F^{-i}\}_{0 \le i \le d}$ is its relationship to the dimension filtration of M. This is answered in the following result.

Theorem 3.4. *Let $\mathcal{M} = \{M_i\}_{0 \le i \le d}$ be the dimension filtration of M. Then it follows $M_i = F^{-i}$ for all $0 \le i \le d$.*

Proof. By the construction of the spectral sequence the term $E_{r+1}^{p,-p}$ is the cohomology at the middle of the following sequence of A-modules

$$E_r^{p-r,-p+r-1} \to E_r^{p,-p} \to E_r^{p+r,-p-r+1}.$$

The term on the left hand side is zero because it is a subquotient of

$$K^{-p+r}(K^{-p+r-1}(M)) = 0.$$

To this end recall that $\dim_A K^{-p+r-1}(M) \le -p + r - 1$, see 3.2.

First of all let us consider the case of $p = 0$. Then also the term on the right hand side is zero since it is a subqoutient of $K^{-r}(K^{r-1}(M)) = 0, r \geq 2$. That is we get a partial degeneration of the spectral sequence to the isomorphisms

$$F^0 \simeq E_\infty^{0,0} \simeq E_2^{0,0} \simeq K^0(K^0(M)).$$

Moreover the local duality theorem implies that $K^0(K^0(M)) \simeq H_{\mathfrak{m}}^0(M) = M_0$. To this end recall that

$$\operatorname{Hom}_A(H^0(\operatorname{Hom}_A(M, D_A^{\cdot})), D_A^{\cdot}) \simeq \operatorname{Hom}_A(H^0(\operatorname{Hom}_A(M, D_A^{\cdot})), E).$$

This proves the claim in the case $p = 0$.

For an arbitrary p the above considerations provide the following chain of inclusions

$$E_\infty^{p,-p} \subseteq E_{r+1}^{p,-p} \subseteq E_r^{p,-p} \subseteq E_2^{p,-p} = K^{-p}(K^{-p}(M))$$

for all $r \geq 2$. By view of 3.2 this implies that either $\dim_A E_\infty^{p,-p} = -p$ or $E_\infty^{p,-p} = 0$. Note that $\operatorname{Ass}_A K^{-p}(K^{-p}(M)) = (\operatorname{Ass}_A K^{-p}(M))_{-p}$ for all $-d \leq p \leq 0$, see 3.2. But now we have

$$E_\infty^{p,-p} \simeq F^p/F^{p+1} \text{ and } E_\infty^{0,0} \simeq F^0 = M_0.$$

Therefore $\dim_A F^p \leq -p$ and $F^p \subseteq M_{-p}$ for all $-d \leq p \leq 0$. In the final part of the proof we have to show equality.

We proceed by an induction on $d = \dim_A M$. As mentioned above the case $d = 0$, i.e. $M_0 = F^0$ is shown to be true. So let $d > 0$. It is known that $(A, \mathfrak{m})$ is a catenary local ring since it possesses a dualizing complex, see [H, Chapter V, Corollary 7.2]. By 2.4 it follows that $(M_{\mathfrak{p}})_{i+\dim A/\mathfrak{p}} = (M_i) \otimes_A A_{\mathfrak{p}}$ for all prime ideals $\mathfrak{p} \in \operatorname{Supp} M$. Next we want to prove a corresponding result for the filtration $\mathcal{F}$, i.e. $F^i(M) \otimes_A A_{\mathfrak{p}} = F^{i+\dim A/\mathfrak{p}}(M_{\mathfrak{p}})$. Here we refer to $F^{\cdot}(M)$ resp. $F^{\cdot}(M_{\mathfrak{p}})$ as the filtration induced by the A-module M resp. by the $A_{\mathfrak{p}}$-module $M_{\mathfrak{p}}$. By 3.3 it turns out that

$$E_2^{pq}(M) \otimes_A A_{\mathfrak{p}} \simeq E_2^{p+t,q-t}(M \otimes_A A_{\mathfrak{p}}), \quad t = \dim A/\mathfrak{p},$$

for all pairs $(p, q) \in \mathbb{Z}^2$. Because of the exactness of the localization functor $\cdot \otimes A_{\mathfrak{p}}$ and because of the functoriality of the spectral sequence this finally shows that $F^p(M) \otimes_A A_{\mathfrak{p}} \simeq F^{p+t}(M_{\mathfrak{p}})$ for all $p \in \mathbb{Z}$.

Now let us finish the proof. Because of the spectral sequence we know that $F^p = M_{-p}$ for $p = -d$. So let us assume statement for p in order to prove it for $p + 1$. To this end consider the injection

$$0 \to M_{-p-1}/F^{p+1} \to F^p/F^{p+1} \simeq E_\infty^{p,-p}.$$

Note that $M_{-p-1} \subseteq M_p = F^p$. Because of the induction hypothesis and the previous considerations we have that $M_{-p-1} \otimes A_{\mathfrak{p}} = F^{p+1} \otimes_A A_{\mathfrak{p}}$ for all non-maximal prime ideals $\mathfrak{p} \in \operatorname{Supp}_A M$. Therefore the module at the left hand side of the above exact sequence has its support contained in $V(\mathfrak{m})$. Moreover by the spectral sequence we get that

$$E_\infty^{p,-p} \subseteq E_2^{p,-p} \simeq K^{-p}(K^{-p}(M)).$$

By virtue of 3.2 we now have the following inclusion

$$\operatorname{Ass}_A E_\infty^{p,-p} \subseteq \{\mathfrak{p} \in \operatorname{Ass}_A M \mid \dim A/\mathfrak{p} = -p\}.$$

Since M_{-p-1}/F^{p+1} is by induction hypothesis an A-module of finite length it is in fact zero, which completes the inductive step. □

It is worth to remark that in general the limit terms $E_\infty^{p,-p}$ of the spectral sequence considered above do not agree with $E_2^{p,-p} \simeq K^{-p}(K^{-p}(M))$. It would be interesting to find an explicit description of these modules.

4. Cohen-Macaulay Filtered Modules

Let M denote a finitely generated A-module, where $(A, \mathfrak{m})$ is a local Noetherian ring. Let $\mathcal{M} = \{M_i\}_{0 \le i \le d}$ denote the dimension filtration.

Definition 4.1. A finitely generated A-module M is called a Cohen-Macaulay filtered module (CMF module), whenever $\mathcal{M}_i = M_i/M_{i-1}$ is either zero or an i-dimensional Cohen-Macaulay module for all $0 \le i \le \dim_A M$.

Note that any Cohen-Macaulay module is also a CMF module. This follows because under this assumption $M_i = 0$ for all $i < \dim_A M$. Conversely an unmixed CMF module is also a Cohen-Macaulay module. Let M be an A-module such that $\operatorname{depth}_A M = 0$ and $M/H_\mathfrak{m}^0(M)$ is a Cohen-Maculay module. Then M is a CMF module as easily seen. For further examples see Section 6.

Related to the definition of a CMF module it will be useful to have the notion of a Cohen-Macaulay filtration.

Definition 4.2. Let M denote a finitely generated A-module with $d = \dim_A M$. An increasing filtration $C = \{C_i\}_{0 \le i \le d}$ of M is called a Cohen-Macaulay filtration whenever $M = C_d, d = \dim_A M$, and $\mathcal{C}_i = C_i/C_{i-1}$ is either zero or an i-dimensional Cohen-Macaulay module for all $1 \le i \le d$.

The following proposition is useful in order to characterize CMF modules. In fact it shows that a Cohen-Macaulay filtration coincides automatically with the dimension filtration.

Proposition 4.3. *Let $C = \{C_i\}_{0 \le i \le d}$ be Cohen-Macaulay filtration of M. Then C coincides with the dimension filtration.*

Proof. First of all it is easily seen that $\dim C_i \le i$ for all $0 \le i \le d$. Moreover it follows that

$$\operatorname{Ass}_A C_i \subseteq \{\mathfrak{p} \in \operatorname{Ass}_A M \mid \dim A/\mathfrak{p} \le i\}.$$

With the definition of $\mathfrak{a}_i$ - as done in Section 2 - this implies that $H_{\mathfrak{a}_i}^0(C_i) = C_i$ for all $0 \le i \le d$. Now fix i and let $j \ge i$ be an integer. Next consider the following short exact sequence

$$0 \to C_j \to C_{j+1} \to \mathcal{C}_{j+1} \to 0.$$

Since $\mathcal{C}_{j+1}$ is either zero or a $(j+1)$-dimensional Cohen-Macaulay module it induces isomorphisms

$$H_{\mathfrak{a}_i}^0(C_j) \simeq H_{\mathfrak{a}_i}^0(C_{j+1}) \text{ for all } j \ge i.$$

Therefore $C_i = H_{\mathfrak{a}_i}^0(C_j)$ for all $j \ge i$. Because of $M = C_d$ this finally proves the claim, see 2.2. □

Let $u_M(0) = \cap_{\dim A/\mathfrak{p}_j = d} N_j$, where $0 = \cap_{j=1}^n N_j$ denotes a reduced primary decomposition of 0 in M as considered in Section 2.

Definition 4.4. A finitely generated A-module $M, d = \dim_A M$, is called an approximately Cohen-Macaulay module whenever $M/u_M(0)$ is a Cohen-Macaulay module and $\operatorname{depth}_A M \geq d - 1$.

This is the extension of the notion of an approximately Cohen-Macaulay ring introduced by S. Gôto, see [G]. Note that a Cohen-Macaulay module is always an approximately Cohen-Macaulay module. Next let us describe the relation of this notion to that of CMF modules.

Proposition 4.5. *Let M be a finitely generated A-module. Then M is approximately Cohen-Macaulay if and only if M is a CMF module and* $\operatorname{depth}_A M \geq \dim_A M - 1$.

Proof. First let M be an approximately Cohen-Macaulay module. Put $d = \dim_A M$. By [M, Theorem 17.2] it follows that

$$d - 1 \leq \operatorname{depth}_A M \leq \dim A/\mathfrak{p} \text{ for all } \mathfrak{p} \in \operatorname{Ass}_A M.$$

Therefore $M_i = 0$ for $i = 0, \dots, d-2$ and $M_{d-1} = u_M(0)$, see 2.2. Now consider the short exact sequence

$$0 \to M_{d-1} \to M \to M/M_{d-1} \to 0.$$

Because M is approximately Cohen-Macaulay it follows that M/M_{d-1} is a d-dimensional Cohen-Macaulay module and $\operatorname{depth}_A M \geq d - 1$. So the short exact sequence implies $\operatorname{depth}_A M_{d-1} \geq d - 1$. Because of $\dim_A M_{d-1} \leq d - 1$ it turns out that M_{d-1} is either zero or a $(d-1)$-dimensional Cohen-Macaulay module.

The reverse statement follows the same line of reasoning. Hence we omit the details. □

Before we shall present in Section 6 a general construction method for CMF modules there are a few results on permanence properties of CMF module. To this end $\hat{A}$ denotes the $\mathfrak{m}$-adic completion of A. Note that there is a natural isomorphism $M \otimes_A \hat{A} \simeq \hat{M}$ for a finitely generated A-module M.

Proposition 4.6. *Let M denote a CMF A-module. Then the following conditions are satisfied:*

a) $\operatorname{Supp}_A M$ *is a catenary subset of* $\operatorname{Spec} A$.
b) *Let* $\mathfrak{p} \in \operatorname{Supp}_A M$. *Then*

$$\dim A/\mathfrak{p} = \dim \hat{A}/\mathfrak{q} \text{ for all } \mathfrak{q} \in \operatorname{Ass}_{\hat{A}} \hat{A}/\mathfrak{p}\hat{A},$$

i.e. $A/\mathfrak{p}$ is formally unmixed for all $\mathfrak{p} \in \operatorname{Supp}_A M$.

Proof. Because M/M_{d-1} is a Cohen-Macaulay module and

$$\operatorname{Supp}_A M = \operatorname{Supp}_A M/M_{d-1}$$

both of the statements follow. For the first statement see [M, §17]. The second is a consequence of [N, (34.9)]. □

Note that a CMF ring A possesses a small Cohen-Macaulay module. That is a Cohen-Macaulay module X such that $\operatorname{depth} X = \dim A$. This follows since $A/A_{d-1}, d = \dim A$, is a d-dimensional Cohen-Macaulay module. Consequently for a CMF ring A all the homological conjectures are true.

Now we start with the permanence properties of the CMF property.

Theorem 4.7. *Let M denote a finitely generated A-module. Let $x \in \mathfrak{m}$ be an M-regular element. Then M is a CMF module if and only if M/xM is a CMF module.*

Proof. First note that whenever $x \in \mathfrak{m}$ is an M-regular element, then $M_0 = 0$ and x is also M/M_i-regular as well as $\mathcal{M}_i$-regular for all $i \geq 1$. Here $\mathcal{M} = \{M_i\}_{0 \leq i \leq d}$ denotes the dimension filtration and $\mathcal{M}_i = M_i/M_{i-1}$. In particular it follows that $M_i \cap xM = xM_i$ for all $1 \leq i \leq d$.

Now suppose that M is a CMF module. Let $i \geq 1$. Then

$$\mathcal{M}_i/x\mathcal{M}_i \simeq ((M_i, xM)/xM)/(M_{i-1}, xM)/xM)$$

is a $(i-1)$-dimensional Cohen-Macaulay module or zero. Therefore by 4.3 it follows that M/xM is a $(d-1)$-dimensional CMF module since

$$\{(M_{i+1}, xM/xM)\}_{0 \leq i < d}$$

is a Cohen-Macaulay filtration.

Conversely let M/xM be a CMF module. Then the dimension filtration $\{M'_i\}_{0 \leq i \leq d}$ of M/xM has the property that $\mathcal{M}'_i = M'_i/M'_{i-1}$ is either zero or an i-dimensional Cohen-Macaulay module. Let $M_{i+1}, 0 \leq i < d$, denote the preimage of M'_i in M and $M_0 = 0$. By the same isomorphism as above it follows now that $\mathcal{M}_i/x\mathcal{M}_i$ is either zero or a $(i-1)$-dimensional Cohen-Macaulay module. Since x is an $\mathcal{M}_i$-regular element, $\mathcal{M}_i$ is either zero or an i-dimensional Cohen-Macaulay module. By view of 4.1 this proves the claim. □

Another permanence poperty of CMF modules is the following result about the localization behaviour.

Proposition 4.8. *Let M denote a CMF module. Then $M_\mathfrak{p}$ is a CMF $A_\mathfrak{p}$-module for any prime ideal $\mathfrak{p} \in \operatorname{Supp}_A M$.*

Proof. Let $\mathcal{M} = \{M_i\}_{0 \leq i \leq d}$ denote the dimension filtration of M. Let $\mathfrak{p} \in \operatorname{Supp}_A M$ and $t = \dim A/\mathfrak{p}$. Then we define $M'_i = (M_{i+t}) \otimes_A A_\mathfrak{p}$, for all $i \geq t$. Then

$$M'_i/M'_{i-1} \simeq (M_{i+t}/M_{i-1+t}) \otimes_A A_\mathfrak{p}$$

is either zero or a Cohen-Macaulay $A_\mathfrak{p}$-module of dimension i, see [M, §17]. By virtue of 4.3 this proves the claim. □

It is worth to note that we do not need 2.4 in order to prove 4.8. The result turns out because of

$$\dim_A X = \dim A/\mathfrak{p} + \dim_{A_\mathfrak{p}} X_\mathfrak{p} \text{ for all } \mathfrak{p} \in \operatorname{Supp}_A X,$$

where X denotes a Cohen-Macaulay A-module. Moreover by 4.6 we know that $\operatorname{Supp}_A M$ is a catenary subset of $\operatorname{Spec} A$.

In the final part of this section consider the behaviour of the CMF property by passing to the completion.

Theorem 4.9. *Let M be a finitely generated A-module. Let M be a CMF A-module. Then $M \otimes_A \hat{A}$ is a CMF $\hat{A}$-module.*

Proof. Let $\mathcal{M} = \{M_i\}_{0 \leq i \leq d}$ denote the Cohen-Macaulay filtration of the CMF A-module M. Then $\{M_i \otimes_A \hat{A}\}_{0 \leq i \leq d}$ is clearly a Cohen-Macaulay filtration of the $\hat{A}$-module $M \otimes_A \hat{A}$. So by 4.3 $M \otimes_A \hat{A}$ is a CMF module over $\hat{A}$. □

The converse of the above statement is not true in general as we will show by an example, see Example 6.1. A more general statement is true for an arbitrary faithful flat extension $(A, \mathfrak{m}) \to (B, \mathfrak{n})$. For the precise statement see the considerations in Section 6.

5. Characterizations of Cohen-Macaulay Filtered Modules

In the first part of this section we consider a few algebraic properties of a CMF module. To this end let M denote a finitely generated A-module M, where $(A, \mathfrak{m})$ is an arbitrary local Noetherian ring. First we need a preliminary result. Here L_A denotes the length function on A-modules.

Lemma 5.1. *Let $\mathcal{M} = \{M_i\}_{0 \le i \le d}$ denote the dimension filtration of M. Then*

$$L_A(M/\underline{x}M) \le \sum_{i=0}^{d} L_A(\mathcal{M}_i/(x_1, \dots, x_i)\mathcal{M}_i)$$

for any distinguished system of parameters $\underline{x} = x_1, \dots, x_d, d = \dim_A M$, of M.

Proof. For $1 \le i \le d$ let us consider the following short exact sequences

$$0 \to M_{i-1} \to M_i \to \mathcal{M}_i \to 0.$$

Tensor it by $A/(x_1, \dots, x_d)A$. Because of $x_i M_{i-1} = 0, 1 \le i \le d$, it induces an exact sequence

$$M_{i-1}/(x_1, \dots, x_{i-1})M_{i-1} \to M_i/(x_1, \dots, x_i)M_i \to \mathcal{M}_i/(x_1, \dots, x_i)\mathcal{M}_i \to 0.$$

Because $\underline{x}$ is a distinguished system of parameters of M the elements $x_1, \dots, x_i$ generate an ideal of definition of M_i. That is, the A-modules

$$M_i/(x_1, \dots, x_i)M_i \text{ and } \mathcal{M}_i/(x_1, \dots, x_i)\mathcal{M}_i, i = 0, \dots, d,$$

are A-modules of finite length. Therefore

$$L_A(M_i/(x_1, \dots, x_i)M_i) \le L_A(M_{i-1}/(x_1, \dots, x_{i-1})M_{i-1}) + L_A(\mathcal{M}_i/(x_1, \dots, x_i)\mathcal{M}_i)$$

for all $i = 1, \dots, d$. Because of $M = M_d$ a recurrence proves the desired inequality. □

Note that the inequality of 5.1 is also true for any system of parameters $\underline{x} = x_1, \dots, x_d$ of M. But in this case it might happen that the modules on the right hand side are not of finite length. In this case the estimate is trivially true.

Theorem 5.2. *Let $\mathcal{M} = \{M_i\}_{0 \le i \le d}$ denote the dimension filtration of a finitely generated A-module M with $d = \dim_A M$ and $t = \operatorname{depth}_A M$. Let $\underline{x} = x_1, \dots, x_d$ be a distinguished system of parameters.*

Suppose that M is a CMF module. Then the following conditions are satisfied:

a) $L_A(M/(x_1, \dots, x_d)M) = \sum_{i=0}^{d} L_A(\mathcal{M}_i/(x_1, \dots, x_i)\mathcal{M}_i)$.

b) *$M/(x_1, \dots, x_{d-t})M$ is a t-dimensional Cohen-Macaulay module.*

The converse is true, i.e. the conditions a) and b) imply that M is a CMF module, provided $\operatorname{depth}_A M \ge d - 1$.

Proof. Suppose that M is a CMF module. Then the factor modules $\mathcal{M}_i, 0 \le i \le d$, are either zero or i-dimensional Cohen-Macaulay modules. Since x_i is an $\mathcal{M}_i$-regular element there are the following short exact sequences

$$0 \to M_{i-1} \to M_i/x_iM_i \to \mathcal{M}_i/x_i\mathcal{M}_i \to 0.$$

Now apply the Koszul homology $H.(x_1,\dots,x_{i-1};\cdot)$ to this sequence. Because of

$$H_1(x_1,\dots,x_{i-1};\mathcal{M}_i/x_i\mathcal{M}_i) = 0,$$

note that $x_1,\dots,x_{i-1}$ is a $\mathcal{M}_i/x_i\mathcal{M}_i$-regular sequence, it induces a short exact sequence

$$0 \to M_{i-1}/(x_1,\dots,x_{i-1})M_{i-1} \to M_i/(x_1,\dots,x_i)M_i \to \mathcal{M}_i/(x_1,\dots,x_i)\mathcal{M}_i \to 0$$

for all $i = 1,\dots,d$. Because of $M = M_d$ a recurrence on i proves the equality of the statement in a).

For the proof of b) first note that $M_i = 0$ for all $i < t = \operatorname{depth}_A M$. Moreover M_t a t-dimensional Cohen-Macaulay module. Now we investigate the above short exact sequence

$$0 \to M_{i-1} \to M_i/x_iM_i \to \mathcal{M}_i/x_i\mathcal{M}_i \to 0.$$

for $i \ge t+1$. Applying the Koszul homology $H.(x_{t+1},\dots,x_{i-1};\cdot)$ to this sequence it induces - by similar arguments as above - a short exact sequence

$$0 \to M_{i-1}/(x_{t+1},...,x_{i-1})M_{i-1} \to M_i/(x_{t+1},...,x_i)M_i \to \mathcal{M}_i/(x_{t+1},...,x_i)\mathcal{M}_i \to 0$$

for all $i = t+1,\dots,d$. Note that $x_{t+1},\dots,x_{i-1}$ forms a regular sequence on the $(i-1)$-dimensional Cohen-Macaulay module $\mathcal{M}_i/x_i\mathcal{M}_i$. Moreover all of the modules considered in these sequences are either zero or of dimension t. By induction we prove now that the module in the middle is a t-dimensional Cohen-Macaulay module. First note that $M_t = \mathcal{M}_t$ is a t-dimensional Cohen-Macaulay module. Because $M_{i-1}/(x_{t+1},\dots,x_{i-1})M_{i-1}$ and $\mathcal{M}_i/(x_{t+1},\dots,x_i)\mathcal{M}_i$ are either zero or t-dimensional Cohen-Macaulay modules the short exact sequence implies that $M_i/(x_{t+1},\dots,x_i)M_i$ is also either zero or a t-dimensional Cohen-Macaulay module. Because of $M = M_d$ this proves the statement of the condition b).

In order to prove the reverse implication let $\operatorname{depth}_A M = d-1$. (In the Cohen-Macaulay case there is nothing to prove.) Therefore we get $M_i = 0$ for all $i < d-1$. Then - as above - there is the short exact sequence

$$0 \to M_{d-1} \to M_d/x_1M_d \to \mathcal{M}_d/x_1\mathcal{M}_d \to 0.$$

Note that x_1 is an $\mathcal{M}_d$-regular element. Next apply the Koszul homology $H.(x_2,\dots,x_d;\cdot)$ to this sequence. Because $x_2,\dots,x_d$ is an M_d/x_1M_d-regular sequence we get the following exact sequence

$$\begin{aligned} 0 \to\ & H_1(x_2,\dots,x_d;\mathcal{M}_d/x_1\mathcal{M}_d) \to \\ & \to M_{d-1}/(x_2,\dots,x_d)M_{d-1} \to M_d/\underline{x}M_d \to \mathcal{M}_d/\underline{x}\mathcal{M}_d \to 0. \end{aligned}$$

By the assumption on the length it turns out that

$$H_1(x_2,\dots,x_d;\mathcal{M}_d/x_1\mathcal{M}_d) = 0.$$

That is, $x_2,\dots,x_d$ forms an $\mathcal{M}_d/x_1\mathcal{M}_d$-regular sequence. So $\mathcal{M}_d$ is a d-dimensional Cohen-Macaulay module. Moreover the above short exact sequence implies that M_{d-1} is a $(d-1)$-dimensional Cohen-Macaulay module. That means, M is a CMF module, as required. □

It would be interesting to generalize the converse of 5.2 to a more general situation. In the case of $\operatorname{depth}_A M = d - 1$ neither a) nor b) will be sufficient for M being a CMF module. For b) see S. Gôto's example [G, Remark 2.9]. For a) consider the ring $A = k[[w,x,y,z]]/(w,x) \cap (y,z) \cap (w^2,x,y^2,z)$, where k denotes a field. Then $\underline{x} = w - y, x - z$ is a distinguished system of parameters satisfying the equality in a) but A is not a CMF ring.

Another partial result in order to prove the converse is the following slight generalization of [G, Lemma 2.1].

Proposition 5.3. *Let M denote a finitely generated A-module with $d = \dim_A M$. Let $r \in \mathbb{N}$ denote an integer. Suppose that there is an element $x \in \mathfrak{m}$ satisfying the following two conditions:*

a) *$M/x^{r+1}M$ is a $(d-1)$-dimensional Cohen-Macaulay module.*
b) $0 :_M x^r = 0 :_M x^{r+1}$.

Then $\operatorname{depth}_A M \geq d - 1$ and M is a CMF module with $M_{d-1} = 0 :_M x^r$.

Proof. Put $N := 0 :_M x^r = 0 :_M x^{r+1}$. We first claim that $\operatorname{depth}_A M/x^r M \geq d - 1$. Suppose the contrary, i.e. $\operatorname{depth}_A M/x^r M =: t < d - 1$. Then the short exact sequence

$$0 \to M/(xM,N) \to M/x^{r+1}M \to M/x^r M \to 0$$

implies that $\operatorname{depth}_A M/(xM,N) = t + 1$. Because x is an M/N-regular element it follows that

$$\operatorname{depth}_A M/N = t + 2 \text{ and } \operatorname{depth}_A M/(x^s M,N) = t + 1 \text{ for all } s \geq 1.$$

Therefore the short exact sequence

$$0 \to N \to M/x^s M \to M/(x^s M,N) \to 0,$$

considered for $s = r+1$, provides that $\operatorname{depth}_A N = t+2$. Then the same sequence considered for $s = r$ yields that $\operatorname{depth}_A M/x^r M \geq t + 1$, a contradiction.

Therefore $M/x^r M$ is a $(d-1)$-dimensional Cohen-Macaulay module. Now the first of the above short exact sequences proves that $M/(xM,N)$ and therefore also M/N is a Cohen-Macaulay module. Moreover the previous exact sequence considered for $s = r$ provides that N is a Cohen-Macaulay module of dimension $d - 1$. By 2.2 this finishes the proof. □

Now we start the cohomological investigation of CMF modules. To this end at first we need a description of the local cohomology modules of a CMF module.

Lemma 5.4. *Let M denote a CMF module with $\mathcal{M} = \{M_i\}_{0 \leq i \leq d}$ its dimension filtration. Let i denote an integer with $0 \leq i \leq d$. Then*

$$H^i_{\mathfrak{m}}(M) \simeq H^i_{\mathfrak{m}}(M_i) \simeq H^i_{\mathfrak{m}}(\mathcal{M}_i).$$

In the case A possesses a dualizing complex it follows that $K^i(M) \simeq K^i(\mathcal{M}_i)$ for all $0 \leq i \leq d$.

Proof. First consider the short excat sequence $0 \to M_{i-1} \to M_i \to \mathcal{M}_i \to 0$. Because of $\dim M_{i-1} \leq i-1$ it induces an isomorphism $H^i_{\mathfrak{m}}(M_i) \simeq H^i_{\mathfrak{m}}(\mathcal{M}_i)$. Second for $j < i$ it yields isomorphims $H^j_{\mathfrak{m}}(M_i) \simeq H^j_{\mathfrak{m}}(M_{i-1})$. Note that $\mathcal{M}_i$ is either zero or an i-dimensional Cohen-Macaulay module. By induction it follows that

$$H^i_{\mathfrak{m}}(M) \simeq H^i_{\mathfrak{m}}(M_d) \simeq H^i_{\mathfrak{m}}(M_{d-1}) \simeq \ldots \simeq H^i_{\mathfrak{m}}(M_{i+1}) \simeq H^i_{\mathfrak{m}}(M_i),$$

which proves the statement about the local cohomology modules. The rest of the claim for $K^i(M)$ follows by similar arguments using the dualizing complex. □

Now we are prepared to prove the main result concerning a characterization of CMF modules in terms of the modules of deficiency $K^i(M), 0 \leq i < d$. Moreover there is an additional information about the canonical module.

Theorem 5.5. *Let* $(A, \mathfrak{m})$ *denote a local ring possessing a dualizing complex* $D^{\cdot}_A$. *Let* M *be a finitely generated* A*-module with* $d = \dim_A M$. *Then the following conditions are equivalent:*

(i) *M is a CMF A-module.*
(ii) *For all $0 \leq i < d$ the module of deficiency $K^i(M)$ is either zero or an i-dimensional Cohen-Macaulay module.*
(iii) *For all $0 \leq i \leq d$ the A-modules $K^i(M)$ are either zero or i-dimensional Cohen-Macaulay modules.*

Proof. First suppose that M is a CMF module. Then the dimension filtration $\mathcal{M} = \{M_i\}_{0 \leq i \leq d}$ has the property that for all $0 \leq i \leq d$ the quotient module $\mathcal{M}_i = M_i/M_{i-1}$ is eiher zero or an i-dimensional Cohen-Macaulay module. By view of 5.4 it follows that $K^i(M) \simeq K^i(\mathcal{M}_i)$ for all $0 \leq i \leq d$. Because $\mathcal{M}_i$ is either zero or an i-dimensional Cohen-Macaulay module we have that $K^i(\mathcal{M}_i)$ is either zero or the canonical module of the i-dimensional Cohen-Macaulay module $\mathcal{M}_i$. But then the canonical module of $\mathcal{M}_i$ is also an i-dimensional Cohen-Macaulay module. So $K^i(M)$ is either zero or an i-dimensional Cohen-Macaulay module. This proves the implication (i) ⇒ (ii) as well as (i) ⇒ (iii).

In order to prove (iii) ⇒ (i) consider the spectral sequence studied in the proof of 3.4. By view of Theorem 3.4 it will be enough to prove that all the quotients $F^p/F^{p+1} \simeq E^{p,-p}_{\infty}$ are either zero or $(-p)$-dimensional Cohen-Macaulay modules. We first claim that $E^{p,-p}_{\infty} \simeq E^{p,-p}_2$ for all $-d \leq p \leq 0$. To this end consider the subsequent stages of the spectral sequence

$$E_r^{p-r,-p+r-1} \to E_r^{p,-p} \to E_r^{p+r,-p-r+1}.$$

The left term is zero because it is a subquotient of $K^{-p+r}(K^{-p+r-1}(M)) = 0$. To this end recall that $\dim_A K^{-p+r-1}(M) \leq -p+r-1$, see 3.2. The term on the right hand side is a subquotient of $K^{-p-r}(K^{-p-r+1}(M))$. By our assumption we have that $K^{-p-r+1}(M)$ is either zero or an $(-p-r+1)$-dimensional Cohen-Macaulay module. But then the $(-p-r)$-th module of deficieny $K^{-p-r}(K^{-p-r+1}(M))$ is zero. That is, the modules at the right are always zero. But this implies that

$$F^p/F^{p+1} \simeq E_2^{p,-p} \simeq K^{-p}(K^{-p}(M))$$

for all $-d \le p \le 0$. We have to finish the proof by showing that $K^{-p}(K^{-p}(M))$ is either zero or a $(-p)$-dimensional Cohen-Macaulay module. By our assumption $K^{-p}(M)$ is either zero or an $(-p)$-dimensional Cohen-Macaulay module. Therefore $K^{-p}(K^{-p}(M))$ is either zero or - as the canonical module of $K^{-p}(M)$ - also a $(-p)$-dimensional Cohen-Macaulay module. By view of 4.3 this proves the claim of (i).

Finally we have to show that (ii) $\Rightarrow$ (iii). That is, we have to show that the canonical module $K(M) = K^d(M)$ is a Cohen-Macaulay module provided for all $0 \le i < d$ the module of deficiency $K^i(M)$ is either zero or an i-dimensional Cohen-Macaulay module. Now we have that $(\mathrm{Hom}_A(M, D_A^{\cdot}))^i = 0$ for all $i < -d$ and $H^{-d}(\mathrm{Hom}_A(M, D_A^{\cdot})) = K(M)$. So there is a short exact sequence of complexes

$$0 \to K(M)[d] \to \mathrm{Hom}_A(M, D_A^{\cdot}) \to C^{\cdot}(M) \to 0,$$

where $C^{\cdot}(M)$ denotes the cokernel of the natural embedding. So the complex $C^{\cdot}(M)$ carries as cohomology modules the modules of deficiencies, i.e.

$$H^i(C^{\cdot}(M)) \simeq \begin{cases} K^{-i}(M) & \text{for } -d < i \le 0, \text{ and} \\ 0 & \text{otherwise.} \end{cases}$$

Now the natural homomorphism of complexes $M \to \mathrm{Hom}_A(\mathrm{Hom}_A(M, D_A^{\cdot}), D_A^{\cdot})$ induces an isomorphism in cohomology for a finitely generated A-module M. Therefore by applying $\mathrm{Hom}_A(\cdot, D_A^{\cdot})$ to the above short exact sequence of complexes it induces a short exact sequence

$$0 \to H^0(\mathrm{Hom}_A(C^{\cdot}(M), D_A^{\cdot})) \to M \to K(K(M)) \to H^1(\mathrm{Hom}_A(C^{\cdot}(M), D_A^{\cdot})) \to 0$$

and isomorphisms $H^i(\mathrm{Hom}_A(C^{\cdot}(M), D_A^{\cdot})) \simeq K^{d-i+1}(K(M))$ for all $i \ge 2$. In order to prove that $K(M)$ is a Cohen-Macaulay module - by local duality - it is enough to prove that $K^{d-i+1}(K(M)) = 0$ for $i \ge 2$. Whence it will be enough to show the vanishing of $H^i := H^i(\mathrm{Hom}_A(C^{\cdot}(M), D_A^{\cdot}))$ for all $i \ge 1$. To this end take the corresponding spectral sequence

$$E_2^{pq} = H^p(\mathrm{Hom}_A(H^{-q}(C^{\cdot}(M)), D_A^{\cdot})) \Rightarrow E_\infty^{p+q} = H^{p+q},$$

derived in the same way as the spectral sequence studied in Section 3. Because of $E_2^{pq} = K^{-p}(K^q(M)) = 0$ for all q and all $p \neq -q$, note that $K^q(M), 0 \le q < d$, is either zero or a q-dimensional Cohen-Macaulay module, there is a partial degeneration to $H^i = 0$ for all $i > 0$. This completes the proof. □

Looking at the second part of Theorem 5.5 there is another sufficient criterion for the canonical module $K(M)$ of M being a Cohen-Macaulay module. Moreover the filtration induced by the spectral sequence for the computation of $H^0(\mathrm{Hom}_A(C^{\cdot}(M), D_A^{\cdot}))$ is just the truncated dimension filtration, i.e. it follows $H^0(\mathrm{Hom}_A(C^{\cdot}(M), D_A^{\cdot})) \simeq M_{d-1}$ and $K(K(M)) \simeq M/M_{d-1}$.

6. Faithful Flat Extensions and Examples

Let M denote a finitely generated A-module, $(A, \mathfrak{m})$ a local Noetherian ring. As mentioned in Section 2 in general M is not a CMF module in case $M \otimes \hat{A}$ is a CMF $\hat{A}$-module. In particular this is not even true for the ring itself as follows by the next example.

Example 6.1. Let $(A, \mathfrak{m})$ denote the 2-dimensional local domain considered by M. Nagata in [N, Example 2]. Clearly it is not a Cohen-Macaulay ring. For the multiplicity $e(\mathfrak{m}, A)$ it is shown that $e(\mathfrak{m}, A) = 1$. Therefore it implies that

$$1 = e(\mathfrak{m}, A) = e(\hat{\mathfrak{m}}, \hat{A}) = e(\hat{\mathfrak{m}}, \hat{A}/u_{\hat{A}}(0)).$$

By the view of [N, (40.6)] it yields that $\hat{A}/u_{\hat{A}}(0)$ is a regular local ring, in particular a 2-dimensional Cohen-Macaulay ring. Moreover since $\operatorname{depth} A = \operatorname{depth} \hat{A} = 1$ the ideal $u_{\hat{A}}(0)$ is - considered as an $\hat{A}$-module - a 1-dimensional Cohen-Macaulay module. But this means that $\hat{A}$ is a CMF ring or equivalently an approximately Cohen-Macaulay ring. But this is not true for A. Otherwise A would be a Cohen-Macaulay ring since it is a domain.

Before we shall formulate our next result let us recall the definition of a Cohen-Macaulay filtration, 4.2. An increasing filtration $C = \{C_i\}_{0 \le i \le d}$ of M is called a Cohen-Macaulay filtration whenever $M = C_d, d = \dim_A M$, and $C_i = C_i/C_{i-1}$ is either zero or an i-dimensional Cohen-Macaulay module for all $1 \le i \le d$. As it was mentioned in 4.3 a Cohen-Macaulay filtration coincides automatically with the dimension filtration.

Now let $(A, \mathfrak{m}) \to (B, \mathfrak{n})$ be a faithful flat homomorphism of local rings. Let M be a finitely generated A-module with $d = \dim_A M$. Let $C = \{C_i\}_{0 \le i \le d}$ denote an increasing filtration of M such that $M = C_d$. Let $C_B = \{(C_B)_i\}_{0 \le i \le n}$ denote the induced filtration defined by $(C_B)_i = C_{i+t} \otimes_A B$, where $t = \dim B/\mathfrak{m}B$ denotes the dimension of the fibre ring.

Theorem 6.2. *Let $(A, \mathfrak{m}) \to (B, \mathfrak{n})$ be a faithful flat homomorphism of local rings. Let M be a finitely generated A-module with $d = \dim_A M$. Then the following conditions are equivalent:*

(i) *The filtration C is a Cohen-Macaulay filtration of M and the fibre ring $B/\mathfrak{m}B$ is a Cohen-Macaulay ring.*
(ii) *The induced filtration C_B is a Cohen-Macaulay filtration of the B-module $M \otimes_A B$.*

Proof. Let X denote an arbitrary finitely generated A-module. By virtue of [M, Theorem 15.1] and [M, Theorem 23.3] the following two equalities are true

$$\dim_B X \otimes_A B = \dim_A X + \dim B/\mathfrak{m}B \text{ and } \operatorname{depth}_B X \otimes_A B = \operatorname{depth}_A X + \operatorname{depth} B/\mathfrak{m}B.$$

First of all this proves that $\dim_B X \otimes_A B = d + t$, i.e. $(C_B)_{d+t} = M \otimes_A B$.

Now suppose that condition (i) is satisfied. Then the above equalities show that each of the B-modules

$$(C_B)_i/(C_B)_{i-1} \simeq (C_{i-t}/C_{i-1-t}) \otimes_A B$$

are either zero or i-dimensional Cohen-Macaulay modules. The converse follows the same line of resoning. Hence we omit it. □

Note that the previuos result 6.2 does not apply to the example considered in 6.1. In the example there does not exist a Cohen-Macaulay filtration in A, while there is one in $\hat{A}$. The Cohen-Macaulay filtration in $\hat{A}$ does not occur as the extension of a Cohen-Macaulay filtration of A.

In the following we want to sum up the examples of CMF modules and rings showing that the occurance of them is quite natural.

Example 6.3. a) Let M be a Cohen-Macaulay module. Then M is also a CMF module.
b) Let $(A, \mathfrak{m})$ be a local ring with $d = \dim A$. Let $N_i, i = 0, \ldots, d$, be a family of A-modules such that either $N_i = 0$ or N_i is an i-dimensional Cohen-Macaulay module. Then $M = \oplus_{i=0}^{d} N_i$ is a CMF module over A. This follows easily by 4.3 since M admitts a filtration $M_i = \oplus_{j=0}^{i} N_j$ such that $M_i/M_{i-1} \simeq N_i, i = 0, \ldots, d$, is either zero or an i-dimenional Cohen-Macaulay module.
c) Let $(A, \mathfrak{m})$ denote a local ring. Let M be a finitely generated A-module. Then consider $A \ltimes M$, the idealization of M over A. That is, the additive group of $A \ltimes M$ coincides with the direct sum of the abelian groups A and M. The muliplication is given by

$$(a, m) \cdot (b, n) := (ab, an + bm).$$

Then $A \ltimes M$ is a d-dimensional local ring, see [N, (1.1)] or [BH, 3.3.22] for these and related facts.

Now suppose that $(A, \mathfrak{m})$ is a d-dimensional Cohen-Macaulay ring. Let M be a CMF module with $\dim M = t < d$. Then $A \ltimes M$ is a d-dimensional CMF ring. To this end let $\mathcal{M} = \{M_i\}_{0 \le i \le t}$ denote the dimension filtration of M. Now put

$$R_i = \begin{cases} A \ltimes M & \text{for} \quad i = d, \\ 0 \ltimes M & \text{for} \quad i = t+1, \ldots, d-1, \text{ and} \\ 0 \ltimes M_i & \text{for} \quad i = 0, \ldots, t. \end{cases}$$

Then $\{R_i\}_{0 \le i \le d}$ is a filtration of $R = A \ltimes M$ such that $R_d = A \ltimes M$ and R_i/R_{i-1} is either zero or an i-dimensional Cohen-Macaulay module. Note that

$$R_i/R_{i-1} \simeq \begin{cases} A & \text{for} \quad i = d, \\ 0 & \text{for} \quad i = t+1, \ldots, d-1, \text{ and} \\ M_i/M_{i-1} & \text{for} \quad i = 1, \ldots, t. \end{cases}$$

By view of 4.3 this proves the claim.
d) Let $A[[x]]$ denote the formal power series ring in one variable x over the local ring $(A, \mathfrak{m})$. Then a finitely A-module M is a CMF module if and only if $M[[x]]$ is a CMF module over the ring $A[[x]]$.
e) Let M be a finitely generated A-module such that $H_{\mathfrak{m}}^i(M), i \neq \dim_A M$, is a finitely generated A-module. Then M is a CMF module if and only if $H_{\mathfrak{m}}^i(M) = 0$ for all $0 < i < \dim_A M$. In particular, under these circumstances M is a Cohen-Macaulay module if and only if M is a CMF module with $\operatorname{depth}_A M > 0$.
f) Every 1-dimensional A-module M is a CMF module. Therefore for any d-dimensional Cohen-Macaulay ring with $d \ge 2$ and a 1-dimensional A-module M the idealization $A \ltimes M$ is a d-dimensional CMF ring.

It would be of some interest to understand the descend of the CMF property from $M \otimes_A \hat{A}$ to M. What are suffficient condition on A? The Example 6.1 does not has Cohen-Macaulay formel fibres. Is it enough to suppose that the homomorphism $A \to \hat{A}$ has Cohen-Macaulay formel fibres ?

References

[BH] W. Bruns, J. Herzog: 'Cohen-Macaulay rings', Cambr. Univ. Press, 1993.
[E] D. Eisenbud: 'Commutative Algebra (with a view towards algebraic geometry)', Springer-Verlag, 1995.

[G] S. GÔTO: *Approximately Cohen-Macaulay rings*, J. Algebra **76** (1981), 214–225.
[H] R. HARTSHORNE: 'Residues and Duality', Lect. Notes in Math., **20**, Springer, 1966.
[HK] J. Herzog, E. Kunz: 'Der kanonische Modul eines Cohen-Macaulay-Ringes', Lect. Notes in Math., **238**, Springer, 1971.
[M] H. MATSUMURA: 'Commutative ring theory', Cambridge University Press, 1986.
[N] M. NAGATA: 'Local rings', Interscience, 1962.
[S1] P. SCHENZEL: 'Dualisierende Komplexe in der lokalen Algebra und Buchsbaum-Ringe', Lect. Notes in Math., **907**, Springer, 1982.
[S2] P. SCHENZEL: 'On the use of local cohomoloy in algebra and geometry', Lectures at the Summmerschool of Commutative Algebra and Algebraic Geometry, Ballaterra, 1996, Birkhäuser Verlag, to appear.
[W] C. WEIBEL: 'An Introduction to Homological Algebra', Cambr. Univ. Press, 1994.

MARTIN-LUTHER-UNIVERSITÄT HALLE-WITTENBERG, FACHBEREICH MATHEMATIK UND INFORMATIK, D — 06 099 HALLE (SAALE), GERMANY
E-mail address: schenzel@mathematik.uni-halle.de

Monomial Conjecture and Auslander's δ-Invariant

Anne-Marie Simon

Free University of Brussels
Brussels, Belgium

Jan R. Strooker

University of Utrecht
Utrecht, The Netherlands

This is a write-up of a lecture delivered by the second author at the conference, which we are pleased to dedicate to our old friend Mario Fiorentini. Full proofs will appear elsewhere.

1 The monomial conjecture

Let $(A, \mathfrak{m}, k)$ be a (commutative) noetherian local ring with its maximal ideal and residue class field. Let A be d-dimensional and let $x = x_1, \ldots, x_d$ be a system of parameters (sop) of A. We say that A satisfies the Monomial Conjecture (MC) provided there is no identity

$$(x_1 \ldots x_d)^t = a_1 x_1^{t+1} + \ldots + a_d x_d^{t+1},$$

where the a_i range over the elements of A, t is a positive integer and x ranges over all systems of parameters of A.

We prefer to think in terms of maps rather than equations. Misusing language, we also write x for the ideal generated by the sop $x_1, \ldots, x_d$ and $x(s)$ for the ideal generated by their s-th powers. Multiplication by the t-th power of the product $x_1 \ldots x_d$ then maps A/x into $A/x(t+1)$, and MC tells us that these maps

$$(x_1 \ldots x_d)^t : A/x \to A/x(t+1)$$

never are null. Since the rings involved are artinian, the maps remain the same when one completes A in its $\mathfrak{m}$-adic topology. Therefore MC need only be proved for complete rings. For ease of exposition we shall from now on assume A to be complete, even though many arguments go through in general.

For each sop, the limit of the above directed system of maps is known to be the d-th local cohomology module of A with respect to the ideal x, but since local cohomology is impervious to taking radicals, we have limit maps

$$\mu_x^A : A/x \to H_{\mathfrak{m}}^d(A)$$

and MC does not allow these maps to be null. Moreover, Grothendieck showed that this local cohomology module is not 0 [17, 10.2.2], so at least there is a chance for MC to be true. Using this result, one can in fact give a fairly straightforward proof of MC when A contains a field of positive characteristic by using the Frobenius endomorphism [17, Cor. 10.3.2]. There is another argument, not too difficult, which works when A contains the field of rational numbers [17, Prop. 10.3.5].

The Monomial Conjecture was introduced by Hochster in the pioneering paper [11], in which many of its fundamental features were already established, and it was shown to hold over a field. The conjecture's chief interest was that it implied the Direct Summand Conjecture, see also section 3. In section 4 we shall find that there are even wider implications. Since that first article, certain special cases of MC have been proved, notably when A is a Buchsbaum ring [8, Cor. 4.8], but the mixed characteristic case is still very much open.

We want in this paper to approach MC by looking at the maps which are induced by μ_x^A on A-modules. We go on to show that MC boils down to questions about ideal theory in Gorenstein rings and that it ties up with notions stemming from the representation theory of modules.

So let M be a finitely generated A-module, and x an sop of A. The maps $M/xM \to M/x(t+1)M$ induced by multiplication with the scalar $(x_1 \ldots x_d)^t$ again have as direct limit the module $H_m^d(M) = H_\mathfrak{m}^d(A) \otimes_A M$, but in [11, Remark 6] there is already an example where the map $\mu_x^M = \mu_x^A \otimes_A M$ is the null map even for a d-dimensional M. Write $Q_x(M)$ for the kernel of the composition of μ_x^M with the residue class map $M \to M/xM$. This is a submodule of M, and MC just means that the ideal $Q_x(A)$ is contained in $\mathfrak{m}$ for every sop x. Notice that μ_x^M and $Q_x(M)$ are functorial in M.

Proposition The following are equivalent for an sop x and a finitely generated module M:

(i) $Q_x(M) = xM$;
(ii) μ_x^M is injective;
(iii) $x_1, \ldots, x_d$ is a regular sequence on M;
(iv) M is a maximal Cohen-Macaulay module (MCM) over A, i.e. depth M = dim M = dim A.

In this case, moreover, im μ_x^M is precisely the submodule of $H_\mathfrak{m}^d(M)$ which is annihilated by the ideal x.

For the proof, see [17, Th. 5.1.12, Cor. 5.2.5, Prop. 5.2.7].

It follows immediately that if A possesses an MCM called C, then A satisfies the Monomial Conjecture, since tensoring a suspected null map μ_x^A with C can never yield an injective map μ_x^C out of the nonnull module C/xC (Nakayama).

Hochster has conjectured that every complete noetherian local ring possesses maximal Cohen-Macaulay modules. This is known for $d \leq 2$, but is wide open even over a field [10, Conj. (E), 3]. However, in [12, Chs. 4, 5] he constructed infinitely generated modules C with the above properties, the so-called Big Cohen-Macaulay modules, first in positive characteristic and then, by general procedures, over a field of characteristic 0, see also [17, Chs. 11, 12]. This then provides another way to prove MC in equal characteristic, and we now also know that a possible counterexample would have to be a ring in mixed characteristic, have dimension at least 3, and not be Buchsbaum.

One way of viewing the present paper is as an attempt to replace Hochster's sweeping conjecture about the existence of MCM's in the complete case by somewhat weaker conditions involving Cohen-Macaulay modules, but which are still strong enough to yield MC. Let us also point out that the representation theoretic aspects of maximal Cohen-Macaulay modules have turned out to be of crucial importance in the theory of singularities. We just refer to the monograph [20] for a survey.

2 Dual form of the monomial conjecture

Let $(A, \mathfrak{m}, k)$ be a d-dimensional complete noetherian local ring and $x = x_1, \ldots, x_d$ an sop. A being a homomorphic image of a complete regular local ring, one can find a complete local ring R which is a complete intersection and which maps onto A. In addition, $\dim R = \dim A = d$ and x can be lifted to an sop $y = y_1, \ldots, y_d$ of R. This is proved in [19, Prop. 1 (i)]. It is also argued that for MC, if $A = R/\mathfrak{a}$, one can restrict to the case where $\mathfrak{a}$ is an unmixed ideal consisting of zerodivisors in R, and that one may even assume that $\mathfrak{a}$ is a component of 0 in R. The proof of [19, Prop. 1 (ii)] is, to say the least, incomplete, but this can be rectified. Furthermore, we shall not insist on R being a complete intersection. We therefore state that it is enough to prove the Monomial Conjecture for $A = R/\mathfrak{a}$ where R is a complete local Gorenstein ring and $\mathfrak{a}$ is an unmixed ideal of height 0 in R. Putting $\mathfrak{b} = \operatorname{Ann} \mathfrak{a}$, we find that $\mathfrak{a} = \operatorname{Ann} \mathfrak{b}$ and $\mathfrak{a}$ and $\mathfrak{b}$ are linked with respect to 0 in R, e.g. [16], [14, Prop. 3.4]. We could even assume geometric linkage, but shall not do so in this paper.

We recall a few facts about the Matlis dual. In the above situation, observe that R and $A = R/\mathfrak{a}$ have the same residue class field k and call the maximal ideal of R by the name $\mathfrak{n}$. Let E be the injective hull of k as an R-module. The functor $\operatorname{Hom}_R(-, E) = -^{\vee}$ is a contravariant functor on R-modules, which is faithfully exact. This Matlis dual preserves the length of modules of finite length, and $M^{\vee\vee} \cong M$ when M is finitely generated. Since the Gorenstein ring R is complete, we have $H^d_{\mathfrak{n}}(R) \cong E$, $E^{\vee} \cong R$, and $H^d_{\mathfrak{n}}(M)^{\vee} \cong \operatorname{Hom}_R(M, R)$ by local duality [17, 3.4, Th. 10.2.1]. .

The idea is now to replace the condition $\mu^A_x \neq 0$ by the equivalent condition $\mu^{A\vee}_x \neq 0$, where we regard μ^A_x not as an A-homomorphism of A/x into the d-th local cohomology module of A with respect to its maximal ideal, but as a homomorphism $\mu^R_y \otimes_R A$ between the R-modules $A/x \cong R/y \otimes_R A$ and $E \otimes_R A$. Using the adjointness between tensor product and Hom, the local duality recalled above and the Proposition in section 1 for the R-module R, we see that $\mu^{A\vee}_x$ is nothing but the map $\operatorname{Hom}_R(A, R) \to \operatorname{Hom}_R(A, R/y)$ induced by the natural projection $R \to R/y$. Hence $\mu^{A\vee}_x \neq 0$ if and only if the ideal $\mathfrak{b}$ is not contained in the ideal y. Thus we obtain the main result of [19]:

Theorem MC is true for $A = R/\mathfrak{a}$ if and only if $\mathfrak{b} = \operatorname{Ann} \mathfrak{a}$ is not contained in any parameter ideal of R.

Thus MC is reduced to a statement about ideals in complete local Gorenstein rings: the "small" ideal $\mathfrak{b}$ consisting of zero divisors is not contained in any "large" parameter ideal. We shall refer to this statement as the "Dual form of MC". This Dual form suggests that there is a strong connection between linkage = "liaison" and MC. In what follows we shall pursue this approach, and it is our hunch that we have barely scratched the surface.

Let us check whether the possession of a maximal Cohen-Macaulay module by the ring $A = R/\mathfrak{a}$, or rather the possession of an MCM by R which is annihilated by the ideal $\mathfrak{a}$, also provides a proof of the Dual form of MC for A. To do this, introduce the notation $\operatorname{ext}^+(X, Y) = \sup i : \operatorname{Ext}^i_R(X, Y) \neq 0$ for any two finitely generated R-modules X and Y. A result in [15, Ch. I, Th. 4.15] tells us that if Y has finite injective dimension, then $\operatorname{ext}^+(X, Y) = \operatorname{depth} R - \operatorname{depth} X$. We obtain

Proposition Suppose $A = R/\mathfrak{a}$ possesses an MCM, and let $\mathfrak{a}$ and $\mathfrak{b}$ be linked w.r.t. 0 in R. Then $\mathfrak{b}$ is not contained in any ideal $\mathfrak{q}$ of R which has finite injective dimension.

Proof Let C be the MCM guaranteed by the hypothesis. Let $\mathfrak{q}$ be an ideal of finite injective dimension in R. By the result recalled above, we have a surjection $\mathrm{Hom}_R(C, R) \to \mathrm{Hom}_R(C, R/\mathfrak{q})$ of nonnull modules. Since C is annihilated by $\mathfrak{a}$, so is its image in R under any homomorphism. This image is therefore contained in $\mathfrak{b}$. If $\mathfrak{b}$ were contained in $\mathfrak{q}$, the given homomorphism would map to 0 in $\mathrm{Hom}_R(C, R/\mathfrak{q})$. This proves the result.

Since an sop in R is a regular sequence, a parameter ideal of R has finite projective dimension. One of the miracles in Gorenstein rings is however that finite injective dimension implies finite projective dimension and vice versa, so the Proposition proves a seemingly stronger statement than the Dual form of MC.

3 Stable modules and the delta invariant

Let $(R, \mathfrak{n}, k)$ again be a complete Gorenstein local ring. Call an R-module stable if it contains no nonnull free direct summand. We elaborate a little on the local duality argument in the previous section to obtain a criterion for such a module M to be stable. This criterion will involve the submodules $Q_x(M)$ which we introduced in connection with MC.

For any sop y of R, denote by v_y an element in R which generates the socle of the artinian local Gorenstein ring R/y. This socle is a one-dimensional k-vector space, whereas v_y is determined modulo y up to a unit of R.

Lemma Let M be a finitely generated module over a complete local Gorenstein ring R. Let $g : R \to M$ be an R-homomorphism. Let $r \in R$. Then $g(r) \notin Q_y(M)$ if and only if there exists an R-linear map $f : M \to R$ with $fg(r) \notin y$.

Proof In the argument of section 2, we did not use that A was a ring, nor that the map from R to A was a ring homomorphism or was surjective.

The following corollary sharpens [17, Prop. 10.3.3]. A more general result is available if one had not decided to stick to complete rings and finitely generated modules.

Splitting lemma A homomorphism $g : R \to M$ admits a splitting if and only if for some sop y the image $g(v_y)$ is not in $Q_y(M)$.

Proof If f is a splitting of g, then $fg = 1$ cannot carry v_y into $y = Q_y(R)$. By functoriality of Q_y, we have $g(v_y) \notin Q_y(M)$. Suppose the latter condition is satisfied. By the Lemma, there is an $f : M \to R$ such that $fg(v_y) \notin y$. Now fg is an endomorphism of R, therefore is multiplication by a scalar $a \in R$. Since $av_y \notin y$, the element a is a unit and $a^{-1}f$ splits g.

We obtain as a consequence that either $g(v_y) \in Q_y(M)$ for every sop y or for none, and a quick proof of the equivalence which started all this off [11, Th. 1]:

Corollary Suppose $R \to S$ is a finite local extension of noetherian local rings, and that R is complete and regular. Then this extension splits if and only if S satisfies MC w.r.t. to a regular sop of R.

Proof Let y be a regular sop of R, then it is an sop of S. Since $v_y = 1 \in R$ is mapped to $1 \in S$, we need only check that $1 \notin Q_y(S)$, and this is tantamount to MC holding for y in S.

The next criterion generalizes [11, Remark 3].

Stability criterion The module M is stable if and only if $v_y M \subset Q_y(M)$ for every sop y.

Since for an MCM, say C, we know that $Q_y(C) = yC$ for every sop y, the module C is stable if and only if $v_y C \subset yC$ for every sop y. An alternative argument might observe that an injection of the selfinjective ring R/y into C/yC would split, and one can lift because $\mathrm{Hom}_R(C, R)$ maps on to $\mathrm{Hom}_R(C, R/y)$ as in the Proof in section 2.

Suppose a stable MCM module surjects onto a module M. Then $v_y M \subset yM$ for every sop y; in particular, M is stable.

Let $\mathfrak{a}$ and $\mathfrak{b}$ again be ideals which are linked in R w.r.t. the null ideal.

Proposition Suppose there is a stable maximal Cohen-Macaulay module which maps onto $R/\mathfrak{b}$. Then $A = R/\mathfrak{a}$ satisfies MC.

Proof By the above, $v_y \cdot R/\mathfrak{b} \subset y \cdot R/\mathfrak{b}$ for each sop y of R. Therefore $(v_y) + \mathfrak{b} \subset y + \mathfrak{b}$, hence $v_y \in y + \mathfrak{b}$. Now $v_y \notin y$, so $\mathfrak{b} \not\subset y$. Via the Dual form of MC, this proves our contention.

Now is the time to recall some more theory. In their important paper [1], Auslander and Buchweitz showed that over a Gorenstein local ring every finitely generated module M admits exact sequences of the type

$$0 \to K \to C \to M \to 0$$

with C an MCM and K finitely generated and of finite injective dimension as an R-module. Let f-rank C denote the rank of a maximal free direct summand of C. One now defines $\delta(M)$ to be the minimum of these f-ranks as we consider all short exact sequences of this type ending in M [2, 1].

One can show that $\delta(M)$ is also the minimum of the f-ranks of the larger class of all MCM's which map onto M [9, Lemma 1.1]. Therefore $\delta(M) = 0$ just means that there exists a stable MCM which maps onto M. This allows us to prove an apparently stronger version of the Dual form of MC, this time by making an assumption on $R/\mathfrak{b}$ rather than on $A = R/\mathfrak{a}$ as in section 2, but again involving Cohen-Macaulay modules.

Proposition Suppose $\delta(R/\mathfrak{b}) = 0$ for an arbitrary ideal $\mathfrak{b}$. Then $\mathfrak{b}$ is not contained in a single ideal $\mathfrak{q}$ of finite injective dimension.

Proof Suppose $\mathfrak{q}$ is an ideal of finite injective dimension containing $\mathfrak{b}$. We show that $\delta(R/\mathfrak{b}) = 1$. Let C be an MCM mapping onto $R/\mathfrak{b}$ and therefore onto $R/\mathfrak{q}$. As in the proof of the Proposition in section 2, we have a surjection of nonnull modules $\mathrm{Hom}_R(C, R) \to \mathrm{Hom}_R(C, R/\mathfrak{q})$. This means that any map from C to $R/\mathfrak{q}$ factors through R. Now there is a surjection from C to $R/\mathfrak{q}$, which needs must factor through a surjection onto the local ring R. This splits, so C contains a free summand of at least rank 1. Since R maps onto $R/\mathfrak{b}$, this proves our contention.

The converse of the Proposition fails in general [3, Ex.], but mark the sequel.

Both Propositions in sections 2 and 3 appear to prove somewhat more than the Dual form of MC. They lead us to ask the

Question In a Gorenstein local ring, possibly complete, is every ideal of finite injective (or projective) dimension contained in a parameter ideal [18]?

Though we do not have lots of calculations to prove this in particular cases, it seems like a natural state of affairs. Note that it is true for a regular local ring...

4 The over-all picture

For at least three years we have been aware of the fact that, in our situation of linkage, $\delta(R/\mathfrak{b}) = 0$ implies that $R/\mathfrak{a} = A$ satisfies MC. But we were, and are, unable to prove that the converse holds. Recently, however, we realized that the prospect changes drastically when we widen our horizon. In this section we sketch these ideas, skimping on proofs.

In a long and involved paper [13], Hochster introduced the Canonical Element Conjecture (CEC). This is again a conjecture for all noetherian local rings, which has been proved in equal characteristic, and for dimension not exceeding 2, for instance because it is enough to have Big Cohen-Macaulay modules. Further special cases have been settled by Dutta in a series of papers [4], [5], [6] and [7]. There are several forms of CEC, which can be shown to be equivalent. We do not propose to state the CEC, but it looks like a somewhat generalized and more flexible MC: certains maps are not allowed to be null. It is therefore not surprising that a ring A which satisfies CEC also satisfies MC. At this conference, it seems worth mentioning that CEC also implies a version of the Evans-Griffith syzygy theorem.

Theorem Let A be a local Cohen-Macaulay domain and let M be a nonfree r-th syzygy with finite projective dimension. Suppose that all local rings of homomorphic image domains of A satisfy CEC. Then M has torsion-free rank at least r [13, Cor. 2.6°].

The syzygy theorem is true in equal characteristic, and experts in the area believe it to be true in general. Which provides yet another reason to wonder about CEC in mixed characteristic.

One of the main results of Hochster's paper is that if all rings in positive characteristic, equal characteristic 0, resp. mixed characteristic, satisfy MC, then they satisfy CEC [13, Th. 2.8].

Now let R be a d-dimensional complete local Gorenstein ring, and the ideals $\mathfrak{a}$ and $\mathfrak{b}$ be linked in R w.r.t. to 0. Hochtster's arguments in [13, Th. 4.3] can be adapted to show the following

Lemma Suppose $A = R/\mathfrak{a}$ satisfies CEC. Then $\operatorname{Ext}^d_R(k, \mathfrak{b})$ maps surjectively onto $\operatorname{Ext}^d_R(k, R)$ by the map stemming from the injection of $\mathfrak{b}$ into R.

One also has the

Proposition Let M be a finitely generated module over a local Gorenstein ring $(R, \mathfrak{n}, k)$. Suppose $F \to M$ is a surjection of a finitely generated free module onto M. Then $\delta(M)$ is the dimension over k of the image of the induced map $\mathrm{Ext}^d_R(k, F) \to \mathrm{Ext}^d_R(k, M)$.

Combining the last two facts for the short exact sequence $0 \to \mathfrak{b} \to R \to R/\mathfrak{b} \to 0$, we find that in the situation of the Lemma, $\delta(R/\mathfrak{b}) = 0$. Using Hochster's major result just mentioned, we find

Theorem Consider all complete noetherian local rings of one of the three types, characteristic wise. The following statements are equivalent:

(i) All these rings satisfy the Canonical Element Conjecture;

(ii) For every unmixed ideal $\mathfrak{b}$ of zero divisors in a Gorenstein local ring of this type, $\delta(R/\mathfrak{b}) = 0$;

(iii) All these rings satisfy the Monomial Conjecture.

In particular, in equal characteristic, always $\delta(R/\mathfrak{b}) = 0$ in (ii).

We think that this result carries some interest. We showed in section 1 that possession of a maximal Cohen-Macaulay module easily implies MC for a particular ring. We believe that this is the first time, however, that MC or CEC or in fact any of the so-called Homological Conjectures which does not specifically involve MCM's, is seen to imply an existence statement for certain MCM's, albeit over Gorenstein rings and through the intervention of linkage.

5 An example

Suppose that the Gorenstein local ring R can be written as $R = S/\mathfrak{q}$, with $(S, \mathfrak{p}, k)$ a Gorenstein local ring and the ideal $\mathfrak{q}$ generated by a regular sequence in S. Remember that this was how we constructed R in section 2, starting from Cohen's structure theorem for complete rings. In this case there is a result of Ding [2, Th. 2.5]:

Proposition Let M be a finitely generated R-module. If the map $\mathfrak{q}/\mathfrak{p}\mathfrak{q} \to \mathrm{Ann}_S M/\mathfrak{p}\mathrm{Ann}_S M$ is not injective, then $\delta(M) = 0$.

Notice that the condition in the Proposition just tells us that a minimal set of generators of the ideal $\mathfrak{q}$ cannot be prolonged to a minimal set of generators of $\mathrm{Ann}_S M$. It is known that the condition is not necessary for $\delta(M)$ to be 0. Just how far away from being necessary, is illustrated by the following familiar example [16, Contr. ex. 1.8]:

Let k be a field and S the ring of power series $k[[X, Y, U, V]]$. Put $Z_1 = XV - YU$, $Z_2 = XU^2 - Y^2V$, and $\mathfrak{q} = (Z_1, Z_2)$. Define $I = (Z_1, Z_2, Y^3 - X^2U, U^3 - YV^2)$. Put $R = S/\mathfrak{q}$ and $\mathfrak{a} = I/\mathfrak{q}$. Then $\mathfrak{a}$, a prime component of 0 in R, and the ideal $\mathfrak{b} = (x, y) \cap (u, v)$ are geometrically linked w.r.t. to 0 in R. The integral domain $R/\mathfrak{a}$ can be parametrized as $k[[s^4, s^3t, st^3, t^4]]$. According to the Proposition, we can conclude that $\delta(R/\mathfrak{b}) = 0$, but we cannot conclude that $\delta(R/\mathfrak{a}) = 0$, because clearly we can prolong Z_1, Z_2 to a minimal set of four generators for the ideal I in S. Nevertheless, of course, $\delta(R/\mathfrak{a}) = 0$ by the Theorem in section 4.

We finish by suggesting that one would like to obtain a workable criterion which should be both necessary and sufficient for a module to have δ-invariant 0.

Note added in proof: The answer to the **Question** in section 3 is "no". J. Stückrad has constructed a host of counter-examples.

References

[1] Auslander, M. & R.-O. Buchweitz, *The homological theory of maximal Cohen-Macaulay approximations*, Memoire 38, Societé Mathématique de France (1989), pp. 5-37.

[2] Ding, S., *Cohen-Macaulay approximation and multiplicity*, J. Algebra 153 (1992), pp. 271-288.

[3] Ding, S., *Auslander's δ-invariant of Gorenstein local rings*, Proc. Amer. Math. Soc. 122 (1994), pp. 649-656.

[4] Dutta, S.P., *On the canonical element conjecture*, Trans. Amer. Math. Soc. 299 (1987), pp. 803-811.

[5] Dutta, S.P., *Dualizing complex and the canonical element conjecture*, J. Lond. Math. Soc. (2) 50 (1994), pp. 477-487.

[6] Dutta, S.P., *On the canonical element conjecture II*, Math. Z. 216 (1994), pp. 379-388.

[7] Dutta, S.P., *Dualizing complex and the canonical element conjecture II*, J. Lond. Math. Soc. (2) 56 (1997), pp. 49-63.

[8] Goto, S., *On the associated graded rings of parameter ideals in Buchsbaum rings*, J. Algebra 85 (1983), pp. 490-534.

[9] Herzog, J., *On the index of a homogeneous Gorenstein ring*, Contemp. Math. 159, Amer. Math. Soc. (1994), pp. 95-102.

[10] Hochster, M., *Cohen-Macaulay modules*, Lect. Notes Math 311, Springer-Verlag (1973), pp. 120-153.

[11] Hochster, M., *Contracted ideals from integral extensions of regular rings*, Nagoya Math. J. 51 (1973), pp. 25-43.

[12] Hochster, M., *Topics in the homological theory of modules over commutative rings*, Reg. Conf. Ser. Math. 24, Amer. Math. Soc. (1975).

[13] Hochster, M., *Canonical elements in local cohomology modules and the direct summand conjecture*, J. Algebra 84 (1983), pp. 503-553.

[14] Kustin, A. & M. Miller, *Structure theory for a class of grade four Gorenstein ideals*, Trans. Amer. Math. Soc. 270 (1982), pp. 287-207.

[15] Peskine, C. & L. Szpiro, *Dimension projective finie et cohomologie locale*, Publ. Math. Inst. Hautes Etud. Sci. 42 (1973), pp. 77-119.

[16] Peskine, C. & L. Szpiro, *Liaisons des variétés algébriques*, Inv. Math. 26 (1974), pp. 271-302.

[17] Strooker, J.R., *Homological questions in local algebra*, Lond. Math. Soc. Lect. Note Ser. 145 (1990).

[18] Strooker, J.R., *Question 2*, Comm. Math. Inst. Utrecht Univ. 19 (1994), p. 36.

[19] Stückrad, J. & J.R. Strooker, *Monomial conjecture and complete intersections*, Manuscr. Math. 79 (1993), pp. 153-159.

[20] Yoshino, Y., *Cohen-Macaulay modules over Cohen-Macaulay rings*, Lond. Math. Soc. Lect. Note Ser. 146 (1990).

Anne-Marie Simon
Département de Mathématique
C.P. 211 Campus Plaine
Boulevard du Triomphe
B-1050 Bruxelles, Belgique
e-mail: ulbmath@ulb.ac.be

Jan R. Strooker
Mathematisch Instituut
Universiteit Utrecht
Postbus 80010
3508 TA Utrecht, Netherlands
e-mail: strooker@math.ruu.nl

Hilbert Function and Numerical Invariants of Closed Subschemes of $\mathbf{P}^n$

MARIO VALENZANO*

Dipartimento di Matematica
Università di Torino
via Carlo Alberto, 10
10123 Torino - Italy
e-mail: valenzano@dm.unito.it

§1 INTRODUCTION

If X is a closed subscheme of $\mathbf{P}^n$, it is useful to associate to it a few integer numbers usually called in the literature "numerical invariants". They are the degree and the arithmetic genus, the least degree of a hypersurface of $\mathbf{P}^n$ which contains X, the index of speciality, the first and the last degree (if defined) of the non-zero components of the intermediate cohomology modules. In addition to these invariants we introduce three numerical functions from $\mathbf{Z}$ to $\mathbf{Z}$ which generalize the characters of Martin-Deschamps and Perrin. Another well known function attached to a subscheme is its Hilbert function. All these invariants come from the cohomology of the ideal sheaf of the subscheme X which, in turn, depends upon the embedding of the scheme in projective space.

Many authors have studied the above numerical invariants in different directions. An old but powerful method to study a closed subscheme, essentially due to Castelnuovo, is to cut the subscheme with hyperplanes or hypersurfaces and deduce some information from the sections so obtained. Thus, in the stream of Castelnuovo Theory, we study the numerical invariants before and after the sections.

We obtain some relations between the numerical invariants of a subscheme and those of its general hypersurface section, both in the case of great generality of an arbitrary subscheme (see Theorem 4.8) and in the case of a k-Buchsbaum subscheme (see Theorem 5.3). These results consist of some inequalities relating, for instance, the least degree of a hypersurface containing the subscheme and the initial degree of the ideal of the section as a subscheme of the hypersurface, or the index of speciality of the subscheme and that of the section. In the case of k-Buchsbaum subschemes we obtain inequalities between the invariants similar to those of Hoa, Miró-Roig and Vogel (see [HMV]), but more general, since in [HMV] the authors consider only hyperplane sections. Furthermore, if we cut a k-Buchsbaum subscheme with a general hypersurface of degree $\geq k$ we obtain new and stronger inequalities, namely Theorem 5.5 and Theorem 5.6.

---

* Member of CNR-GNSAGA.

In section 3 we introduce, starting from the cohomology of the ideal sheaf of a subscheme X, three numerical functions, namely the functions γ_X, R_X and σ_X, which are extensions to an arbitrary subscheme in $\mathbf{P}^n$ of the characters introduced in [MDP] by Martin-Deschamps and Perrin for curves in $\mathbf{P}^3$. We show that there exists a relation between these three functions, then we describe how γ_X and σ_X are related to the postulation and the speciality of the subscheme, and also how to derive the degree and the arithmetic genus of X from γ_X (see Proposition 3.7). Furthermore, we relate the three functions of a hypersurface section to the same functions of the subscheme X, both for arbitrary X (see Theorem 4.10) and for a k-Buchsbaum subscheme (see Theorem 5.7).

We also generalize two results of Geramita and Migliore. Firstly, we establish a formula relating the Hilbert function of any closed subscheme of $\mathbf{P}^n$ with the Hilbert function of a hypersurface section. We prove that the Hilbert function of the hypersurface section depends upon two terms: a suitable difference of the Hilbert function of the subscheme and a term related to the module structure of the first deficiency module of the subscheme (see Theorem 4.2). Geramita and Migliore prove a similar formula only for a hyperplane section of a curve in $\mathbf{P}^n$, i.e. of a 1-dimensional locally Cohen-Macaulay and equidimensional subscheme (see [GM] Proposition 2.5). Secondly, we generalize the formula in [GM], Proposition 2.7, concerning the behaviour of Hilbert functions under liaison for arithmetically Buchsbaum curves in $\mathbf{P}^n$. Indeed, we prove a formula in which the Hilbert functions of a pair of subschemes directly linked by a complete intersection are related with the Hilbert function of the complete intersection and the function R of one of the two subschemes (see Theorem 6.3).

§2 Notations and Preliminaries

Let $\mathbf{k}$ be an algebraically closed field of any characteristic, $S = \mathbf{k}[X_0, \ldots, X_n]$ the polynomial ring in $n+1$ variables over $\mathbf{k}$ with $n \geq 3$, $\mathfrak{m}$ the maximal homogenous ideal $(X_0, \ldots, X_n)$ in S and $\mathbf{P}^n = \operatorname{Proj} S$. For every closed subscheme X of $\mathbf{P}^n$ we will denote with $\mathcal{I}_X$ the ideal sheaf of X in $\mathcal{O}_{\mathbf{P}^n}$, with $\mathcal{O}_X$ the structure sheaf of X, with $I(X)$ the homogeneous (saturated) ideal in S defining X and with $H(X, -)$ the Hilbert function of X, i.e.

$$H(X,t) = \dim_{\mathbf{k}} (S/I(X))_t = \dim_{\mathbf{k}} H^0(\mathbf{P}^n, \mathcal{O}_{\mathbf{P}^n}(t)) - \dim_{\mathbf{k}} H^0(\mathbf{P}^n, \mathcal{I}_X(t)) \quad \forall t \in \mathbf{Z}$$

where $(S/I(X))_t$ denotes the t-th graded component of $S/I(X)$.

For a sheaf $\mathcal{F}$ of $\mathcal{O}_{\mathbf{P}^n}$-modules we will use the following notations:

$$H^i\mathcal{F}(t) = H^i(\mathbf{P}^n, \mathcal{F}(t))$$
$$H^i_*\mathcal{F} = \bigoplus_{t \in \mathbf{Z}} H^i(\mathbf{P}^n, \mathcal{F}(t))$$
$$h^i\mathcal{F}(t) = \dim_{\mathbf{k}} H^i(\mathbf{P}^n, \mathcal{F}(t))$$

for $0 \leq i \leq n$ and $t \in \mathbf{Z}$. Notice that, for all i, $H^i_*\mathcal{F}$ is a graded S-module. In particular we have the cohomology modules $H^i_*\mathcal{I}_X$, where $H^0_*\mathcal{I}_X = I(X)$ is the defining ideal of X, while for $1 \leq i \leq \dim X$ the modules $H^i_*\mathcal{I}_X$ are the so called "deficiency modules" for they measure the failure of X to be arithmetically

Cohen-Macaulay (briefly aCM), where a closed subscheme X of $\mathbf{P}^n$ is aCM if its homogeneous coordinate ring $S/I(X)$ is Cohen-Macaulay (i.e. a ring with Krull dimension equal to its depth) which is equivalent to the condition $H^i_*\mathcal{I}_X = 0$ for $i = 1, \ldots, \dim X$. In particular the first deficiency module $H^1_*\mathcal{I}_X$ measures the failure of X to be projectively normal, where we call projectively normal a subscheme X such that the restriction map $H^0\mathcal{O}_{\mathbf{P}^n}(t) \to H^0\mathcal{O}_X(t)$ is surjective for every t, which is obviously equivalent to the condition $H^1\mathcal{I}_X(t) = 0$ for all t.

Given a form $F \in S_a$, we have the homomorphism "multiplication by F" ϕ^i_F on the graded module $H^i_*\mathcal{I}_X$, so we get for each t a linear map of $\mathbf{k}$-vector spaces $\phi^i_{F,t}: H^i\mathcal{I}_X(t) \to H^i\mathcal{I}_X(t+a)$; then we put

$$\mathcal{K}^i_{X,F}(t) = \dim_{\mathbf{k}} \ker \phi^i_{F,t}$$

for all $t \in \mathbf{Z}$ and $1 \leq i \leq \dim X$. Therefore given any closed subscheme X of $\mathbf{P}^n$ we have the cohomology functions $t \mapsto h^i\mathcal{I}_X(t)$ for $0 \leq i \leq n$ and also the functions $t \mapsto \mathcal{K}^i_{X,F}(t)$ for $1 \leq i \leq \dim X$ and for each $F \in S_a$ $(a \geq 1)$. We note that the cohomology function $t \mapsto h^i\mathcal{I}_X(t)$ is the null function for $\dim X + 2 \leq i \leq n-1$, $h^n\mathcal{I}_X(t) = h^n\mathcal{O}_{\mathbf{P}^n}(t)$ for every t if $\operatorname{codim} X \geq 2$, and the function $t \mapsto \mathcal{K}^1_{X,F}(t)$ is of finite support for each $F \in S_a$ such that the corresponding hypersurface contains no component of X, since $\ker \phi^1_F$ is isomorphic, up to a shift, to a quotient of the homogenous ideal defining the section in the hypersurface, so $\mathcal{K}^1_{X,F}(t) = 0$ for $t \ll 0$ and, moreover, by Serre, $\mathcal{K}^1_{X,F}(t) = 0$ for $t \gg 0$.

Let M be a S-graded module, we define the dual module of M over $\mathbf{k}$ as $M^* = \operatorname{Hom}_{\mathbf{k}}(M, \mathbf{k})$ with the following structure as a graded S-module: $[M^*]_t = \operatorname{Hom}_{\mathbf{k}}(M_{-t}, \mathbf{k})$ (the dual vector space) and $\phi^*_{F,t} = {}^t\phi_{F,-t-a}$ (with ϕ^*_F multiplication by $F \in S_a$ on M^* and ${}^t\phi_{F,s}$ the dual map of $\phi_{F,s}$).

If $Z \subset Y \subset \mathbf{P}^n$ are closed subschemes, then we will denote with $\mathcal{I}_Z$ the ideal sheaf of Z as a subscheme of $\mathbf{P}^n$ and with $\mathcal{I}_{Z|Y}$ the ideal sheaf of Z as a subscheme of Y, so we have the exact sequence of $\mathcal{O}_{\mathbf{P}^n}$-modules

$$0 \to \mathcal{I}_Y \to \mathcal{I}_Z \to \iota_*\mathcal{I}_{Z|Y} \to 0$$

where $\iota: Y \hookrightarrow \mathbf{P}^n$ is the closed embedding of Y in $\mathbf{P}^n$, but for the sake of simplicity we will write

$$0 \to \mathcal{I}_Y \to \mathcal{I}_Z \to \mathcal{I}_{Z|Y} \to 0.$$

Given a function $f: \mathbf{Z} \to \mathbf{Z}$ we define its first difference Δf and its successive differences $\Delta^m f$ putting for all t

$$\Delta f(t) = \Delta^1 f(t) = f(t) - f(t-1)$$
$$\Delta^m f(t) = \Delta^{m-1} f(t) - \Delta^{m-1} f(t-1).$$

Notice that if the function g is given by $g(t) = f(a-t)$, then we have

$$\Delta^m g(t) = (-1)^m \Delta^m f(a - t + m).$$

We define also the operator ${}_a\Delta$ (a a positive integer)

$$_a\Delta f(t) = f(t) - f(t-a) \qquad \forall t.$$

For the binomial coefficients $\binom{n}{p}$, with $n, p \in \mathbf{Z}$, we will use the following convention:

$$\binom{n}{p} = \begin{cases} \frac{n!}{p!(n-p)!} & \text{if } p \geq 0 \text{ and } n \geq p \\ (-1)^{-(n+p)}\binom{-p-1}{-n-1} & \text{if } p < 0 \text{ and } p \leq n \leq -1 \\ 0 & \text{otherwise} \end{cases}$$

where obviously $0! = 1$.

By abuse of notation we will denote with the same symbol F a hypersurface of $\mathbf{P}^n$ and any homogeneous polynomial defining it.

For all general facts not explicitly mentioned we refer to Hartshorne's book [H].

§3 Numerical Invariants and Characters

Given a closed subscheme X of $\mathbf{P}^n$ of dimension r, with $0 \leq r \leq n$, we consider the following numerical invariants:

$$\begin{aligned} d(X) &= \text{degree of } X \\ p_a(X) &= \text{arithmetic genus of } X \\ s(X) &= \min\{t \mid h^0\mathcal{I}_X(t) \neq 0\} \\ e(X) &= \max\{t \mid h^r\mathcal{O}_X(t) \neq 0\} \\ p_i(X) &= \max\{t \mid h^i\mathcal{I}_X(t) \neq 0\} \quad \text{for} \quad 1 \leq i \leq r \end{aligned}$$

(if $H^i_*\mathcal{I}_X = 0$ we don't define $p_i(X)$).

We note that, if $H^i_*\mathcal{I}_X \neq 0$, the integers $p_i(X)$ are well defined by the vanishing theorem of Serre (see [H] Theorem III.5.2), in fact there exists an integer t_0 such that $h^i\mathcal{I}_X(t) = 0$ for every $t \geq t_0$ and $i > 0$; $s(X)$ is the least degree of a hypersurface of $\mathbf{P}^n$ containing X, while $e(X)$ is the so called "index of speciality".

If the deficiency module $H^i_*\mathcal{I}_X$ is of finite length, then we can define the invariant

$$q_i(X) = \min\{t \mid h^i\mathcal{I}_X(t) \neq 0\}$$

with $1 \leq i \leq r$, but if $H^i_*\mathcal{I}_X = 0$ we don't define $q_i(X)$.

Notice that a scheme X has all its deficiency modules of finite length if and only if it is locally Cohen-Macaulay and equidimensional (see [M] Theorem 1.2.2). In this case also the integers $q_i(X)$ are well defined (if $H^i_*\mathcal{I}_X \neq 0$) and we have obviously $q_i(X) \leq p_i(X)$ for $i = 1, \ldots, r$.

Now we want to introduce other invariants related to a closed subscheme of $\mathbf{P}^n$, which extend to higher dimension the characters used by Martin-Deschamps and Perrin for curves in $\mathbf{P}^3$ (see [MDP]).

Definition 3.1. Given a closed subscheme $X \subseteq \mathbf{P}^n$ of dimension r, with $0 \leq r \leq n$, we define the following functions from $\mathbf{Z}$ to $\mathbf{Z}$

$$\gamma_X(t) = -\Delta^{r+2}H(X,t) = \Delta^{r+2}h^0\mathcal{I}_X(t) - \binom{t+n-r-2}{n-r-2}$$

$$R_X(t) = \sum_{i=1}^{r}(-1)^{i+1}h^i\mathcal{I}_X(t)$$

$$\sigma_X(t) = \Delta^{r+2}h^r\mathcal{O}_X(t).$$

Notice that we have

$$\gamma_X(t) = 0 \quad \text{for} \quad t < 0 \quad \text{and} \quad \gamma_X(0) = -1$$

$$R_X(t) = 0 \quad \text{and} \quad \sigma_X(t) = 0 \quad \text{for} \quad t \gg 0.$$

Remark 3.2. The Hilbert function of X, $H(X,t) = h^0\mathcal{O}_{\mathbf{P}^n}(t) - h^0\mathcal{I}_X(t)$, is zero for $t < 0$ and agrees for $t \gg 0$ with the Hilbert polynomial of X, i.e. with

$$P(X,t) = \chi(\mathcal{O}_X(t)) = \sum_{i=0}^{r}(-1)^i h^i\mathcal{O}_X(t)$$

(which is a polynomial in t of degree r), hence the $(r+1)$-th difference of $H(X,t)$ is a function of finite support and therefore γ_X is a character in the sense of Martin-Deschamps and Perrin (see [MDP] page 29), that is γ_X is a function of finite support such that $\sum_{t\in\mathbf{Z}}\gamma_X(t) = 0$. We will call γ_X the "postulation character" of X and σ_X the "speciality function" of X.

Proposition 3.3. *For all closed subscheme $X \subseteq \mathbf{P}^n$ of dimension r we have*

$$\gamma_X(t) = \Delta^{r+2}R_X(t) + (-1)^r\sigma_X(t)$$

for every $t \in \mathbf{Z}$.

Proof. From the exact sequence

$$0 \to \mathcal{I}_X \to \mathcal{O}_{\mathbf{P}^n} \to \mathcal{O}_X \to 0$$

we get the equalities (for all $t \in \mathbf{Z}$)

$$H(X,t) = h^0\mathcal{O}_X(t) - h^1\mathcal{I}_X(t)$$
$$h^i\mathcal{O}_X(t) = h^{i+1}\mathcal{I}_X(t) \qquad 1 \le i \le r-1$$

so we obtain

$$\begin{aligned} P(X,t) &= \chi(\mathcal{O}_X(t)) \\ &= h^0\mathcal{O}_X(t) + \sum_{i=1}^{r-1}(-1)^i h^i\mathcal{O}_X(t) + (-1)^r h^r\mathcal{O}_X(t) \\ &= H(X,t) + h^1\mathcal{I}_X(t) + \sum_{i=2}^{r}(-1)^{i+1}h^i\mathcal{I}_X(t) + (-1)^r h^r\mathcal{O}_X(t) \\ &= H(X,t) + R_X(t) + (-1)^r h^r\mathcal{O}_X(t) \end{aligned}$$

from which taking the $(r+2)$-th difference we get the statement, since we have $\Delta^{r+2}P(X,t) \equiv 0$.

Remark 3.4. If X is locally Cohen-Macaulay and equidimensional, then R_X is a function of finite support and therefore $\Delta^{r+2}R_X$ is a character. In this case, it follows that also σ_X is a character in the sense of Martin-Deschamps and Perrin.

Corollary 3.5. *If X is aCM, then $\gamma_X(t) = (-1)^r \sigma_X(t)$ for every $t \in \mathbf{Z}$.*

Remark 3.6. Notice that, if X is a 1-dimensional subscheme of $\mathbf{P}^3$, we recover the relation $\gamma_X = \Delta^3 \rho_X - \sigma_X$, points out by Martin-Deschamps and Perrin (see [MDP] page 30), where ρ_X is the "Rao function" of the curve, i.e. $\rho_X : t \mapsto h^1 \mathcal{I}_X(t)$. We note also that, in dimension greater then 1, the right function to study is R_X, which is the alternated sum of the dimensions of the intermediate cohomology groups, and not just the Hilbert function of a graded module.

Proposition 3.7. *Let $X \subseteq \mathbf{P}^n$ be a closed subscheme of dimension r. Then:*

(1)

$$s(X) = \min \left\{ t > 0 \mid \gamma_X(t) \neq -\binom{t+n-r-2}{n-r-2} \right\}.$$

(2)

$$e(X) = \max\{t \mid \sigma_X(t) \neq 0\} - r - 2.$$

(3) *The character γ_X determines the postulation of X:*

$$h^0 \mathcal{I}_X(t) = \binom{t+n}{n} + \sum_{j \leq t} \binom{t-j+r+1}{r+1} \gamma_X(j) \qquad \forall t \in \mathbf{Z}.$$

(4) *The function σ_X determines the speciality of X:*

$$h^r \mathcal{O}_X(t) = (-1)^r \sum_{j \geq t+r+2} \binom{j-t-1}{r+1} \sigma_X(j) \qquad \forall t \in \mathbf{Z}.$$

(5) *The character γ_X determines the degree and the arithmetic genus of X:*

$$d(X) = \sum_{j \geq 1} j \gamma_X(j)$$

$$p_a(X) = \sum_{j \geq r+2} \binom{j-1}{r+1} \gamma_X(j).$$

(6) *If X is locally Cohen-Macaulay and equidimensional, then*

$$d(X) = (-1)^r \sum_{j \in \mathbf{Z}} j \sigma_X(j)$$

$$p_a(X) = (-1)^r \left\{ \sum_{j \in \mathbf{Z}} \frac{(j-1)(j-2)\cdots(j-r-1)}{(r+1)!} \sigma_X(j) - 1 \right\}.$$

Proof. (1) Obvious from the definition of γ_X.
(2) Obvious from the definition of σ_X and the properties of the difference of a numerical function vanishing for $t \gg 0$.
(3) It follows from the formula for the primitive of a numerical function with the vanishing condition for $t \ll 0$ (see [MDP] page 29).
(4) It follows from the formula for the primitive of a numerical function with the

vanishing condition for $t \gg 0$ (see [MDP] page 29).
(5) By (3) we have

$$H(X,t) = -\sum_{j \le t} \binom{t-j+r+1}{r+1} \gamma_X(j).$$

Now, by a straightforward computation we get

$$\binom{t-j+r+1}{r+1} = \frac{1}{(r+1)!}\left\{t^{r+1} + \left[\frac{1}{2}(r+1)(r+2) - (r+1)j\right] t^r + \right.$$
$$\left. + \cdots + (-1)^{r+1}\prod_{l=1}^{r+1}(j-l)\right\}$$

hence

$$H(X,t) = -\frac{t^{r+1}}{(r+1)!}\sum_{j \le t}\gamma_X(j) - \frac{(r+1)(r+2)t^r}{2(r+1)!}\sum_{j\le t}\gamma_X(j) +$$
$$+ \frac{t^r}{r!}\sum_{j\le t} j\,\gamma_X(j) + \cdots + (-1)^r \sum_{j \le t}\frac{\prod_{l=1}^{r+1}(j-l)}{(r+1)!}\gamma_X(j).$$

For $t \gg 0$ it results

$$\gamma_X(j) = 0 \quad \forall j > t$$
$$\sum_{j \le t}\gamma_X(j) = \sum_{j \in \mathbf{Z}}\gamma_X(j) = 0$$

and the Hilbert function equals the Hilbert polynomial

$$H(X,t) = P(X,t) = \frac{d(X)}{r!}t^r + \cdots + 1 + (-1)^r p_a(X),$$

so, comparing the two expressions of $H(X,t)$, we obtain

$$d(X) = \sum_{j \in \mathbf{Z}} j\,\gamma_X(j) = \sum_{j \ge 1} j\,\gamma_X(j)$$

and

$$p_a(X) + (-1)^r = \sum_{j \in \mathbf{Z}}\frac{(j-1)(j-2)\cdots(j-r-1)}{(r+1)!}\gamma_X(j).$$

Put $A(j) = (j-1)(j-2)\cdots(j-r-1)/(r+1)!$, then we have

$$A(j) = \binom{j-1}{r+1} \qquad \text{for} \quad j \ge r+2$$
$$A(j) = \binom{j-1}{r+1} = 0 \qquad \text{for} \quad j = 1, \ldots, r+1$$
$$A(0)\gamma_X(0) = \frac{(-1)(-2)\cdots(-r-1)}{(r+1)!}(-1) = (-1)^r$$
$$\binom{-1}{r+1}\gamma_X(0) = 0$$

and therefore

$$p_a(X) = \sum_{j\in\mathbf{Z}} \binom{j-1}{r+1} \gamma_X(j) = \sum_{j\geq r+2} \binom{j-1}{r+1} \gamma_X(j).$$

(6) If X is locally Cohen-Macaulay and equidimensional, then the function R_X is of finite support, hence we have

$$\sum_{j\in\mathbf{Z}} j^m \, \Delta^{r+2} R_X(j) = 0 \quad \text{for} \quad 0 \leq m \leq r+1$$

(see [MDP] page 29). Therefore by Proposition 3.3 we obtain

$$\begin{aligned} d(X) &= \sum_{j\geq 1} j\,\gamma_X(j) = \sum_{j\in\mathbf{Z}} j\,\gamma_X(j) \\ &= \sum_{j\in\mathbf{Z}} j\,\Delta^{r+2} R_X(j) + (-1)^r \sum_{j\in\mathbf{Z}} j\,\sigma_X(j) \\ &= (-1)^r \sum_{j\in\mathbf{Z}} j\,\sigma_X(j) \end{aligned}$$

and

$$\begin{aligned} p_a(X) + (-1)^r &= \sum_{j\in\mathbf{Z}} \frac{(j-1)(j-2)\cdots(j-r-1)}{(r+1)!} \gamma_X(j) \\ &= \sum_{j\in\mathbf{Z}} A(j) \Delta^{r+2} R_X(j) + (-1)^r \sum_{j\in\mathbf{Z}} A(j)\sigma_X(j) \\ &= (-1)^r \sum_{j\in\mathbf{Z}} \frac{(j-1)(j-2)\cdots(j-r-1)}{(r+1)!} \sigma_X(j) \end{aligned}$$

since $A(j) = (j-1)\cdots(j-r-1)/(r+1)!$ is a polynomial in j of degree $r+1$.

Corollary 3.8. *Let $X \subseteq \mathbf{P}^n$ be a closed subscheme of dimension r. If $\gamma_X(t) = 0$ for every $t \geq r+2$, then X has arithmetic genus $p_a(X) = 0$.*

Proof. It follows directly from Proposition 3.7(5).

§4 Numerical Invariants and Hypersurface Sections

Let X be a closed subscheme of $\mathbf{P}^n$ of dimension r, $1 \leq r \leq n-2$, and let F be a hypersurface of degree a. If F contains no component of X, then, taking into account that $\mathcal{I}_F \simeq \mathcal{O}_{\mathbf{P}^n}(-a)$, we have the following exact sequences of $\mathcal{O}_{\mathbf{P}^n}$-modules

$$0 \to \mathcal{I}_X(-a) \xrightarrow{\cdot F} \mathcal{I}_X \to \mathcal{I}_{X\cap F|F} \to 0 \tag{†}$$

and

$$0 \to \mathcal{O}_{\mathbf{P}^n}(-a) \xrightarrow{\cdot F} \mathcal{I}_{X\cap F} \to \mathcal{I}_{X\cap F|F} \to 0 \tag{††}$$

where the morphisms denoted with $\cdot F$ are "multiplication by F" and $X \cap F$ is the scheme-theoretic intersection of X and F, i.e. $\mathcal{I}_{X\cap F} = \mathcal{I}_X + \mathcal{I}_F$. We observe that, twisting (††) by t and passing to cohomology, we get

$$\begin{aligned} h^0\mathcal{I}_{X\cap F}(t) &= h^0\mathcal{O}_{\mathbf{P}^n}(t-a) + h^0\mathcal{I}_{X\cap F|F}(t) \\ h^i\mathcal{I}_{X\cap F}(t) &= h^i\mathcal{I}_{X\cap F|F}(t) \end{aligned} \tag{‡}$$

for $1 \le i \le n-2$ and $t \in \mathbf{Z}$; actually (††) implies

$$H^i_*\mathcal{I}_{X\cap F} \cong H^i_*\mathcal{I}_{X\cap F|F} \quad \text{for} \quad 1 \le i \le n-2$$

(isomorphism of S-modules).

Proposition 4.1. *Let X be a closed subscheme of $\mathbf{P}^n$ and F a hypersurface of degree a not containing any component of X. Then*

$$\begin{aligned} h^0\mathcal{I}_{X\cap F|F}(t) &= {}_a\Delta h^0\mathcal{I}_X(t) + \mathcal{K}^1_{X,F}(t-a) \\ h^i\mathcal{I}_{X\cap F|F}(t) &= {}_a\Delta h^i\mathcal{I}_X(t) + \mathcal{K}^i_{X,F}(t-a) + \mathcal{K}^{i+1}_{X,F}(t-a) \quad (1 \le i \le r-1) \\ h^r\mathcal{I}_{X\cap F|F}(t) &= {}_a\Delta h^r\mathcal{I}_X(t) + \mathcal{K}^r_{X,F}(t-a) - {}_a\Delta h^{r+1}\mathcal{I}_X(t) \end{aligned}$$

and

$$\begin{aligned} h^0\mathcal{I}_{X\cap F}(t) &= \binom{t-a+n}{n} + {}_a\Delta h^0\mathcal{I}_X(t) + \mathcal{K}^1_{X,F}(t-a) \\ h^i\mathcal{I}_{X\cap F}(t) &= {}_a\Delta h^i\mathcal{I}_X(t) + \mathcal{K}^i_{X,F}(t-a) + \mathcal{K}^{i+1}_{X,F}(t-a) \quad (1 \le i \le r-1) \\ h^r\mathcal{I}_{X\cap F}(t) &= {}_a\Delta h^r\mathcal{I}_X(t) + \mathcal{K}^r_{X,F}(t-a) - {}_a\Delta h^{r+1}\mathcal{I}_X(t) \end{aligned}$$

for all $t \in \mathbf{Z}$.

Proof. Twisting (†) by t and passing to cohomology we get the exact sequences

$$0 \to H^0\mathcal{I}_X(t-a) \to H^0\mathcal{I}_X(t) \to H^0\mathcal{I}_{X\cap F|F}(t) \to \ker\phi^1_{F,t-a} \to 0$$

$$0 \to \ker\phi^i_{F,t-a} \to H^i\mathcal{I}_X(t-a) \to H^i\mathcal{I}_X(t) \to H^i\mathcal{I}_{X\cap F|F}(t) \to \ker\phi^{i+1}_{F,t-a} \to 0$$

for $1 \le i \le r-1$ and

$$\begin{aligned} 0 \to \ker\phi^r_{F,t-a} \to H^r\mathcal{I}_X(t-a) \to H^r\mathcal{I}_X(t) \to H^r\mathcal{I}_{X\cap F|F}(t) \to \\ \to H^{r+1}\mathcal{I}_X(t-a) \to H^{r+1}\mathcal{I}_X(t) \to 0 \end{aligned}$$

where ϕ^i_F is "multiplication by F" on the cohomology module $H^i_*\mathcal{I}_X$. So, taking into account (‡) we obtain the claim.

The following Theorem extends [GM] Proposition 2.5.

Theorem 4.2. *Let X be a closed subscheme of $\mathbf{P}^n$ and F a hypersurface of degree a not containing any component of X. Then*

$$H(X \cap F, t) = {}_a\Delta H(X, t) - \mathcal{K}^1_{X,F}(t-a).$$

Proof. Using the definition of Hilbert function and Proposition 4.1 we get

$$\begin{aligned} H(X \cap F, t) &= h^0\mathcal{O}_{\mathbf{P}^n}(t) - h^0\mathcal{I}_{X\cap F}(t) \\ &= \binom{t+n}{n} - \binom{t-a+n}{n} - {}_a\Delta h^0\mathcal{I}_X(t) - \mathcal{K}^1_{X,F}(t-a) \\ &= {}_a\Delta H(X,t) - \mathcal{K}^1_{X,F}(t-a). \end{aligned}$$

Remark 4.3. In [GM, Prop. 2.5] the above result is proved only for hyperplane sections of a curve in $\mathbf{P}^n$ (i.e. a 1-dimensional subscheme without embedded or isolated points). Actually the formula holds for hypersurface sections of any closed subscheme of $\mathbf{P}^n$.

Remark 4.4. From Theorem 4.2 it follows that

$$H(X \cap F, t) = {}_a\Delta H(X,t) \quad \text{for} \quad 0 \le t \le \sigma - 1 \quad \text{and} \quad t \ge p_1(X) + a + 1$$

where $\sigma = \min\{t \mid h^0\mathcal{I}_{X\cap F|F}(t) \neq 0\}$ and also that for a projectively normal subscheme $X \subset \mathbf{P}^n$ it results $H(X \cap F, t) = {}_a\Delta H(X,t)$ for all t and for every general hypersurface F of degree a. In particular we recover the well known relation between the Hilbert function of an aCM scheme and its general hyperplane section, i.e. $H(X \cap H, t) = \Delta H(X,t)$ for all t.

Now to establish some relations between the numerical invariants of a subscheme and those of its general hypersurface section we will give the following technical lemma needed in the sequel (basically it is [GM] Lemma 3.1 with a slight modification).

Lemma 4.5. *Let $X \subset \mathbf{P}^n$ be a closed subscheme, F a hypersurface of degree a not containing any component of X, and G any form of degree b. Then $\ker\phi^1_{F,t} \subseteq \ker\phi^1_{G,t}$ if and only if for all $x \in H^0\mathcal{I}_{X\cap F|F}(t+a)$, $G\cdot x$ lifts to $H^0\mathcal{I}_X(t+a+b)$.*

Proof. From the following commutative diagram with exact rows

$$\begin{array}{ccccccc} I_{t+a} & \longrightarrow & J_{t+a} & \xrightarrow{f} & M_t & \xrightarrow{\phi^1_{F,t}} & M_{t+a} \\ \downarrow & & \downarrow{\scriptstyle \psi_{G,t+a}} & & \downarrow{\scriptstyle \phi^1_{G,t}} & & \downarrow \\ I_{t+a+b} & \xrightarrow[h]{} & J_{t+a+b} & \xrightarrow[g]{} & M_{t+b} & \longrightarrow & M_{t+a+b} \end{array}$$

where $I = H^0_*\mathcal{I}_X$, $J = H^0_*\mathcal{I}_{X\cap F|F}$, $M = H^1_*\mathcal{I}_X$ and the vertical arrows are "multiplication by G", we get

$$\begin{aligned} &\operatorname{im} f = \ker\phi^1_{F,t} \subseteq \ker\phi^1_{G,t} \\ \Leftrightarrow \quad & g \circ \psi_{G,t+a} = \phi^1_{G,t} \circ f = 0 \\ \Leftrightarrow \quad & \operatorname{im}\psi_{G,t+a} \subseteq \ker g = \operatorname{im} h \end{aligned}$$

and therefore the claim.

Corollary 4.6. *Let $X \subset \mathbf{P}^n$ be a closed subscheme, F a hypersurface of degree a not containing any component of X, and G any form of degree b. If, for some integer t, $h^0\mathcal{I}_{X\cap F|F}(t+a) \neq 0$ and $\phi^1_{G,t} = 0$, then $h^0\mathcal{I}_X(t+a+b) \neq 0$.*

Proof. It follows directly from the above lemma, taking into account that we have $\ker \phi^1_{F,t} \subseteq \ker \phi^1_{G,t} = H^1\mathcal{I}_X(t)$.

Definition 4.7. Given a closed subscheme $X \subset \mathbf{P}^n$ and a hypersurface F not containing any component of X we put

$$s(X \cap F|F) = \min\{t \mid h^0\mathcal{I}_{X\cap F|F}(t) \neq 0\}$$

while

$$s(X \cap F) = \min\{t \mid h^0\mathcal{I}_{X\cap F}(t) \neq 0\}$$

so by an easy computation we obtain

$$s(X \cap F) = \min\{s(X \cap F|F), \deg F\}.$$

We note that, thanks to (‡), for the other invariants it results:

$$\begin{aligned} e(X \cap F) &= \max\{t \mid h^r\mathcal{I}_{X\cap F}(t) \neq 0\} = \max\{t \mid h^r\mathcal{I}_{X\cap F|F}(t) \neq 0\} \\ p_i(X \cap F) &= \max\{t \mid h^i\mathcal{I}_{X\cap F}(t) \neq 0\} = \max\{t \mid h^i\mathcal{I}_{X\cap F|F}(t) \neq 0\} \\ q_i(X \cap F) &= \min\{t \mid h^i\mathcal{I}_{X\cap F}(t) \neq 0\} = \min\{t \mid h^i\mathcal{I}_{X\cap F|F}(t) \neq 0\} \end{aligned}$$

$(1 \leq i \leq r-1)$.

Theorem 4.8. *Let X be a closed subscheme of $\mathbf{P}^n$ and F a general hypersurface. Then:*

(1) *if $H^1_*\mathcal{I}_X \neq 0$, then $s(X) \leq p_1(X) + 2$;*
(2) *if $H^1_*\mathcal{I}_X \neq 0$, then $p_1(X) \geq 0$;*
(3) *$s(X \cap F|F) \leq s(X)$;*
(4) *$p_i(X \cap F) \geq p_{i+1}(X) + \deg F$ for $i = 1, \ldots, r-1$, and if $H^i_*\mathcal{I}_X = 0$, then $p_i(X \cap F) = p_{i+1}(X) + \deg F$;*
(5) *$e(X \cap F) \geq e(X) + \deg F$;*
(6) *if $H^r_*\mathcal{I}_X = 0$, then $e(X \cap F) = e(X) + \deg F$, if $H^r_*\mathcal{I}_X \neq 0$, then $e(X \cap F) \leq \max\{e(X) + \deg F, p_r(X)\}$.*

Proof. (1) Let H, H' be two general hyperplanes. Put $p = p_1(X)$, then we have $\phi^1_{H',p} = 0$ and from the exact sequence

$$H^0\mathcal{I}_{X\cap H|H}(p+1) \to H^1\mathcal{I}_X(p) \to 0$$

we get $h^0\mathcal{I}_{X\cap H|H}(p+1) \neq 0$, so by Corollary 4.6 we obtain $h^0\mathcal{I}_X(p+2) \neq 0$, hence $s(X) \leq p_1(X) + 2$.
(2) Let H be a general hyperplane. It results $h^0\mathcal{I}_{X\cap H|H}(t) = 0$ for every $t \leq 0$, so we have $h^1\mathcal{I}_X(t-1) \leq h^1\mathcal{I}_X(t)$ for $t \leq 0$ and therefore the last non-zero component of the first deficiency module must be in non-negative degree.

(3) Obvious, since it results $h^0\mathcal{I}_{X\cap F|F}(s) \geq h^0\mathcal{I}_X(s) > 0$, where $s = s(X)$.
(4) Put $p = p_{i+1}(X)$ and $a = \deg F$, then we have the surjection

$$H^i\mathcal{I}_{X\cap F}(p+a) \to H^{i+1}\mathcal{I}_X(p) \to 0$$

so $h^i\mathcal{I}_{X\cap F}(p+a) \geq h^{i+1}\mathcal{I}_X(p) \neq 0$, i.e. $p_i(X \cap F) \geq p_{i+1}(X) + \deg F$.
Moreover, if $H^i_*\mathcal{I}_X = 0$, then $h^i\mathcal{I}_{X\cap F}(t+a) \leq h^{i+1}\mathcal{I}_X(t)$ for every t, and so $p_i(X \cap F) = p_{i+1}(X) + \deg F$.
(5) Put $e = e(X)$ and $a = \deg F$, then we have the surjection

$$H^r\mathcal{I}_{X\cap F}(e+a) \to H^{r+1}\mathcal{I}_X(e) \to 0$$

so $h^r\mathcal{I}_{X\cap F}(e+a) \geq h^{r+1}\mathcal{I}_X(e) \neq 0$ and therefore $e(X \cap F) \geq e(X) + \deg F$.
(6) If $H^r_*\mathcal{I}_X = 0$, then we have $h^r\mathcal{I}_{X\cap F}(t+a) = h^{r+1}\mathcal{I}_X(t) = 0$ for every $t \geq e(X)+1$ and $h^r\mathcal{I}_{X\cap F}(e+a) = h^{r+1}\mathcal{I}_X(e) \neq 0$ ($e = e(X)$), and so $e(X \cap F) = e(X) + \deg F$.
If $H^r_*\mathcal{I}_X \neq 0$, put $a = \deg F$. From the exact sequence

$$H^r\mathcal{I}_X(t) \to H^r\mathcal{I}_{X\cap F}(t) \to H^{r+1}\mathcal{I}_X(t-a)$$

we get $h^r\mathcal{I}_{X\cap F}(t) = 0$ for every $t > \max\{e(X)+a, p_r(X)\}$, that is $e(X \cap F) \leq \max\{e(X) + \deg F, p_r(X)\}$.

Proposition 4.9. *Let $X \subset \mathbf{P}^n$ be a closed subscheme of dimension r and F a hypersurface of degree a containing no component of X. Then*

$$\begin{aligned}\gamma_{X\cap F}(t) &= \Delta^{r+1}h^0\mathcal{I}_{X\cap F}(t) - \binom{t+n-r-1}{n-r-1}\\ &= \Delta^{r+1}h^0\mathcal{I}_{X\cap F|F}(t) + \binom{t-a+n-r-1}{n-r-1} - \binom{t+n-r-1}{n-r-1}.\end{aligned}$$

If $a = 1$

$$\begin{aligned}\gamma_{X\cap F}(t) &= \Delta^{r+1}h^0\mathcal{I}_{X\cap F}(t) - \binom{t+n-r-1}{n-r-1}\\ &= \Delta^{r+1}h^0\mathcal{I}_{X\cap F|F}(t) - \binom{t+n-r-2}{n-r-2}.\end{aligned}$$

Proof. By definition and (‡) it results

$$\begin{aligned}H(X \cap F, t) &= h^0\mathcal{O}_{\mathbf{P}^n}(t) - h^0\mathcal{I}_{X\cap F}(t)\\ &= h^0\mathcal{O}_{\mathbf{P}^n}(t) - h^0\mathcal{O}_{\mathbf{P}^n}(t-a) - h^0\mathcal{I}_{X\cap F|F}(t)\end{aligned}$$

so, being $\gamma_{X\cap F}(t) = -\Delta^{r+1}H(X \cap F, t)$, we get the claim.

Theorem 4.10. *Let $X \subset \mathbf{P}^n$ be a closed subscheme of dimension r and F a hypersurface of degree a not containing any component of X. Then:*
(1) *For the postulation character we have*

$$\gamma_{X\cap F}(t) = \sum_{j=0}^{a-1} \gamma_X(t-j) + \Delta^{r+1}\mathcal{K}^1_{X,F}(t-a)$$

and if $a = 1$

$$\gamma_{X\cap F}(t) = \gamma_X(t) + \Delta^{r+1}\mathcal{K}^1_{X,F}(t-1).$$

(2) *For the function R we have*

$$R_{X\cap F}(t) = {}_a\Delta R_X(t) + (-1)^r{}_a\Delta h^r\mathcal{I}_X(t) + \mathcal{K}^1_{X,F}(t-a) + (-1)^r\mathcal{K}^r_{X,F}(t-a).$$

(3) *For the speciality function we have*

$$\sigma_{X\cap F}(t) = \sum_{j=0}^{a-1} \Delta^{r+2}h^r\mathcal{I}_X(t-j) + \Delta^{r+1}\mathcal{K}^r_{X,F}(t-a) - \sum_{j=0}^{a-1}\sigma_X(t-j)$$

and if $a = 1$

$$\sigma_{X\cap F}(t) = \Delta^{r+2}h^r\mathcal{I}_X(t) + \Delta^{r+1}\mathcal{K}^r_{X,F}(t-1) - \sigma_X(t).$$

Proof. (1) By Theorem 4.2 we have

$$\begin{aligned} H(X\cap F, t) &= {}_a\Delta H(X,t) - \mathcal{K}^1_{X,F}(t-a) \\ &= \sum_{j=0}^{a-1} \Delta H(X, t-j) - \mathcal{K}^1_{X,F}(t-a) \end{aligned}$$

and therefore $\gamma_{X\cap F}(t) = \sum_{j=0}^{a-1}\gamma_X(t-j) + \Delta^{r+1}\mathcal{K}^1_{X,F}(t-a)$.
(2) By Proposition 4.1 we get

$$\begin{aligned} R_{X\cap F}(t) &= \sum_{i=1}^{r-1}(-1)^{i+1}h^i\mathcal{I}_{X\cap F}(t) \\ &= \sum_{i=1}^{r-1}(-1)^{i+1}\left\{{}_a\Delta h^i\mathcal{I}_X(t) + \mathcal{K}^i_{X,F}(t-a) + \mathcal{K}^{i+1}_{X,F}(t-a)\right\} \\ &= \sum_{i=1}^{r-1}(-1)^{i+1}{}_a\Delta h^i\mathcal{I}_X(t) + \mathcal{K}^1_{X,F}(t-a) + (-1)^r\mathcal{K}^r_{X,F}(t-a) \\ &= {}_a\Delta R_X(t) + (-1)^r{}_a\Delta h^r\mathcal{I}_X(t) + \mathcal{K}^1_{X,F}(t-a) + (-1)^r\mathcal{K}^r_{X,F}(t-a). \end{aligned}$$

(3) By Proposition 4.1 we have

$$\begin{aligned} h^r\mathcal{I}_{X\cap F}(t) &= {}_a\Delta h^r\mathcal{I}_X(t) + \mathcal{K}^r_{X,F}(t-a) - {}_a\Delta h^{r+1}\mathcal{I}_X(t) \\ &= \sum_{j=0}^{a-1}\Delta h^r\mathcal{I}_X(t-j) + \mathcal{K}^r_{X,F}(t-a) - \sum_{j=0}^{a-1}\Delta h^{r+1}\mathcal{I}_X(t-j) \end{aligned}$$

so, taking the $(r+1)$-th difference, we get the claim.

From now on we restrict our attention to closed subschemes of $\mathbf{P}^n$ which are locally Cohen-Macaulay and equidimensional. For such a scheme all its deficiency modules have finite length (see [M] Theorem 1.2.2), then there exists a positive integer t such that $\mathfrak{m}^t$ annihilates all the deficiency modules, so it makes sense to give the following definition.

Definition 5.1. Let X be a closed subscheme of dimension r $(1 \leq r \leq n-2)$, locally Cohen-Macaulay and equidimensional, then X is called k-Buchsbaum if and only if

$$k = \min\{t \mid \mathfrak{m}^t \cdot H^i_* \mathcal{I}_X = 0 \text{ for } 1 \leq i \leq r\}.$$

Notice that X is 0-Buchsbaum if and only if it is aCM.

Proposition 5.2. *Let X be a k-Buchsbaum subscheme of $\mathbf{P}^n$. Then the cohomology (of the ideal sheaf) of the degree a hypersurface section $X \cap F$ is dimensionally independent of the hypersurface F, provided that $a \geq k$ and F contains no component of X. In fact*

$$\begin{aligned} h^0\mathcal{I}_{X\cap F}(t) &= \binom{t-a+n}{n} + {}_a\Delta h^0\mathcal{I}_X(t) + h^1\mathcal{I}_X(t-a) \\ h^i\mathcal{I}_{X\cap F}(t) &= h^i\mathcal{I}_X(t) + h^{i+1}\mathcal{I}_X(t-a) \quad 1 \leq i \leq r-1 \\ h^r\mathcal{I}_{X\cap F}(t) &= h^r\mathcal{I}_X(t) - {}_a\Delta h^{r+1}\mathcal{I}_X(t). \end{aligned} \qquad (*)$$

Proof. Let F be a hypersurface of degree $a \geq k$ not containing any component of X. Since $a \geq k$ the map $\phi^i_{F,t}$ is the zero map for all t and $1 \leq i \leq r$, and so $\mathcal{K}^i_{X,F}(t) = h^i\mathcal{I}_X(t)$ for $1 \leq i \leq r$ and $t \in \mathbf{Z}$. Therefore by Proposition 4.1 we get $(*)$, from which we deduce that dimensionally the cohomology of the hypersuface section $X \cap F$ is independent of the hypersurface F, provided that F has degree at least k and meets X properly.

Now we give some relations between the numerical invariants of a k-Buchsbaum subscheme and those of its general hypersurface section. Naturally all the relations of Theorem 4.8 still hold, but the k-Buchsbaum hypothesis make possible to prove the following ones, which generalize Prop. 2.2 and Prop. 2.4 of [HMV].

Theorem 5.3. *Let X be a k-Buchsbaum subscheme of dimension r and $k \geq 1$, and let F be a general hypersurface. Then:*

(1) $p_r(X) \leq e(X) + k + 1$;
(2) $s(X) - k \leq s(X \cap F|F) \leq s(X)$;
(3) $q_i(X \cap F) \leq q_i(X)$ *for* $i = 1, \ldots, r-1$;
(4) $q_i(X \cap F) \leq p_{i+1}(X) + \deg F \leq p_i(X \cap F)$ *for* $i = 1, \ldots, r-1$;
(5) $p_i(X \cap F) \geq q_{i+1}(X \cap F) + \deg F$ *for* $i = 1, \ldots, r-2$;
(6) $e(X) + \deg F \leq e(X \cap F) \leq e(X) + \deg F + k$.

Proof. (1) See [HMV] Prop. 2.4(a).
(2) Obviously $s(X \cap F|F) \leq s(X)$. Put $\sigma = s(X \cap F|F)$ and let G be a general form of degree k. Then we have $h^0\mathcal{I}_{X\cap F|F}(\sigma) \neq 0$ and $\phi^1_{G,\sigma} = 0$, so by Corollary 4.6 we

get $s(X) \leq s(X \cap F|F) + k$.
(3) Put $q = q_i(X)$, then we have the injection

$$0 \to H^i\mathcal{I}_X(q) \to H^i\mathcal{I}_{X\cap F}(q)$$

so $h^i\mathcal{I}_{X\cap F}(q) \geq h^i\mathcal{I}_X(q) \neq 0$ and therefore $q_i(X \cap F) \leq q_i(X)$.
(4) It results $h^i\mathcal{I}_{X\cap F}(p_{i+1}(X) + \deg F) \neq 0$ as in the proof of Theorem 4.8(4), and therefore the claim.
(5) By (3) and Theorem 4.8(4) we have $q_{i+1}(X \cap F) \leq q_{i+1}(X) \leq p_{i+1}(X) \leq p_i(X \cap F) - \deg F$, that is $p_i(X \cap F) \geq q_{i+1}(X \cap F) + \deg F$.
(6) By Theorem 4.8(5) we have $e(X) + \deg F \leq e(X \cap F)$. Put $e = e(X)$ and $a = \deg F$. Suppose $h^r\mathcal{I}_{X\cap F}(e + a + k + 1) \neq 0$, then $h^r\mathcal{I}_X(e + a + k + 1) \neq 0$ and hence $p_r(X) \geq e + a + k + 1 > e(X) + k + 1$ which contradicts the relation $p_r(X) \leq e(X) + k + 1$, therefore $e(X \cap F) \leq e(X) + \deg F + k$.

Remark 5.4. In [HMV] the authors have considered only general hyperplane sections of a k-Buchsbaum subscheme, getting some inequalities between the numerical invariants of the scheme and those of the section. In the above theorem we have generalized these inequalities to all general hypersurface sections, obtaining in particular (2), (3) and (6) which are the analogous of Prop. 2.2 and Prop. 2.4(c) in [HMV].

Theorem 5.5. *Let X be a k-Buchsbaum subscheme of dimension r and $k \geq 1$, and let F be a general hypersurface of degree $\geq k$. Then:*

(1) *if $H^1_*\mathcal{I}_X \neq 0$, then $s(X \cap F|F) \leq q_1(X) + \deg F$;*
(2) $q_i(X \cap F) \leq q_i(X) \leq p_i(X) \leq p_i(X \cap F)$ *for* $i = 1, \ldots, r-1$;
(3) $q_i(X \cap F) \leq q_{i+1}(X) + \deg F \leq p_{i+1}(X) + \deg F \leq p_i(X \cap F)$ *for $i = 1, \ldots, r-1$;*
(4) *if $H^r_*\mathcal{I}_X \neq 0$, then $e(X \cap F) = \max\{e(X) + \deg F, p_r(X)\}$.*

Proof. (1) Put $q = q_1(X)$ and $a = \deg F$, then we have the exact sequence

$$H^0\mathcal{I}_{X\cap F|F}(q + a) \to H^1\mathcal{I}_X(q) \to 0$$

from which $s(X \cap F|F) \leq q_1(X) + \deg F$.
(2) & (3) Put $a = \deg F$. For $1 \leq i \leq r - 1$ we have the exact sequence

$$0 \to H^i\mathcal{I}_X(t) \to H^i\mathcal{I}_{X\cap F}(t) \to H^{i+1}\mathcal{I}_X(t - a) \to 0$$

for every t. Since it results

$$h^i\mathcal{I}_{X\cap F}(t) = 0 \quad \forall t \geq p_i(X \cap F) + 1 \quad \text{and} \quad \forall t \leq q_i(X \cap F) - 1$$

we get $p_i(X) \leq p_i(X \cap F)$ and $q_{i+1}(X) + a \geq q_i(X \cap F)$, then, together with Theorem 5.3(3) and Theorem 4.8(4), we obtain the claim.
(4) For $t = p_r(X)$ we have the injection

$$0 \to H^r\mathcal{I}_X(t) \to H^r\mathcal{I}_{X\cap F}(t)$$

hence $e(X \cap F) \geq p_r(X)$, and by Theorem 4.8(5) we have $e(X \cap F) \geq e(X) + \deg F$, therefore

$$e(X \cap F) \geq \max\{e(X) + \deg F, p_r(X)\}.$$

Moreover, by Theorem 4.8(6) it holds also the opposite inequality, so the claim.

Theorem 5.6. *Let X be a k-Buchsbaum subscheme of dimension r and $k \geq 1$, and let F be a general hypersurface of degree $\geq k$.*

(1) *If $p_i(X \cap F) = p_{i+1}(X) + \deg F$, then $p_i(X) \leq p_{i+1}(X) + \deg F$.*
If $p_i(X \cap F) > p_{i+1}(X) + \deg F$, then $p_i(X) > p_{i+1}(X) + \deg F$.

(2) *If $q_i(X \cap F) = q_{i+1}(X) + \deg F$, then $q_i(X) \geq q_{i+1}(X) + \deg F$.*
If $q_i(X \cap F) < q_{i+1}(X) + \deg F$, then $q_i(X) < q_{i+1}(X) + \deg F$.

Proof. Put $a = \deg F$. Being $a \geq k$, we have the exact sequence

$$0 \to H^i\mathcal{I}_X(t) \to H^i\mathcal{I}_{X\cap F}(t) \to H^{i+1}\mathcal{I}_X(t-a) \to 0$$

for every t.
(1) If $p_i(X \cap F) = p_{i+1}(X) + a$, then we have

$$h^i\mathcal{I}_X(t) = 0 \qquad \forall t \geq p_i(X \cap F) + 1 = p_{i+1}(X) + a + 1$$

and therefore $p_i(X) \leq p_{i+1}(X) + a$.
If $p_i(X \cap F) > p_{i+1}(X) + a$, then we have

$$h^i\mathcal{I}_X(t) = 0 \qquad \forall t \geq p_i(X \cap F) + 1$$

and

$$h^i\mathcal{I}_X(p_i(X \cap F)) = h^i\mathcal{I}_{X\cap F}(p_i(X \cap F)) \neq 0$$

so $p_i(X) = p_i(X \cap F) > p_{i+1}(X) + a$.
(2) If $q_i(X \cap F) = q_{i+1}(X) + a$, then we have

$$h^i\mathcal{I}_X(t) = 0 \qquad \forall t \leq q_i(X \cap F) - 1 = q_{i+1}(X) + a - 1$$

and therefore $q_i(X) \geq q_{i+1}(X) + a$.
If $q_i(X \cap F) < q_{i+1}(X) + a$, then we have

$$h^i\mathcal{I}_X(t) = 0 \qquad \forall t \leq q_i(X \cap F) - 1$$

and

$$h^i\mathcal{I}_X(q_i(X \cap F)) = h^i\mathcal{I}_{X\cap F}(q_i(X \cap F)) \neq 0$$

so $q_i(X) = q_i(X \cap F) < q_{i+1}(X) + a$.

Theorem 5.7. *Let X be a k-Buchsbaum subscheme with $k \geq 1$ and F a hypersurface of degree $a \geq k$ containing no component of X. Then:*
(1) *For the postulation character we have*

$$\gamma_{X\cap F}(t) = \sum_{j=0}^{a-1} \gamma_X(t-j) + \Delta^{r+1}h^1\mathcal{I}_X(t-a)$$

and if $k = a = 1$

$$\gamma_{X\cap F}(t) = \gamma_X(t) + \Delta^{r+1}h^1\mathcal{I}_X(t-1).$$

(2) *For the function R we have*

$$R_{X\cap F}(t) = {}_a\Delta R_X(t) + h^1\mathcal{I}_X(t-a) + (-1)^r h^r\mathcal{I}_X(t).$$

(3) *For the speciality function we have*

$$\sigma_{X\cap F}(t) = \Delta^{r+1}h^r\mathcal{I}_X(t) - \sum_{j=0}^{a-1}\sigma_X(t-j)$$

and if $k = a = 1$

$$\sigma_{X\cap F}(t) = \Delta^{r+1}h^r\mathcal{I}_X(t) - \sigma_X(t).$$

Proof. (1) It follows from Theorem 4.10(1), since for $a \geq k$ we have $\mathcal{K}^1_{X,F}(t) = h^1\mathcal{I}_X(t)$ for every t.
(2) It follows by Theorem 4.10(2), since for $a \geq k$ we have $\mathcal{K}^1_{X,F}(t) = h^1\mathcal{I}_X(t)$ and $\mathcal{K}^r_{X,F}(t) = h^r\mathcal{I}_X(t)$ for every t.
(3) Since $a \geq k$, by Proposition 5.2 we have

$$\begin{aligned} h^r\mathcal{I}_{X\cap F}(t) &= h^r\mathcal{I}_X(t) - {}_a\Delta h^{r+1}\mathcal{I}_X(t) \\ &= h^r\mathcal{I}_X(t) - \sum_{j=0}^{a-1}\Delta h^{r+1}\mathcal{I}_X(t-j) \end{aligned}$$

so, taking the $(r+1)$-th difference, we get the claim.

§6 Linked Subschemes

Definition 6.1. Let X, X' be closed subschemes of $\mathbf{P}^n$ of dimension r, locally Cohen-Macaulay and equidimensional, then X is (algebraically) directly linked to X' by the complete intersection V if and only if it holds $[I(V) : I(X)] = I(X')$ and $[I(V) : I(X')] = I(X)$ (where $[I : J]$ denotes the ideal quotient of the ideals I and J in the ring S), or equivalently if and only if $\mathcal{I}_{X'|V} \simeq \mathcal{H}om_{\mathcal{O}_{\mathbf{P}^n}}(\mathcal{O}_X, \mathcal{O}_V)$ and $\mathcal{I}_{X|V} \simeq \mathcal{H}om_{\mathcal{O}_{\mathbf{P}^n}}(\mathcal{O}_{X'}, \mathcal{O}_V)$ (see [PS] page 277).

Lemma 6.2. *Let X, $X' \subset \mathbf{P}^n$ be closed subschemes of dimension r, locally Cohen-Macaulay and equidimensional, directly linked by the complete intersection V of hypersurfaces of degree $a_1, \ldots, a_{n-r}$. Let $\nu = \sum a_i - n - 1$. Then for every homogeneous polynomial F of degree a we have*

$$\mathcal{K}^i_{X',F}(j) = {}_a\Delta h^{r-i+1}\mathcal{I}_X(\nu - j) + \mathcal{K}^{r-i+1}_{X,F}(\nu - j - a)$$

for $1 \leq i \leq r$ and $j \in \mathbf{Z}$.

Proof. By hypothesis X is linked to X' by the complete intersection V, hence we have the isomorphism of S-graded modules

$$H^i_*\mathcal{I}_{X'} \cong (H^{r-i+1}_*\mathcal{I}_X)^*(-\nu) \quad \text{for} \quad 1 \leq i \leq r$$

(see [M] Theorem 4.3.1) where $\nu = \sum a_i - n - 1$, therefore it results

$$h^i \mathcal{I}_{X'}(j) = h^{r-i+1} \mathcal{I}_X(\nu - j) \qquad \forall j \in \mathbf{Z}.$$

Moreover for all $F \in S_a$ the multiplication map

$$\hat{\phi}^i_{F,j} \colon H^i \mathcal{I}_{X'}(j) \to H^i \mathcal{I}_{X'}(j + a)$$

is the dual of

$$\phi^{r-i+1}_{F,\nu-j-a} \colon H^{r-i+1} \mathcal{I}_X(\nu - j - a) \to H^{r-i+1} \mathcal{I}_X(\nu - j)$$

so

$$\operatorname{rk} \hat{\phi}^i_{F,j} = \operatorname{rk} \phi^{r-i+1}_{F,\nu-j-a} = h^{r-i+1} \mathcal{I}_X(\nu - j - a) - \mathcal{K}^{r-i+1}_{X,F}(\nu - j - a)$$

and therefore

$$\begin{aligned} \mathcal{K}^i_{X',F}(j) &= h^i \mathcal{I}_{X'}(j) - \operatorname{rk} \hat{\phi}^i_{F,j} \\ &= h^{r-i+1} \mathcal{I}_X(\nu - j) - h^{r-i+1} \mathcal{I}_X(\nu - j - a) + \mathcal{K}^{r-i+1}_{X,F}(\nu - j - a) \\ &= {}_a\Delta h^{r-i+1} \mathcal{I}_X(\nu - j) + \mathcal{K}^{r-i+1}_{X,F}(\nu - j - a). \end{aligned}$$

Theorem 6.3. *Let X, $X' \subset \mathbf{P}^n$ be closed subschemes of dimension r, locally Cohen-Macaulay and equidimensional, directly linked by the complete intersection V of hypersurfaces of degree $a_1, \ldots, a_{n-r}$. Then*

$$\Delta^{r+1} H(V,t) = \Delta^{r+1} H(X,t) + \Delta^{r+1} H(X', N - t) + \Delta^{r+1} R_X(t)$$

where $N = \max\{t \mid \Delta^{r+1} H(V,t) \neq 0\} = \sum a_i - n + r$.

Proof. By induction on the dimension r. Assume $\dim X = \dim X' = \dim V = 1$. Let H be a general hyperplane, then the 0-dimensional scheme $X \cap H$ is linked to $X' \cap H$ by the complete intersection $V \cap H$ (see [M] Prop. 4.2.13), so we have

$$\Delta H(V \cap H, t) = \Delta H(X \cap H, t) + \Delta H(X' \cap H, N - t)$$

where $N = \sum_{i=1}^{n-1} a_i - n + 1$ (see [DGO] Theorem 3). By Theorem 4.2 it results (since V is aCM)

$$\begin{aligned} H(V \cap H, t) &= \Delta H(V, t) \\ H(X \cap H, t) &= \Delta H(X, t) - \mathcal{K}^1_{X,H}(t - 1) \\ H(X' \cap H, N - t) &= \Delta H(X', N - t) - \mathcal{K}^1_{X',H}(N - t - 1) \end{aligned}$$

so we have

$$\Delta^2 H(V,t) = \Delta^2 H(X,t) - \Delta \mathcal{K}^1_{X,H}(t-1) + \Delta^2 H(X', N - t) - \Delta \mathcal{K}^1_{X',H}(N - t - 1).$$

Put $\nu = \sum_{i=1}^{n-1} a_i - n - 1$, then using Lemma 6.2 we get (since $\nu - N = -2$)

$$\begin{aligned}
-\Delta\mathcal{K}^1_{X',H}(N-t-1) &= \mathcal{K}^1_{X',H}(N-t-2) - \mathcal{K}^1_{X',H}(N-t-1) \\
&= \Delta h^1\mathcal{I}_X(\nu - N + t + 2) + \mathcal{K}^1_{X,H}(\nu - N + t + 2 - 1) - \\
&\quad - \Delta h^1\mathcal{I}_X(\nu - N + t + 1) - \mathcal{K}^1_{X,H}(\nu - N + t + 1 - 1) \\
&= \Delta h^1\mathcal{I}_X(t) - \Delta h^1\mathcal{I}_X(t-1) + \mathcal{K}^1_{X,H}(t-1) - \mathcal{K}^1_{X,H}(t-2) \\
&= \Delta^2 h^1\mathcal{I}_X(t) + \Delta\mathcal{K}^1_{X,H}(t-1)
\end{aligned}$$

and therefore we prove the claim for 1-dimensional subschemes

$$\Delta^2 H(V,t) = \Delta^2 H(X,t) + \Delta^2 H(X', N-t) + \Delta^2 h^1\mathcal{I}_X(t).$$

Now, let assume true the statement for subschemes of dimension $r-1$. Let X, X' be linked subschemes of dimension r. Let H be a general hyperplane, then $X \cap H$ is linked to $X' \cap H$ by the complete intersection $V \cap H$ of hypersurfaces of degree $a_1, \ldots, a_{n-r}, 1$ (see [M] Prop. 4.2.13). By induction hypothesis we have

$$\Delta^r H(V \cap H, t) = \Delta^r H(X \cap H, t) + \Delta^r H(X' \cap H, N - t) + \Delta^r R_{X\cap H}(t)$$

where $R_{X\cap H}(t) = \sum_{i=1}^{r-1}(-1)^{i+1}h^i\mathcal{I}_{X\cap H}(t)$ and $N = \sum_{i=1}^{n-r} a_i + 1 - n + r - 1 = \sum a_i - n + r$. By Theorem 4.2 it results (since V is aCM)

$$\begin{aligned}
H(V \cap H, t) &= \Delta H(V,t) \\
H(X \cap H, t) &= \Delta H(X,t) - \mathcal{K}^1_{X,H}(t-1) \\
H(X' \cap H, N-t) &= \Delta H(X', N-t) - \mathcal{K}^1_{X',H}(N-t-1).
\end{aligned}$$

Put $\nu = \sum_{i=1}^{n-r} a_i - n - 1$, then by Lemma 6.2 we have

$$\mathcal{K}^1_{X',H}(N-t-1) = \Delta h^r\mathcal{I}_X(t-r) + \mathcal{K}^r_{X,H}(t-r-1)$$

(since $\nu - N + t + 1 = t - r$), therefore

$$\Delta^r\mathcal{K}^1_{X',H}(N-t-1) = (-1)^r\Delta^{r+1}h^r\mathcal{I}_X(t) + (-1)^r\Delta^r\mathcal{K}^r_{X,H}(t-1).$$

By Theorem 4.10(2) we have

$$R_{X\cap H}(t) = \Delta R_X(t) + (-1)^r\Delta h^r\mathcal{I}_X(t) + \mathcal{K}^1_{X,H}(t-1) + (-1)^r\mathcal{K}^r_{X,H}(t-1)$$

so we obtain

$$\begin{aligned}
\Delta^r R_{X\cap H}(t) =& \Delta^{r+1}R_X(t) + (-1)^r\Delta^{r+1}h^r\mathcal{I}_X(t) + \\
&+ \Delta^r\mathcal{K}^1_{X,H}(t-1) + (-1)^r\Delta^r\mathcal{K}^r_{X,H}(t-1)
\end{aligned}$$

and therefore

$$\begin{aligned}
\Delta^{r+1}H(V,t) &= \Delta^{r+1}H(X,t) - \Delta^r\mathcal{K}^1_{X,H}(t-1)+\\
&+\Delta^{r+1}H(X',N-t) - \Delta^r\mathcal{K}^1_{X',H}(N-t-1)+\\
&+\Delta^{r+1}R_X(t) + (-1)^r\Delta^{r+1}h^r\mathcal{I}_X(t)+\\
&+\Delta^r\mathcal{K}^1_{X,H}(t-1) + (-1)^r\Delta^r\mathcal{K}^r_{X,H}(t-1)\\
&= \Delta^{r+1}H(X,t) - \Delta^r\mathcal{K}^1_{X,H}(t-1)+\\
&+\Delta^{r+1}H(X',N-t) - (-1)^r\Delta^{r+1}h^r\mathcal{I}_X(t) - (-1)^r\Delta^r\mathcal{K}^r_{X,H}(t-1)+\\
&+\Delta^{r+1}R_X(t) + (-1)^r\Delta^{r+1}h^r\mathcal{I}_X(t)+\\
&+\Delta^r\mathcal{K}^1_{X,H}(t-1) + (-1)^r\Delta^r\mathcal{K}^r_{X,H}(t-1)\\
&= \Delta^{r+1}H(X,t) + \Delta^{r+1}H(X',N-t) + \Delta^{r+1}R_X(t).
\end{aligned}$$

Remark 6.4. The theorem above generalizes an analogous result for linked arithmetically Buchsabum curves proved in [GM] Proposition 2.7.

Remark 6.5. Notice that the Hilbert function of a complete intersection depends only on the degree of the hypersurfaces and is completely computable using the Koszul complex. For instance, if $V = F \cap G$ is a two-codimensional complete intersection in $\mathbf{P}^n$ with $\deg F = a$ and $\deg G = b$, then we have the following Koszul resolution for V

$$0 \to \mathcal{O}_{\mathbf{P}^n}(-a-b) \to \mathcal{O}_{\mathbf{P}^n}(-a) \oplus \mathcal{O}_{\mathbf{P}^n}(-b) \to \mathcal{O}_{\mathbf{P}^n} \to \mathcal{O}_V \to 0$$

therefore

$$H(V,t) = \binom{t+n}{n} - \binom{t-a+n}{n} - \binom{t-b+n}{n} + \binom{t-a-b+n}{n}$$

and

$$\Delta^{n-1}H(V,t) = \binom{t+1}{1} - \binom{t-a+1}{1} - \binom{t-b+1}{1} + \binom{t-a-b+1}{1}.$$

The case of higher codimension is analogous.

References

[DGO] E. Davis, A.V. Geramita, F. Orecchia, *Gorestein algebras and the Cayley-Bacharach theorem*, Proc. Amer. Math. Soc. **93** (1985), 593–597.

[GM] A.V. Geramita, J.C. Migliore, *On the ideal of an arithmetically Buchsbaum curve*, J. Pure App. Alg. **54** (1988), 215–247.

[H] R. Hartshorne, *Algebraic Geometry*, Graduate Texts in Mathematics 52, Springer-Verlag, 1977.

[HMV] L.T. Hoa, R.M. Miró-Roig, W. Vogel, *On numerical invariants of locally Cohen-Macaulay schemes in* $\mathbf{P}^n$, Hiroshima Math. J. **24** (1994), 299–316.

[M] J.C. Migliore, *An Introduction to Deficiency Modules and Liaison Theory for Subschemes of Projective Space*, Lecture Notes Series 24, Seoul National University, 1994.

[MDP] M. Martin-Deschamps, D. Perrin, *Sur la Classification des Courbes Gauches*, Astérisque 184–185, Soc. Math. de France, 1990.

[PS] C. Peskine, L. Szpiro, *Liaison des variétés algébriques*, Invent. Math. **26** (1974), 271–302.

A Deformation of Projective Schemes

Freddy Van Oystaeyen
Dept. of Mathematics and Computer Science
University of Antwerp, UIA
B.2610-Antwerp, Belgium
e-mail : voyst@uia.ua.ac.be

Dedicated to Mario Fiorentini, a commutative friend

0 Introduction

Recently there is a growing interest in the construction of an algebraic geometry for a class of noncommutative algebras said to be "quantized" algebras. In one approach, probably the most abstract one, those noncommutative algebras are viewed as a ring of functions on a kind "variety" that remains **virtual** throughout the whole theory. Then the methods used can only be of categorical nature (derived categories,...) and focus eventually on cohomological methods (quantum cohomology). There is another approach dealing with more ring theoretical objects, i.e. starting from a nice class of algebras, called **schematic algebras**, it is possible to define a space, a Zariski topology, points, lines etc..., cf. [VOW1], [AZ], [ATV]. then it becomes possible to develop a scheme theory, completely extending the theory of schemes in classical algebraic geometry.

This scheme theory, including Čhech cohomology calculations for sheaf cohomology, Serre duality, Serre's global section theorems, local-global results on affine coverings,... may be combined with more arithmetical aspects like : divisors (e.g. on curves) and the relation to valuation theory, regularity conditions in terms of homological algebra... . All of this may be applied to a class of algebras containing most if not all algebras recently receiving extra attention in connection with the theory of quantum-groups and quantum spaces e.g. Weyl algebras, rings of differential operators on smooth varieties, enveloping algebras of Lie algebras, Sklyanin algebras, quantum matrix rings, quantized Weyl algebras, Witten algebras and generalized gauge algebras, quantized enveloping algebras like the quantum group $U_q(sl_2)$, rings of microlocal fractions of the forementioned e.g. Kashiwara's rings of pseudo-differential operators, prime algebras satisfying polynomial identities,..., commutative algebras.

In this paper attention is restricted to rings R with a Zariakian filtration FR such that the associated graded ring $G(R)$ is a positive graded affine K-algebra, $G(R) = K \oplus G(R)_1 \oplus G(R)_2 \oplus, \ldots$, which is **commutative** and generated by $G(R)_1$ as a K-algebra. We consider the scheme on Proj$G(R) = Y$ given by the usual structure sheaf $\underline{O}_Y$ of $G(R)$. Then we construct two deformations of $(Y, \underline{O}_Y)$ obtaining a first one by using classical localizations at Ore sets, and another one by using the rings of quantum sections which can be obtained from the first one by locally completing. In fact the sheaf of quantum sections of R appears as a degeneration of a formal scheme. The "geometric" properties of the deformed schemes are controlled by the algebraic theory relating "filtered properties" of filtered R-modules to "graded properties" of the associated graded $G(R)$-modules via the graded properties of the Rees modules over the Rees ring (Blow-up ring).

For full detail on Zariskian filtrations and the basic techniques concerning the use of graded ring theory on the level of the Rees ring we refer to [LVO].

The scheme theory for schematic algebras has been introduced in [VOW1], [VOW2]. Quantum sections stem from [VOR], examples and specific applications appeared in [VOW3], [LEV0]. Microstructure sheaves were studied further in [RVO]. A good survey of structure sheaves on almost commutative algebras was given in [LH], however the author ignored the fact that those structure sheaves are indeed given by localizations everywhere locally. In Section 2 of this paper we present the theory completely in terms of real localizations. Not only does this provide a complete analogue for the commutative case, it also confirms that we are really dealing with a "deformation" that respects localization; this serves as a basis for the structure sheaves defined in terms of localizations, microlocalizations and quantum sections.

That the "deformed geometry" of R can to some extent be studied by the methods we propose is suggested in special cases, in particular for rings of differential oprators, by [BB], [G], [LEVO]. The methods of noncommutative geometry, now available for the surprizingly wide class of schematic algebras, show that the basic ingredients of classical algebraic geometry are all present in the noncommutative theory. We hope to establish in this paper that the case of filtered rings with commutative associated graded ring **really** features a "non-commutative geometry" that can be understood to be just a natural deformation of the projective geometry at the associared graded ring level.

1 Preliminaries and Notation

All rings are associative with unit. Module will mean left module but ideal will mean two-sided ideal.

Throughout, R is a K-algebra over a field K and R has a positive filtration FR given by $F_0R = K \subset F_1R \subset \ldots F_nR \subset \ldots$ such that $\cup_n F_nR = R$ and $G(R) = \oplus_{n \in \mathbb{N}} F_nR/F_{n-1}R$ is a commutative domain such that $G(R)$ is generated by $G(R)_1$ as a K-algebra. The filtration FR is Zariskian exactly then when $G(R)$ is Noetherian i.e. when $G(R)_1$ is a finite dimensional κ-vectorspace. In this paper we always assume the foregoing properties of $G(R)$, i.e. we assume that $G(R)$ is an affine K-algebra, positively graded; we put $Y = \text{Proj}G(R)$, $\underline{O}_Y$ the projective structure sheaf of $G(R)$ and $\underline{O}_Y^g$ the graded structure sheaf obtained by associating to a Zariskian open basis set $Y(\overline{f})$, $\overline{f}$ homogeneous in $G(R)$, the graded ring $< \overline{f} >^{-1} G(R)$, where $< \overline{f} >$ stands for the multiplicatively closed set $\{1, \overline{f}, \overline{f}^2, \ldots\}$.

We put $A = \widetilde{R} = \oplus_{n \in \mathbb{N}} F_nR$ for the Rees ring of R with respect to FR. The element $1 \in F_0R = K$ viewed as an element of F_1R defines an element $T \in A_1$ which is a central regular element of A and moreover homogeneous of degree 1. The properties : $A/TA \simeq G(R)$, $A/(1-T)A = R$, state that we may view R as a "deformation" of $G(R)$ in a sense that can be made precise, in terms of the T-adic completion of A (that will contain $K[[T]]$), but may remain intuitive here.

A filtration on an R-module M will be given by an ascending chain of additive subgroups F_nM such that : $\cup_n F_nM = M$; $m, n \in \mathbb{Z}$, $F_nRF_mM \subset F_{n+m}M$. The associated graded module $G(M)$, $G(M) = \oplus_{n \in \mathbb{N}} F_nM/F_{n-1}M$ is a graded $G(R)$-module and the Rees module $\widetilde{M} = \oplus_{n \in \mathbb{Z}} F_nM$ is a graded $\widetilde{R}$-module.

For a filtered R-module M we have $\widetilde{M}/(1-T)\widetilde{M} = M$ and $\widetilde{M}/T\widetilde{M} = G(M)$. The category R-filt consists of filtered R-modules and R-linear morphisms preserving filtrations. A filtered morphism is **strict** if $f : M \to N$ is such that $F_nN \cap f(M) = f(F_nM)$ for every $n \in \mathbb{Z}$. The functor $\sim$ given by $\sim$: R-filt $\to \widetilde{R}$-gr defines an equivalence of R-filt and the full subcategory $\mathcal{F}_T$ in the category of graded $\widetilde{R}$-modules defined as the class of T-torsion free graded $\widetilde{R}$-modules. If $f : M \to \mathbb{N}$ is a morphism in R-filt then f is strict if and only if $\widetilde{f} : \widetilde{M} \to \widetilde{N}$ has T-torsionfree cokernel.

Recall that FM is called a **good filtration** if there exist generators $m_1, \ldots, m_t$ for M such that for every $n \in \mathbb{Z}$ we have : $F_nM = \sum_{i=1}^{t} F_{n=d_i}Rm_i$ for some $d_1, \ldots, d_t \in \mathbb{Z}$. It is well-known that FM is good exactly then when $\widetilde{M}$ is finitely generated !

Unless otherwise stated all filtrations considered will be **separated** i.e. $\cap_{n \in \mathbb{Z}} F_nM = 0$. For FR we already assumed this property because $F_mR = 0$ when $m < 0$. If $r \in R$ then there is a unique $n \in \mathbb{N}$ such that $r \in F_nR - F_{n-1}R$ and we define the principal symbol map $\sigma : R \to G(R)$ by $\sigma(r) = r\text{-mod } F_{n-1}R$, i.e. $\sigma(r) \in G(R)_n$. Since we assume $G(R)$ to be a **domain**, it follows that σ **is a multiplicative map**, meaning that $\sigma(xy) = \sigma(x)\sigma(y)$ for nonzero x and y. In a similar way we may define $\sigma_M : M \to G(M)$ but this is in general just a set map.

For basic facts concerning filtrations and associated graded structures we refer to [LVO]; we shall use results from this book freely in the sequel.

A next section of preliminaries is concerned with Ore sets of a noncommutative ring and the related schematic property in terms of coverings.

A multiplicatively closed set S of a noncommutative ring R, say $0 \notin S$ and $1 \in S$, is said to be a **left Ore set** if

i) Given $r \in R$, $s \in S$ there are $r' \in R$, $s' \in S$, such that $s'r = r's$.

ii) If $rs = 0$ for $r \in R, s \in S$ then $s'r = 0$ for some $s' \in S$.

An **Ore set** is a left Ore set that is at the same time a right Ore set. To a (left) Ore set there corresponds a left **Gabriel filter** on R gven by $\mathcal{L}(S) = \{L$ left(right) ideal of $R, L \cap S \neq \emptyset\}$. For a (left) Ore set one can form the (left) ring of fractions $S^{-1}R$; a fraction is then an equivalence class of pairs $(s,r) \in S \times R$ with respect to $(s,r) \sim (s',r')$ if and only if $s''(s'r - sr') = 0$ for some $s'' \in S$. For an R-module M, $Q_S(M) = S^{-1}R \otimes_R M$ is the S-localization of M. This defines an exact localization functor : $Q_S : R\text{-mod} \to R\text{-mod}$ called the **localization functor at** S. For $M \in R$-mod, the S-torsion part $\kappa_S M$ is defined to be the R-submodule $\kappa_S M = \{m \in M, sm = 0 \text{ for some } s \in S\}$. Clearly $M/\kappa_S M$ is S-torsionbfree and $Q_S(M)\kappa_S M) = Q_S(M)$. We say that κ_S is the **kernel functor** corresponding to S. To any ideal $\overline{I}$ of a commutative ring, say $G(R)$; we may associate a Gabriel filter $\mathcal{L}(\overline{I}) = \{\overline{L}, \overline{L} \supset (\overline{I})^n, n \in \mathbb{N}\}$ a kernel functor $\kappa_{\overline{I}}$ defined by $\kappa_{\overline{I}}\overline{M} = \{\overline{m} \in \overline{M}; \overline{L}, \overline{m} = 0 \text{ for some } \overline{I} \in \mathcal{L}(\overline{I})\}$. General localization theory allows to define the localizations at $\overline{I}$ as $Q_{\overline{I}}(\overline{M}) = \varinjlim_{n \in \mathbb{N}} \mathrm{Hom}_{G(R)}(\overline{I}^n, \overline{M}/\kappa_{\overline{I}}R)$ and we obtain a localization functor $Q_{\overline{I}} : G(R)\text{-mod} \to G(R)\text{-mod}$ which need not be exact in general. For generalities concerning localization theory, cf. [Go], [Gold], ...

A noncommutative ring R is said to be **affine schematic** if there exist finitely many Ore sets $S_1, \ldots, S_n$ such that for every choice of $s_i \in S_i, i = 1, \ldots, n$ we have $\sum_{i=1}^n Rs_i = R$.

Now look at a positively graded K-algebra A, $A = K \oplus A_1 \oplus A_2 \oplus \ldots$. The positive part $A_+ = \oplus_{n>0} A_n$ is called the **irrelevant maximal ideal of** A. The ideal A_+ defines a (graded) Gabriel filter $\mathcal{L}_+$ having powers of A_+ for a filter basis; the corresponding κ_+ may be defined by $\kappa_+ M = \{m \in M, A_+^n m = 0,$ some $n \in \mathbb{N}\}$. The corresponding localization functor $Q_{\kappa_+} : A\text{-gr} \to A\text{-gr}$ is very usefull e.g. [NVO], [VO1]. The full subcategory of $Q_{\kappa_+}(A)$-gr consisting of $Q_{\kappa_+}(M)$ for graded A-modules M is called the **quotient category** of A-gr with respect to κ_+, denoted by (A, κ_+)-gr. For homogeneous Ore sets we have $(A, K_S)\text{-gr} = (S^{-1}A)\text{-gr}$. However Q_{κ_+} is almost never exact ! Note that for a commutative A, the functor Q_{κ_+} is **nothing but the global section**

functor for Proj$A$: $Q_{\kappa_+}(M) = \Gamma_*(M) \cong \oplus_{n\in\mathbb{Z}}\Gamma(\text{Proj}A, \widetilde{M(n)})$, the latter in the notation of [HART].

A graded algebra A is said to be **schematic** if there is a finite set of homogeneous Ore sets $S_1, \ldots, S_n$ of A that are nontrivial in the sense that $S_i \cap A_+ \neq \emptyset$ for $i = 1, \ldots, n$ such that $\cap_i \mathcal{L}(S_i) = \mathcal{L}_+$, or equivalently : $\cap_i \kappa_{S_i}(M) = \kappa_+ M$ for all graded A-modules M, or equivalently : for every choice of $s_i \in S_i$ there exists an $m \in \mathbb{N}$ such that

$$(*) \qquad (A_+)^m \subset \sum_i A s_i$$

In case A is commutative and $S_i =< f_i >$ the multiplicative set generated by an $f_i \in A$ then the schematic condition (*) just means that $Y = \text{Proj}A = \cup_i Y(f_i)$, i.e. the **covering property**.

The interesting fact for $A = \widetilde{R}$ for an R of the type we are considering here, is that A **is necessarily schematic** (cf. [VOW]), moreover if $G(R)$ is normal, i.e. integrally closed, then R **is weakly affine schematic** (cf. loc. cit). The Ore sets used in A and R are lifted from multiplicative sets in $G(R)$ in a special way as described hereafter.

Given R as before, $\overline{S}$ an Ore set of $G(R)$ (hence this is just a multiplicative set because $G(R)$ is assumed to be commutative). Consider $S = \{r \in R, \sigma(r) \in \overline{S}\}$; since σ is multiplicative here, S is clearly multiplicatively closed and $0 \notin S$, $1 \in S$. It is known (cf. [AVV]) that S is an Ore set (left and right !) and it is saturated in the terminology of [AVV]. If $s \in S$ is such that $\sigma(s) \in G(R)_n$ then we look at $\widetilde{s} = sT^n \in A$ and define $\widetilde{S} = \{\widetilde{s}, s \in S\}$. It is not hard to see that $\widetilde{S}$ is an Ore set of A consisting of homogeneous elements. Let $\pi_n : A \twoheadrightarrow A/T^nA$ be the canonical epimorphisms; then $\pi_1 : A \twoheadrightarrow G(R)$ maps $\widetilde{S}$ exactly to $\overline{S}$.

The localized ring $S^{-1}R$ has a quotient filtration such that $G(S^{-1}R) = \overline{S}^{-1}G(R)$ and $(S^{-1}R)^\sim = \widetilde{S}^{-1}A$. The **microlocalization of** R at S may be defined by first defining the microlocalization of A at $\widetilde{S}$ by : $\varprojlim_n \pi_n(\widetilde{S})^{-1}A/T^nA$, say $Q^\mu_{\widetilde{S}}(\widetilde{A})$, then putting $Q^\mu_S(R) = Q^\mu_{\widetilde{S}}(\widetilde{R})/TQ^\mu_{\widetilde{S}}(\widetilde{R})$.

From [AVV] we recall that $G(Q^\mu_S(R)) = \overline{S}^{-1}G(R)$ and $(Q^\mu_S(R))^\sim = Q^\mu_{\widetilde{S}}(A)$. The ring of **quantum sections** at $\overline{S}$ is then defined as $F_0Q^\mu_S(R) = Q^\mu_{\widetilde{S}}(A)_0$; it has an induced negative filtration $F_-Q^\mu_S(R)$ and **it is Zariskian** with respect to this. Note that $F_0Q^\mu_S(R) \supset F_0(S^{-1}R)$ and $F_0Q^\mu_S(R) = F_0(S^{-1}R)^\wedge$ where completion is with respect to $F_-(S^{-1}R)$. The advantage of working with $F_0Q^\mu_S(R)$ is that $F_-(S^{-1}R)$ need not be Zariskian on $F_0(S^{-1}R)$. We denote $Q^q_S(R) = F_0Q^\mu_S(R)$. In a similar way, we introduce $Q^q_S(M)$.

Let us recall, for the reader's convenience, some more important properties stemming from [VOR], [AVV]; these results have been proved in different generality in loc. cit. but all of them apply in the case we are considering here.

For a saturated Ore set S we have $Q^{\mu}_{S}(R) = (S^{-1}R)^{\wedge}$ where $\wedge$ stands for completion in the quotient filtration. If FM is good, i.e. $\widetilde{M}$ is finitely generated, then we have $Q^{\mu}_{\widetilde{S}}(\widetilde{M}) = Q^{\mu}_{\widetilde{S}}(\widetilde{R}) \otimes_{\widetilde{R}} \widetilde{M}$ and $Q^{\mu}_{\widetilde{S}}(\widetilde{R})$ is a flat right $\widetilde{R}$-module.

The covering property forced by the schematic property is in fact a covering by **affine** sets. The graded ring $\overline{S}^{-1}G(R)$ is strongly graded for any Ore set $\overline{S}$ of $G(R)$ because $G(R)$ is generated by $G(R)_1$ over $G(R)_0 = K$. Hence $\overline{S}^{-1}G(R)$-gr $\cong (\overline{S}^{-1}G(R))_0$-mod. In particular for $\overline{S} =< \overline{f} >$ we have that $< \overline{f} >^{-1} G(R)$-gr $= (< \overline{f} >^{-1} G(R))_0$-mod (proving $Y(\overline{f}) \cong \mathrm{Spec}\underline{O}_Y(Y(\overline{f}))$ as schemes). It follows from this that $Q^{\mu}_{\widetilde{S}}(\overline{M}) = Q^{\mu}_{\widetilde{S}}(\widetilde{R}) \otimes_{Q^{q}_{\widetilde{S}}(\widetilde{R})} Q^{q}_{\widetilde{S}}(\widetilde{M})$.

The foregoing surveys most basic facts we need in the sequel, for full detail the reader may consult [VOR], [AVV] and [NVO], [LVO] as well as further references given there.

2 The Saturated Deformation of Projective Schemes

The first deformation of $\underline{O}_Y, Y = \mathrm{Proj}G(R)$, we want to present is obtained by using only classical localizations in the construction of the structure sheaves. The reason why we will go on to the quantum-sections later on is that for saturated Ore sets S and T, the localization functor Q_S and Q_T do not commute (whereas $Q_{\sigma(S)}$ and $Q_{\sigma(T)}$ do commute !) hence $Q_S Q_T(R)$ need not be a ring.

Consider $Y = \mathrm{Proj}G(R)$ as before. For the graded ideals $\overline{I}$ of $G(R)$ the $Y(\overline{I}) = \{P \in Y, P \not\supset \overline{I}\}$ are the Zariski open sets in Y. The set $\mathcal{B} = \{Y(\overline{f}), \overline{f} \in \mathcal{L}G(R)\}$ is a basis for the Zariski topology. The graded structure sheaf of $G(R)$, say $\underline{O}^{g}_{Y}$ is obtained by associating $G(R)[\overline{f}^{-1}]$ to $Y(\overline{f})$. The stalk at $P \in Y$, denoted by $\underline{O}^{g}_{Y,P}$ is the graded ring $\overline{S}_P^{-1}G(R)$ where $\overline{S}_P = \mathcal{L}(G(P) - P)$. Associating $G(R)[\overline{f}^{-1}]_0$ to $Y(\overline{f})$ defines the structure sheaf $\underline{O}_Y$ and we write $G(R)_{(\overline{f})}$ for its sections over $Y(\overline{f})$, $G(R)_{(P)}$ for the stalk at $P \in Y$, or respectively : $T(Y(\overline{f}), \underline{O}_Y) = G(R)_{(\overline{f})}, \underline{O}_{Y,P} = G(R)_{(P)}$.

As before we put $A = \widetilde{R}$ and let $\pi_n : A \to A/T^n A$ be the canonical epimorphism, $\pi_0 : A \to G(R) = A/TA$.

If $\widetilde{f}$ represents $\overline{f}$ then we write $S(\widetilde{f})$ for the homogeneous multiplicative set in A obtained as $h(\pi_0^{-1}S(\overline{f}))$ where $S(\overline{f}) = \{1, \overline{f}, \overline{f}^2, \ldots\}$. The image of $S(\widetilde{f})$ in R is obtained as $S(\widetilde{f})\mathrm{mod}(1-T)A$ and we denote it by $< f >$ where $f = \widetilde{f}$-mod $(1-T)A$. Consequently we have $\sigma(f) = \overline{f}$. It is clear from earlier observations that $< f >$ is a saturated left and right Ore set. Note that $Y(\overline{f}) \cap Y(\overline{g}) = Y(\overline{fg})$ but $< fg >$ is only contained in the saturated Ore set generated by $\{f, g\}$

in R, the latter is just $< f > \vee < g >$. The localizations at $< fg >$ and $< f > \vee < g >$ may be different (but the microlocalizations will coincide and that is one good reason to consider them later on). For $P \in Y$ we have $\overline{S}_P = \cup\{S(\overline{f}), \overline{f}$ homogeneous in $G(R)$ such that $\overline{f} \notin P\}$ and we obtain $S_P = \vee\{< f >, \overline{f} \notin P\}$. For a graded ideal $\overline{I}$ of $G(R)$ the localization corresponding to $Y(\overline{I})$ is given by the Gabriel filter $\mathcal{L}(\overline{I})$ having powers of $\overline{I}$ as a filterbasis. In case $\overline{I} = G(R)\overline{f}_1 + \ldots + \overline{f}_d$ then $\mathcal{L}(\overline{I}) = \cap_{i=1}^d \mathcal{L}(\overline{f}_j)$ corresponds to the kernel functor (torsion theory) : $\kappa_{\overline{I}} = \Lambda_{j=1}^d \kappa_{S(\overline{f}_j)}$ where the $\overline{f}_j$ may be taken to be homogeneous.

Then $\wedge_{j=1}^d \kappa_{<f_j>}$ is a (localization) kernel functor on R-mod which we denote by $\kappa_{<I>}$, where $\mathcal{L} < I >= \cap_{j=1}^d \mathcal{L} < f_j >$. Hence a left ideal L of R is in $\mathcal{L} < I >$ if and only if $L \supset \sum_{j=1}^d Rs_j$ with $s_j \in < f_j >$.

Similarly S_P corresponds to a kernel functor $\kappa_{<P>} = \vee\{\kappa_{<f>}, \overline{f} \notin P\}$. The latter being the supremum of perfect torsion theories it does itself define a perfect torsion theory (cf. [VO]), meaning that the corresponding localization functor $Q_P : R$-mod $\to R$-mod is exact and commutes with direct sums.

2.1 Lemma

With notation as before : $\kappa_{<f>} = \wedge\{\kappa_{<P>}, \overline{f} \notin P\}$.

Proof For P such that $\overline{f} \notin P$ we have $\kappa_{<f>} < \kappa_P$. Conversely if $L \in \mathcal{L}(\wedge\{\kappa_{<P>}, \overline{f} \notin P\})$ then there is a $z_P \in L$ such that $\sigma(z_p) \notin P$ for every P such that $\overline{f} \notin P$. Since $\sigma(z_p)$ is not a zerodivisor in $G(R)$ it follows that $G(R)\sigma(z_p) = \sigma(Rz_p)$. Put $\overline{J} = \sum_p G(R)\sigma(z_p)$. If some prime ideal Q of $G(R)$ contains $\overline{J}$ then Q must contain $\overline{f}$, hence $\overline{f} \in \text{rad}\overline{J}$ or $\overline{f}^m \in \overline{J}$. Write $\overline{f}^m = \overline{g}_{P_1} + \ldots + \overline{g}_{P_t}$ with $\overline{g}_{P_i} \in G(R)\sigma(z_{p_i})$ being all of the of the same degree $\mu = m\text{deg}\overline{f}$. Let $q_{p_i} \in Rz_{P_i}$ represent $\overline{g}_{P_i}$, $i = 1, \ldots, t$, i.e. $g_{P_i} \in F_\mu R - F_{\mu-1}R$. We may assume that the expression of $\overline{f}^m$ has minimal length. If then $t_0 < t$ is such that $\sum_{i=1}^{t_0} g_{P_i} \in F_\mu R - F_{\mu-1}R$ but $g_{P_{t_0+1}} + \sum_{i=1}^{t_0} g_{P_i} \in F_{\mu-1}R$ then $\sigma(\sum_{i=1}^{t_0} g_{P_i}) = \sum_{i=1}^{t_0} \sigma(g_{P_i})$ and $\sigma(g_{P_{t_0+1}}) = \sigma(-\sum_{i=1}^{t_0} g_{P_i})$ yields $\sigma(\sum_{i=1}^{t_0} g_{P_i}) = \sum_{i=1}^{t_0} \sigma(g_{P_i}) = -\sigma(g_{P_{t_0+1}})$, hence $\sum_{i=0}^{t_0+1} \overline{g}_{P_i} = 0$ contradicting the minimality assumption on the expression $\overline{f}^m = \sum_{i=1}^t \overline{g}_{P_i}$. Consequently, $s = \sum_{i=1}^t g_{P_i}$ is in $F_\mu R - F_{\mu-1}R$ and $\sigma(s) = \sigma(\sum_{i=1}^t g_{P_i}) = \sum_{i=1}^t \overline{g}_{P_i} = f^m$ follows then. This yields $s \in < f >$ because the multiplicative set $< f >$ is saturated. Then $s \in L$ entails $L \in \mathcal{L} < f >$, as desired. □

2.2 Proposition

With notation as before : $\wedge\{\kappa_{<P>}, \overline{I} \not\subset P\} = \wedge_{j=1}^d \kappa_{<f_j>} = \kappa_{<I>}$, where $\overline{I} = G(R)\overline{f}_1 + \ldots + G(R)\overline{f}_d$.

Proof First observe :

$$\wedge_{i=1}^{d}(\wedge\{\kappa_{<P>}, \overline{f}_i \notin P\}) = \wedge\{\kappa_{<P>}, \overline{I} \not\subset P\}.$$

Then apply the lemma and the claim follows. □

The foregoing reduces all localization theory we need to $\wedge$ and $\vee$ of Ore set localizations (in fact saturated ones). Therefore the following result may be summarized as stating that the Zariskian property is preserved under saturated localization. Note that in general, as we observed earlier $S^{-1}R$ need not be Zariskian filtered for an arbitrary Ore set S.

2.3 Proposition

Let R be a filtered ring such that $A = \widetilde{R}$ is Noetherian, let S be a saturated Ore set of R then the quotient filtration of $S^{-1}R$ is a Zariskian filtration.

Proof The definition of $FS^{-1}R$ implies $(S^{-1}R)^{\sim} = \widetilde{S}^{-1}A = \widetilde{B}$ and this is a Noetherian ring. The Zariskian property follows if we establish : $F_{-1}(S^{-1}R) \subset J(F_0S^{-1}R)$ or equivalently $J \in J^g(\widetilde{B})$. By definition, the graded Jacobson radical of $\widetilde{B}$ is the largest graded ideal such that its intersection with $\widetilde{B}_0$ is in $J(\widetilde{B}_0)$. If $T\widetilde{s}^{-1}\widetilde{r} \in (T\widetilde{B})_0$, then the fact that S is saturated entails that $\widetilde{s} - T\widetilde{r}b \in \widetilde{S}$ for any $b \in \widetilde{B}_0$ (note : $\deg\widetilde{s} = \deg T\widetilde{r}b$). Therefore $(\widetilde{s} - T\widetilde{r}b)$ is invertible in $\widetilde{B}$; or $1 - T\widetilde{s}^{-1}\widetilde{r}b$ is a unit in $\widetilde{B}_0$. It follows that $T\widetilde{s}^{-1}\widetilde{r} \in J(\widetilde{B}_0)$.□

2.4 Corollary

The assumptions on R as set forth at the beginning of Section 1 entail that for a saturated Ore set S of R the localized ring $S^{-1}R$ has a Zariskian quotient filtration. In particular the localized ring with respect to $< f >$ and $\kappa_{<P>}$ are Zariski filtered rings.

The relations between $\kappa_{<P>}$ and $\kappa_{<f>}$ or $\kappa_{<I>}$ established before and in Lemma 2.1. and in Proposition 2.2. lead to :

2.5 Theorem

Associating to a Zariski open $Y(\overline{I})$ of Proj$G(R)$ the localized ring $R_{<I>} = Q_{\kappa<I>}(R)$ defines a sheaf of noncommutative rings denoted by $\underline{R}_Y$. If $\overline{I} = G(R)\overline{f}$ for some homogeneous $\overline{f} \in G(R)$ then $R_{<f>}$ is the localization at a saturated Ore set $< f >$ of R. For $P \in Y$ the stalk $\underline{R}_{Y,P}$ is exactly $R_{<P>} = Q_{\kappa<P>}(R)$. All $R_{<f>}$ and $R_{<f>}$ are Zariski filtered rings.

Proof It suffices to add a remark about the stalk property. Since $\kappa_{<P>} = \vee\{\kappa_{<f>}, \overline{f} \notin P\}$ it is indeed clear that $\underline{R}_{Y,P} = \varinjlim_{\overline{f}\notin P} R_{<f>} = R_{<P>}$. □

The advantage of having derived the sheaf conditions at the level of torsion theories (this may be translated into localizing Serre subcategories and quotient categories), i.e.

$$(*)\qquad \begin{cases} \kappa_{<P>} = \vee\{\kappa_{<f>}, \overline{f} \notin P\} = \vee\{\kappa_{<I>}, \overline{I} \not\subset P\} \\ \kappa_{<f>} = \wedge\{\kappa_{<P>}, \overline{f} \notin P\} \\ \kappa_{<I>} = \wedge\{\kappa_{<P>}, \overline{I} \not\subset P\} = \wedge_{i=1}^{d} \kappa_{<f_i>} \text{ if } \overline{I} = (\overline{f}_1, \ldots, \overline{f}_d) \end{cases}$$

is that results about structure sheaves of modules follow also from this without extra work.

2.6 Theorem

Let M be an R-module. Associating the localization $M_{<I>}$ to $Y(\overline{I})$ defines a sheaf of R-modules, $\underline{M}_Y$, say. For a homogeneous $\overline{f} \in G(R)$, $M_{<f>} = R_{<f>} \otimes_R M$. For $P \in Y$ the stalk of $\underline{M}_Y$ at P is given by : $\underline{M}_{Y,P} = M_{<P>} = R_{<P>} \otimes_R M$.

If M is a filtered R-module then $M_{<f>}$, resp. $M_{<P>}$ have the localized filtrations and $\underline{M}_Y$ is a sheaf of filtered R-modules.

Proof Let us just point out that we did not (have to) restrict to absolutely torsionfree M, exactly because we have derived the "sheaf-properties" on the torsion theoretic level (see (*)). More precisely $\underline{\kappa M}$ given by $\kappa, \underline{M}(Y(\overline{I})) = \kappa_{<I>} M$ defines a sheaf $\underline{M/\kappa M}$ given by $\underline{M/\kappa M}(Y(\overline{I})) = M/\kappa_{<I>} M$ and we obtain a sheaf morphism $\underline{M/\kappa M} \hookrightarrow \underline{M}_Y$ locally given by : $M/\kappa_{<I>} M \hookrightarrow M_{<I>}$. It is even possible to verify that $\underline{M}_Y$ is the localization of $\underline{M/\kappa M}$ in the category of sheaves of modules over Y. □

$G(R)$ is schematic with respect to all $S(f)$ (or a selected finite set of such) and $A = \widetilde{R}$ is schematic with respect to $S(\widetilde{f})$ (or a corresponding finite set of such ones). In case $G(R) = Q_{\kappa_+}(G(R)) = \Gamma_*(G(R))$ then R is **weakly affine schematic** in the sense that $\underline{R}_Y(Y) = R$. This leads to :

2.7 Proposition

If $G(R)$ is integrally closed with $\dim G(R) > 1$ then $\underline{R}_Y(Y) = R$.

Proof From the fact that $G(R)$ is normal of dimension more that 1 it follows that $G(R) = \cap_{P \in Y} G(R)_P$, hence we also have $G(R) = \cap_{\overline{f} \in \mathcal{L}(G(R))} G(R)_{\overline{f}} = Q_{\kappa_+}(G(R))$. Then it follows easily (passing via $\widetilde{R} = A$ and checking that $A = Q_{\kappa_+}(A)$) that $R = \underline{R}_Y(Y)$. □

A filtered Ring (or a sheaf of filtered rings), $\underline{R}$ say, is a sheaf of rings endowed with a family of subsheaves of additive groups $\mathcal{F}_n\mathcal{R}$ such that :

1. $\ldots \subset \mathcal{F}_n\mathcal{R} \subset \mathcal{F}_{n+1}\mathcal{R} \subset \ldots$, and $\mathcal{R} = \cap_n^s \mathcal{F}_n\mathcal{R}$

2. $\mathcal{F}_n\mathcal{R}.\mathcal{F}_m\mathcal{R} \subset \mathcal{F}_{n+m}\mathcal{R}$

3. The unit section $\underline{1}$ is in $\mathcal{F}_0\mathcal{R}$

where $\cup_n^s$ is the sheaf associated to the presheaf obtained by taking unions of sections locally.

A lot of the theory of filtered and associated graded rings may be translated directly to the case of sheaves of filtered rings. We do not develop this almost purely linguistic excercise further but phrase the following theorem is that language because it makes the deformation aspect clear.

2.8 Theorem

$\mathcal{F}_0\underline{R}_Y$ is a Noetherian and coherent sheaf; the stalk at $P \in Y$ equals $F_0Q_{<P>}(R)$.

The Ideal (sheaf of ideals) $\mathcal{F}_{-1}\underline{R}_Y$ of $\mathcal{F}_0\underline{R}_Y$ is coherent and $\mathcal{F}_0\underline{R}_Y/\mathcal{F}_{-1}\underline{R}_Y - \underline{O}_Y$, the structure sheaf of $\text{Proj}(G(R)) = Y$.

We call $\mathcal{F}_0\underline{R}_Y$ the **saturated deformation** of $\underline{O}_Y$. In Section 3 we shall return to some properties of a more algebraic nature, now let us end this section by passing to the microlocal and quantum versions. The reason for wanting to do this is that for Ore sets S and T of R, even when $\sigma(S)$ and $\sigma(T)$ commute, we do **not** have $Q_SQ_T(R) = Q_TQ_S(R)$. Recall from [VOW 2] that, under the hypotheses put on $G(R)$, we do have $Q_S^\mu Q_T^\mu(R) = Q_T^\mu Q_S^\mu(R)$ and also $Q_S^q Q_T^q(R) = Q_T^q Q_S^q(R)$. In particular we obtain :

2.9 Lemma

1. For a saturated Ore set S of R we have $Q_S^\mu(R) = (S^{-1}R)^\wedge$, $\wedge$ standing for completion with respect to the quotient filtration; moreover $Q_S^\mu(R) = Q_{S_{\text{sat}}}^\mu(R)$ holds for any Ore set S, S_{sat} denoting the saturation of S.

2. $Q^\mu_{<fg>} = Q^\mu_{<f>}Q^\mu_{<g>} = Q^\mu_{<g>}Q^\mu_{<f>} = Q^\mu_{<f>\vee<g>}$

3. For any Ore set S of R, $Q_S^\mu(R)$ is Zariski filtered (because $Q_S^\mu(R) = (S_{\text{sat}}^{-1}R)^\wedge$ follows from part 1, so we may apply Corollary 2.4.).

If $Y(\overline{I}) \subset Y(\overline{J})$ then $\kappa_{<J>} < \kappa_{<I>}$ and the canonical ring morphism $Q_{<J>}(R) \to Q_{<I>}(R)$ extends to a ring morphism $\mu_{\overline{I}}^{\overline{J}} : Q^\mu_{<J>}(R) \to Q^\mu_{<I>}(R)$, defined by passing to the completion.

In a similar way, for a filtered R-module M, we obtain a morphism $\mu_{\overline{I}}^{\overline{J}}$: $Q^\mu_{<J>}(M) \to Q^\mu_{<I>}(M)$, which is a **strict** filtered morphism.

Let us write $< \widetilde{I} >$, resp. $< \widetilde{P} >$ for the homogeneous Ore sets of $A = \widetilde{R}$ corresponding to the saturated Ore sets $< I >$, resp. $< P >$ of R. On the level of Rees rings, we have :

$$(Q^{\mu}_{<I>}(R))^{\sim} = Q^{\mu}_{<\widetilde{I}>}(A), \left(Q^{\mu}_{<P>}(R)\right)^{\sim} = Q^{\mu}_{<\widetilde{P}>}(A)$$

and similarly for a filtered R-module M :

$$(Q^{\mu}_{<I>}(M))^{\sim} = Q^{\mu}_{<\widetilde{I}>}(\widetilde{M}), (Q^{\mu}_{<P>}(M))^{\sim} = Q^{\mu}_{<\widetilde{P}>}(\widetilde{M})$$

2.10 Theorem

1. Associating $R^{\mu}_{<I>} = Q^{\mu}_{<I>}(R)$ to $Y(\overline{I})$ defines a sheaf of complete Zariski rings, denoted $\underline{R}^{\mu}_{Y}$. Associating $M^{\mu}_{<I>} = Q^{\mu}_{<I>}(M)$ to $Y(\overline{I})$ defines a sheaf, $\underline{M}^{\mu}_{Y}$, of $\underline{R}^{\mu}_{Y}$-modules.

2. At $P \in Y$ the stalk $\underline{R}^{\mu}_{Y,P}$ is a Zariski filtered ring such that $(\underline{R}^{\mu}_{Y,P})^{\wedge} = Q^{\mu}_{<P>}(R)$, where $\wedge$ now stands for completion with respect to the microlocalized filtration of $\underline{R}^{\mu}_{Y,P}$. For a filtered R-module M we obtain :

$$\underline{M}^{\mu}_{Y,P} = \underline{R}^{\mu}_{Y.P} \otimes_R M, (\underline{M}^{\mu}_{Y,P})^{\wedge} = Q^{\mu}_{<P>}(M)$$

3. The modified versions of 1. and 2. hold with respect to

$$\widetilde{R} = A, \widetilde{M}, \widetilde{\underline{M}^{\mu}_{Y}}, \underline{A}^{\mu}_{Y}, < \widetilde{I} >, < \widetilde{P} >, \text{ etc...}$$

Proof **1.** Assumptions on $G(R)$ entail that A is schematic. Since Ore sets of A are weakly compatible in the sense of [VOW2], i.e. for $\widetilde{S}$ and $\widetilde{T}$ the $Q_{\widetilde{S}}$ and $Q_{\widetilde{T}}$ on $G(R)$-gr commute, it follows that the following equalizer diagram is exact :

$$Q^{\mu}_{\widetilde{ST}}(\widetilde{M}) \longrightarrow Q^{\mu}_{\widetilde{S}}(\widetilde{M}) \times Q^{\mu}_{\widetilde{T}}(\widetilde{M}) \rightrightarrows Q^{\mu}_{\widetilde{S}\vee\widetilde{T}}(\widetilde{M})$$

and same for diagrams like this corresponding to finite words in Ore sets (cf. loc. cit). The torsion theoretic condition (*) before Theorem 2.6. (and in fact using the theorem too) allow to establish the claim in a straightforward way.

2. Put $S_p = \varinjlim_{f \notin P} Q^{\mu}_{<f>}(R)$, pick $x \in F_{-1}S_p$, $a \in F_0 S_p$. For some $\overline{f}$ homogeneous, $\overline{f} \notin P$ we have : $x \in F_{-1}Q^{\mu}_{<f>}(R), a \in F_0 Q^{\mu}_{<f>}(R)$.

The Zariskian property of $Q^{\mu}_{<f>}(R)$ yields $(1 - ax)^{-1} \in F_0 Q^{\mu}_{<f>}(R)$, hence $(1 - ax)^{-1} \in F_0 S_P$ or $F_{-1}S_P \subset J(F_0 S_P)$.

Let us check whether $\widetilde{S}_P$ is Noetherian. For homogeneous $\overline{f} \notin P$, $Q^{\mu}_{<f>}(R) = (Q_{<f>}(R))^{\wedge}$, where this completion is flat because $(Q_{<f>}(R))^{\sim}$ is Noetherian ! For a graded left ideal $\widetilde{L} \subset \widetilde{S}_P$ (similar argument will work for right ideals) : $\widetilde{L}(f) = \widetilde{L} \cap Q^{\mu}_{<\widetilde{f}>}(A)$ is finitely generated.

For another element $\overline{g} \notin P$ we clearly have :

$$\widetilde{L}(fg) \cap Q_{<\widetilde{fg}>}(A) = Q_{<\widetilde{fg}>}(A)(\widetilde{L}(f) \cap Q_{<\widetilde{f}>}(A))$$

The flatness of $Q_{<\widetilde{fg}>}(A)^\wedge$ as a right $Q_{<\widetilde{fg}>}(A)$-module yields :

$$\begin{aligned} \widetilde{L}(fg) &= Q_{<\widetilde{fg}>}(A)^\wedge(\widetilde{L}(fg) \cap Q_{<\widetilde{fg}>}(A)) \\ &= Q_{<\widetilde{fg}>}(A)^\wedge \left(\widetilde{L}(f) \cap Q_{<\widetilde{f}>}(A)\right) = Q^\mu_{<\widetilde{fg}>}(A)\widetilde{L}(f) \end{aligned}$$

(note : $Q_{<\widetilde{fg}>}(A)^\wedge$ is also flat as a $Q_{<\widetilde{f}>}(A)$-module).

It follows that a finite set of generators, $\lambda_1, \ldots, \lambda_t$ say, for $\widetilde{L}(f)$ still generates $\widetilde{L}(fg)$ as a left $Q^\mu_{<\widetilde{fg}>}(A)$-modules. Since $\widetilde{L}$ is a union of images of $\widetilde{L}(f)$ for all $\overline{f} \notin P$ homogeneous, $\widetilde{L}$ must be generated by the images of $\lambda_1, \ldots, \lambda_t$ over $\widetilde{S}_P$. It is then clear that $\widetilde{S}_P$ is Noetherian and consequently $\underline{R}^\mu_{Y,P} = S_P$ is Zariski filtered. From (*) we retain that $Q_{<P>}(R) = \varinjlim Q_{<f>}(R)$, hence $Q_{<P>}(R)$ is a filtered subring of S_P. The canonical localization morphism $Q_{<f>}(R) \to Q_{<P>}(R)$ is a strict filtered morphism defining a strict filtered morphism $Q^\mu_{<f>}(R) \to Q^\mu_{<P>}(R)$ (since we restrict attention to domains R usually, all maps are then injective). The universal property of $\varinjlim$ yields an injection $S_p \to Q^\mu_{<P>}(R)$, but $Q_{<P>}(R) = \varinjlim_{\overline{f} \notin P} Q_{<f>}(R)$ entails $Q_{<P>}(R) \subset S_P$. Thus $Q^\mu_{<P>}(P)$ must be the completion of S_P.

The final claim in 2. follows again by the flatness of completion at a Zariskian filtration. For a filtered R-module

$$\begin{aligned} M : \quad & \underline{M}^\mu_Y(Y(\overline{f})) = \underline{R}^\mu_Y(Y(\overline{f})) \otimes_R M \\ & \underline{M}^\mu_{Y,P} = \varinjlim_{\overline{f} \notin P}(Q^\mu_{<f>}(R) \otimes_R M) = \underline{R}^\mu_{Y,P} \otimes_R M \end{aligned}$$

Consequently $(\underline{M}^\mu_{Y,P})^\wedge = (\underline{R}^\mu_{Y,P} \otimes_R M)^\wedge = Q^\mu_{<P>}(R) \otimes_R M$, second equality following from flatness of $Q^\mu_{<P>}(R)$ as a right $\underline{R}^\mu_{Y,P}$-module (because $F\underline{R}^\mu_{Y,P}$ is Zariskian as was established earlier).

3. Follows from **1.** and **2.** in the obvious way. □

In sheaf-language we may summarize things as follows :

2.11 Theorem

1. $\mathcal{F}_0\underline{R}^\mu_Y$ is a Noetherian and coherent sheaf, the Ideal $\mathcal{F}_{-1}\underline{R}^\mu_Y$ is coherent and we have : $\underline{O}_Y = \mathcal{F}_0\underline{R}^\mu_Y / \mathcal{F}_{-1}\underline{R}^\mu_Y$.

2. If M is a filtered R-module with good filtration FM then the filtered sheaf $\underline{M}^\mu_Y$ of $\underline{R}^\mu_Y$-modules has associated graded sheaf $\mathcal{G}(\underline{M}^\mu_Y)$ equal to the graded structure sheaf $\underline{O}^g_{G(M)}$ of $G(R)$ over Y. Since the latter is coherent and $\underline{M}^\mu_Y$ is locally finite, it follows that $\underline{M}^\mu_Y$ is coherent.

Moreover, the sheaf $\mathcal{F}_0\underline{M}^\mu_Y = \underline{M}^q_Y$ is coherent and it has an $\mathcal{F}_0\underline{R}^\mu_Y = \underline{R}^q_Y$ - submodule $\mathcal{F}_{-1}\underline{M}^\mu_Y$ such that : $\mathcal{F}_0\underline{M}^\mu_Y/\mathcal{F}_{-1}\underline{M}^\mu_Y = \underline{O}_{G(M)}$.

The foregoing just states that the sheaves of quantum sections $\underline{R}^q_Y, \underline{M}^q_Y$ are deformations (sheafwise) of the structure sheaves of Proj$G(R)$, resp. $\underline{\underline{O}}_{G(M)}$ for $G(M) \in G(R)$-gr.

For some easy examples, e.g. the noncommutative scheme structure, we may refer to [LVO].

Let us point out that a deformation of Noetherian formal schemes may be defined, leading to the definition of formal quantum sections. The new results in the foregoing part of this section, e.g. those related to $< I >$ and $< f >$ and $< P >$ have formal versions in the sense that I-adic completions in the classical theory of formal schemes are to be replaced by $< I >$-completions (i.e. completion in the linear topology of $R^\mu_{<f>}$ defined by the $Q^\mu_{<f>}(L), L \in \mathcal{L}(< I >)$).

The quantized formal scheme theory then depends on some compatibility properties between $< I >^\mu$-completions and completions with respect to the localized filtrations deriving from FR on $Q^\mu_{<f>}(R)$. Detail is included in a forthcoming work [VO2].

Summarizing, we may state that the sheaf of formal quantum sections of a filtered module M with good filtration is the deformation of the formal structure sheaf of $G(M)$ over the formal scheme $(\widehat{Y}, \underline{O}_{\widehat{Y}})$. If we put $\overline{I} = G(R)$ i.e. $\widehat{Y} = Y$ the we rediscover the theory of quantum sections as mentioned before, now appearing as a "degenerations" of the theory of formal "schemes" of noncommutative objects.

3 Some Lifting Properties

Notation and assumption on $R, G(R)$ are as in the foregoing section, usually we assume that $G(R)$ is a domain i.e. Y is a connected scheme. Combining the results of [VVO], [VOW], [LVO] applied to this special situation, we obtain a list of "lifting" as well as "local global" properties showing that the noncommutative deformation of Proj$G(R) = Y$ has most of the properties one would expect for a "geometrical" space.

3.1 Theorem

1. If $\underline{O}_{Y,P}$ is a maximal order then $\underline{R}_{Y,P}$ is so too.
2. If $\underline{O}_{Y,P}$ has finite injective dimension then $\underline{R}_{Y,P}$ too.

3. If $\underline{O}_{Y,P}$ is regular, i.e. has finite global dimension, then $\underline{R}_{Y,P}$ has finite global dimension.

4. Every finitely generated projective $\underline{R}_{Y,P}$-module is stably free.

5. If $\underline{O}_{Y,P}$ is regular then every finitely generated projective $\underline{R}_{Y,P}$-module is free.

Proof **1.** cf. [VVO]. **2.** and **3.** may be found in [LVO 1,2], **4.** stems from [LVO 3], [LH].

5. Since $\underline{R}_{Y,P}$ is Noetherian, is having finite global dimension and is faithfully flat over $F_0\underline{R}_{Y,P}$ it follows that a finitely generated $\underline{R}_{Y,P}$-module M with filtration FM satisfies :

$$\mathrm{Ext}^k_{F_0\underline{R}_{Y,P}}(F_0M, F_0\underline{R}_{Y,P}) \otimes_{F_0\underline{R}_{Y,P}} \underline{R}_{Y,P} = \mathrm{Ext}^k_{\underline{R}_{Y,P}}(M, \underline{R}_{Y,P})$$

If M is finitely generated projective then by the foregoing and some homological algebra, F_0M is projective as a $F_0\underline{R}_{Y,P}$-module and hence free. It follows that M is free.

Recall that a (left and right) Noetherian ring R is said to have the **Auslander condition** satisfied if for every finitely generated R-module M and any $k \geq 0$; for every R-submodule $N \subset \mathrm{Ext}^k_R(M, R)$ we have $j_R(N) \geq k$, where $j_R(-)$ stands for the grade number over R. If R has finite injective dimension and satisfies the Auslander condition then R is said to be **Auslander regular**. If R has finite global dimension and satisfies the Auslander condition then R is said to be **Auslander-Gorenstein regular** (or R is an Auslander-Gorenstein ring).

3.2 Theorem

1. If $\underline{O}_{Y,P}$ is Auslander-Gorenstein then so is $\underline{R}_{Y,P}$.

2. If $\underline{O}_{Y,P}$ is Auslander regular then so is $\underline{R}_{Y,P}$.

For an R-module M filtered by FM, the **characteristic ideal**, denoted by $J(M)$, is defined by $J(M) = \mathrm{rad}(\mathrm{ann}_{G(R)}G(M))$.

The **characteristic variety** $V(M)$ is defined as the set of minimal prime divisors of $J(M)$ in Y.

3.3 Proposition

(cf. [LH] Proposition 5.1., for this formulation). Let M have a good filtration FM, then the following are equivalent :

1. $G(M)$ is κ_+-torsion, i.e. $Q_{\kappa_+}(G(M)) = 0$.

2. $\underline{O}^g_{GM,P} = 0$ for all $P \in Y$.
3. $V(M) = \emptyset$
4. $\text{Kdim}_{G(R)}G(M) = 0$, where Kdim stands for all Krull dimension
5. $GK\text{dim}G(M) = 0$, where GKdim stands for Gelfand-Kirillov dimension.
6. $GK\text{dim}(M) = 0$.
7. M is finite dimensional over K the base field of R.
8. $G(M)$ is finite dimensional over K.

3.4 Theorem

(cf. [LH]) The following statements are equivalent :

1. $\oplus_{P \in Y} \underline{R}_{Y,P}$ is a faithfully flat left and right R-module.
2. For every $M \neq 0$ with good filtration FM, $\text{Kdim}G(M) > 0$.
3. Every nonzero finitely generated R-module M has $\text{GKdim}(M) > 0$.

3.5 Theorem

(Local-Global properties)

1. $\text{inj.dim}R = \sup(\text{injdim}\underline{R}_{Y,P}, P \in Y\}$. If $G(R)$ has finite dimension and $\underline{O}_{Y,P}$ is a Gorenstein ring for every $P \in Y$ then $\text{injdim}R < \infty$.
2. If R has finite global dimension and M has good filtration FM then M is projective if and only if $\underline{M}_{Y,P}$ is a projective $\underline{R}_{Y,P}$-module for every $P \in Y$. In this case $\text{gldim}R = \sup\{\text{gldim}\underline{R}_{Y,P} P \in Y\}$. If $G(R)$ has finite dimension and every $P \in Y$ is regular i.e. $G(R)$ is a regular commutative ring, then $\text{gldim}R < \infty$.
3. If $\underline{O}_{Y,P}$ is Gorenstein for every $Y \in P$ then R is Auslander-Gorenstein. If $G(R)$ is a regular domain then R is Auslander regular.
4. If $\underline{R}_{Y,P}$ is a maximal order for every $P \in Y$ then R is too, if $\underline{O}_{Y,P}$ is a maximal order for every P, then R is a maximal order.

Proof Direct consequences of foregoing results and those cited ([VVO], [LVO 1], [LH], ...) □

An interesting feature, desirable for applications in cohomological considerations and calculations using Čhech cohomology, is condition χ introduced by M. Artin, J. Zhang in [AZ], i.e. $\text{Ext}^n_A({}_AK, M)$ is finite dimensional for each finitely generated graded A-module, $A = \widetilde{R}$.

3.6 Theorem

(**[VOW 3]**) If A is schematic of finite global dimension then condition χ holds.

Now in the situation we consider here $\widetilde{R} = A$ is schematic, if $G(R)$ is regular then A is Auslander regular ([LVO], [LVO1]) and so it certainly has finite global dimension. Therefore Theorem 3.6. may be applied to $\widetilde{R}$ in our situation and one derives from this :

3.7 Theorem

If $G(R)$ is regular then for every graded finitely generated $\widetilde{R}$-module N we have :

1. $Q_{\kappa_+}(N) = \mathrm{Hom}_A(A_{\geq n}, N)$ for n sufficiently large.
2. $Q_{\kappa_+}(N) \cong N_{\geq n}$ for n sufficiently large.
3. $Q_{\kappa_+}(N)_{\geq n}$ is a finitely generated A-module for all n.
4. For $j \geq 1$, $\left(\underline{H}^j(\pi(M))\right)_n$ is finite dimensional for all n
5. For $j \geq 1$, $\left(\underline{H}^j(\pi(N))\right)_n = 0$ for n sufficiently large.
6. $\mathrm{Ext}^n_{\widetilde{R}}(N_1, N)$ is finite dimensional for all n, if N_1 is finitely generated.

In the foregoing $\underline{H}^j(\pi(M))$ stands for the graded cohomology, cf. [VOW]; however it is one of the main results of [VOW3] that this cohomology may be calculated via the Čhech cohomology corresponding to the Grothendieck topology defined on Proj$A$($\cong$ Proj$G(R)$" $\coprod$ "Spec$R$ in some well defined sense, viewing Spec$R$ as the category R-mod, Proj$G(R)$ as the usual quotient category of $G(R)$- gr and similar for proj$A$). Statement **5.** is then an interesting **"vanishing theorem"**.

The noncommutative geometry of Proj$A$ in the sense of [VOW 1, 2, 3], [VO2] is phrased in terms of a Grothendieck topology providing a noncommutative equivalent of the Zariski topology (covers are given in terms of "words of Ore sets of A''). It turns out that a noncommutative version of Grothendieck's scheme theory does indeed extend to the class of schematic algebras, containing most if not all recently popular algebras of quantized type. In this paper we have deformed the structure sheaf of proj$G(R)$ to a noncommutative sheaf over the same topological space Y and have shown that most of the basic results of scheme theory remain available. This may be extended to Proj$A$ and Proj$A$ = Proj$G(R) \coprod$ Spec$R$ may be given a schematically correct meaning (expressing fully the relation between filtered structures over R and graded structures over $G(R)$ and $\widetilde{R}$). For a detailed treatment of this kind of noncommutative geometry and many concrete examples related to quantum groups, Weyl algebras, gauge algebras,..., we refer to the forthcoming work [VO2].

References

[**VOW1**] F. Van Oystaeyen, L. Willaert, *Grothendieck Topology, Coherent Sheaves and Serre's Theorem for Schematic Algebras*, J. Pure Applied Algebra, 104, 1995, 109-122.

[**VOW2**] F. Van Oystaeyen, L. Willaert, *Examples and Quantum Sections of Schematic Algebras*, J. Pure Applied Algebra.

[**VOW3**] F. Van Oystaeyen, L. Willaert, *Cohomology of Schematic Algebras*, J. of Algebra 183, 2, 1996, 359-364.

[**AZ**] M. Artin, J. J. Zhang, *Noncommutative Projective Schemes*, Adv. in Math. 109, 1994, 228-287.

[**ATV**] M. Artin, J. Tate, M. Van den Bergh, *Modules over Regular Algebras of Dimension 3*, Invent. Math. 106, 1991, 335-388.

[**LVO**] Li Huishi, F. Van Oystaeyen, *Zariskian Filtrations*, Kluwer Monographs, Vol. 2., Kluwer Acad. Publ., 1995.

[**VOR**] F. Van Oystaeyen, R. Sallam, *A Microstructure Sheaf and Quantum Sections over a Projective Scheme*, J. of Algebra, 1993, 201-225.

[**LEVO**] L. Le Bruyn, F. Van Oystaeyen, *Quantum Sections and Gauge Algebras*, Publ. Mathematiques, 36(2A), 1992, 693-714.

[**RVO**] A. Radwan, F. Van Oystaeyen, *Micro-Structure Sheaves and Quantum Sections over Formal Schemes*, Bull. Soc. Math. Belge, 1993.

[**LH**] Li Huishi, *Structure Sheaves on Almost Commutative Algebras*, Comm. in Algebra, to appear.

[**BB**] W. Borho, J.-L. Brylinski, *Differential Operators in Homogeneous Spaces I. Irreducibility of the Associated Variety*, Invent. Math. 69, 1982, 437-476.

[**G**] V. Ginsburg, *Characteristic Varieties and Vanishing Cycles*, Invent. Math., 84, 1986, 327-402.

[**GO**] J. Golan, *Localization of Noncommutative Rings*, Marcel Dekker, New York, 1975.

[**GOLD**] O. Goldman, *Rings and Modules of Quotients*, J. of Algebra, 13, 1969, 10-47.

[**NVO**] C. Năstăsescu, F. Van Oystaeyen, *Graded Ring Theory*, Library of Math., North Holland, Amsterdam, 1982.

[**VO1**] F. Van Oystaeyen, *Prime Spectra in Noncommutative Algebra*, Lect. Notes in Math., 444, Springer-Verlag, Berlin, 1978.

[**AVV**] J.M. Asensio, F. Van Oystaeyen, M. Van den Bergh, *A New Algebraic Approach to Microlocalization of Filtered Rings*, Trans. Am. Math. Soc., 316, 1989, 15-25.

[**HART**] R. Hartshorne, *Algebraic Geometry*, Springer Verlag, Berlin, New York, 1977.

[**VO2**] F. Van Oystaeyen, *Geometric Methods in Noncommutative Algebra*, Monograph, to appear.

[**VVO**] M. Van den Bergh, F. Van Oystaeyen, *Lifting Maximal Orders*, Comm. in Algebra, 17, 1989, 341-349.

[**LVO1**] Li Huishi, F. Van Oystaeyen, *Filtrations on Simple Artinian Rings*, J. of Algebra, Vol. 132, 1990, 361-376.

[**LVO2**] Li Huishi, F. Van Oystaeyen, *Global Dimension and Auslander Regularity of Rees Rings*, Bull. Soc. Math. Belg.

[**JPS**] J.-P. Serre, *Faisceaux Algébriques Cohérents*, Ann. of Math. 61, 1955, 197-278.

[**BJ**] J.-E. Björk, *Rings of Differential Operators*, Math. Library, 21, North-Holland, Amsterdam, 1979.